Current Topics in Microbiology and Immunology

173

Editors

Function and Specificity of γ/δ T Cells

International Workshop,
Schloß Elmau, Bavaria, FRG
October 14-16, 1990

Edited by
K. Pfeffer, K. Heeg,
H. Wagner, and G. Riethmüller

With 41 Figures

Springer-Verlag
Berlin Heidelberg New York
London Paris Tokyo
Hong Kong Barcelona
Budapest

Dr. Klaus Pfeffer
Prof. Dr. Klaus Heeg
Prof. Dr. Hermann Wagner
Institut für Medizinische
Mirobiologie und Hygiene
TU München
Trogerstraße 9
8000 München 8

Prof. Dr. Gert Riethmüller
Institut für Immunologie
University of Munich
Goethestraße 11
8000 München

ISBN-13: 978-3-642-76494-3 e-ISBN-13: 978-3-642-76492-9
DOI: 10.1007/ 978-3-642-76492-9

Library of Congress Catalog Card Number 15-12910
Softcover reprint of the hardcover 1st edition 1991

23/3020-543210 – Printed on acid-free paper

Preface

Our current understanding of α/β T cell receptor (TCR) expressing T cells advanced from function and specificity to the molecular organization of the TCR. We now know that the TCR α and β chains together express specificity for (antigenic) peptides presented by the "responder" MHC allele, thus explaining the phenomenon of MHC restriction at a molecular level.

Surprisingly even though our perception of the molecular organization of the γδ TCR is well advanced, current knowledge of function and specificity of the γδ T cell subset is poor. Therefore it appeared rather timely to bring together scientists pioneering research on γδ T cells for the International Workshop on Function and Specificity of γδ T cells, held October 11-14, 1990 at Schloss Elmau/Bavaria, FRG. Besides offering a scientific forum for open discussions, it was also hoped that such a workshop would be seminal for collaborative interactions and personal relationships among scientists "addicted" to γ/δ T cells.

This volume of Current Topics in Microbiology and Immunology details the workshop reports und thus aims at covering the state of the art of γδ T cell research in autumn 1990. The organizers of the workshop believe that the issue provides an actual survey about what is presently known about the biological role of γδ T cells not only in man and mouse, but also in other species such as chicken, sheep, rat and swine. The individual chapters deal with the genomic organization of the γδ T cell receptor and present the most recent data on maturation and differentiation of γδ T cells in vivo and in vitro, including γδ T cells in transgenic mice. Putative antigens and the specificity and probable restriction elements of γδ T cell receptors are discussed, as is involvement of γδ T cells in mucosal and dermal immunity. A number of old questions have been answered. Yet more new questions have arisen during the workshop discussions which now await clarification.

We are just beginning to discern the outlines of the function and specificity of γδ T cells at their various localizations in the body. It is obvious, however, that progress will depend on more direct exchange of data and ideas among scientists viewing different perspectives. The workshop and this issue of CTMI serve this purpose. The organizers thank the Deutsche Forschungsgemeinschaft for its generous funding.

KLAUS PFEFFER
KLAUS HEEG
HERMANN WAGNER
GERT RIETHMÜLLER

Contents

Other Species

Transgenic Mice/Manipulation In Vivo

Antigen/Heat Shock Proteins

Restriction

CD5 B Cells

Mucosal and Dermal Immunity

Contributors

(Their adresses can be found at the beginning of the respective chapters)

Genomic Organization

Genetic Organization of the Human T Cell Receptor *Gamma* Locus

MARIE-PAULE LEFRANC[1] and T.H. RABBITTS[2]

[1] Laboratoire d'Immunogénétique Moléculaire, URA CNRS 1191, Université Montpellier II, Sciences et Techniques du Languedoc, CP012, Place E. Bataillon, 34095 Montpellier Cedex 5, France

[2] Laboratory of Molecular Biology, Medical Research Council, Hills Road, Cambridge CB2 2QH, England

The T-cell gamma/delta receptor is expressed on about 3-5 % of the circulating T lymphocytes in human. In this report, we will review the genetic organization of the human T cell receptor gamma locus.

THE HUMAN T CELL RECEPTOR GAMMA LOCUS

The human T cell receptor gamma (TCRG or TRG) locus consists of genes which are rearranged and joined during T cell differentiation. The human TRG gene locus has been mapped to chromosome 7 (Rabbitts et al., 1985) at band 7p14-p15 (Murre et al., 1985 ; Bensmana et al., 1990). We have extensively studied its organization by phage and cosmid clone analysis and gene deletion mapping (for review, see Lefranc et al 1987 ; Lefranc 1988 ; Lefranc and Rabbitts 1989) and we have linked the variable and constant regions by pulse field gel electrophoresis (PFGE) (Lefranc et al., 1989). The gamma locus comprises two constant regions (TRGC) linked to each other at a distance of 16 kilobases (Lefranc and Rabbitts, 1985 ; Lefranc et al., 1986 a,b), five joining segments (TRGJ) (reviewed in Huck and Lefranc 1987) and, in most cases, 14 variable gamma genes (TRGV) belonging to four subgroups and located upstream of the two C gamma genes (Lefranc et al. 1986a, c ; Forster et al. 1987 ; Huck et al. 1988) (Fig. 1). The human T cell receptor gamma locus spans 160 kb of genomic DNA, with only 16 kb separating the most 3'V gene from the most 5' J segment (Lefranc et al. 1989).

Fourteen Variable Gamma Genes (TRGV)

Fourteen TRGV genes have been identified in human DNA (Lefranc et al 1986a, c ; Forster et al. 1987 ; Huck et al. 1988). The group of TRGV genes includes six pseudogenes and eight potentially active genes. These active genes fall into four distinct subgroups, designated VγI - VγIV. Nine Vγ genes, five of them functional (V2, V3, V4, V5 and V8), and four pseudogenes (V1, V5P, V6 and V7), belong to subgroup I, whereas subgroups II, III and IV each consists of a single gene, designated V9, V10 and V11, respectively (Lefranc et al. 1986a,c ; Forster et al. 1987 ; Huck et al. 1988). Two pseudogenes, VA and VB, located upstream of V9 and V11, respectively, belong to none of these subgroups (Forster et al. 1987, Huck et al. 1988). A haplotypic variation of the number of the VγI genes, from 7 to 10, can be observed due to the deletion of the V4 and V5 genes (Font et al. 1988 ; Ghanem et al. 1989) or the insertion of an additional gene, V3P, between V3 and V4 (Ghanem et al. 1989, 1990) (Fig. 2A). The frequency of the VγI subgroup gene

Current Topics in Microbiology and Immunology, Vol. 173

haplotypes has been studied in five different populations using the VγI probe pV3S (Ghanem et al. 1989, 1990). As an example, the frequency of the 7-gene haplotype (with deletion of V4 and V5) is 0.21 in the French population, 0.13 in the Black African population (Ghanem et al. 1989), and 0.17 in the Chinese population (Ghanem et al. 1990).

Five Joining Gamma Segments (TRGJ)

Five joining gamma segments (TRGJ) have been identified : J1, J2 (Lefranc et al. 1986a), JP (Lefranc et al. 1986c), JP1 and JP2 (Huck and Lefranc, 1987 ; Quertermous et al. 1987 ; Tighe et al. 1987) (Fig 1). JP1, JP and J1 are located upstream of TRGC1 whereas JP2 and J2 are upstream of TRGC2. These segments encode 16-20 aminoacids of the variable region of the γ-chain (Huck and Lefranc 1987), the major part of the variable region being encoded by one of the TRGV genes located further upstream (Lefranc et al. 1986c).

Two Constant Gamma Genes (TRGC)

The two human constant gamma genes TRGC1 and TRGC2 are separated by a distance of 16 kb (Lefranc and Rabbitts, 1985). Structural differences exist between the two C genes : TRGC1 consists of three exons (ex1, ex2 and ex3) (Lefranc et al. 1986b), whereas the TRGC2 gene contains in some cases two (Lefranc et al. 1986b) and in some others three copies of exon 2 in addition to exon 1 and exon 3 (Lefranc et al. 1986b ; Littman et al. 1987 ; Buresi et al. 1989) (Fig.2B). Therefore the TRGC2 gene, spanning 9.5 kb or 12 kb of genomic DNA, respectively (Lefranc et al. 1986b ; Buresi et al. 1989), is longer than the TRGC1 gene (6 kb only) (Lefranc et al. 1986b) and it displays an allelic polymorphism due to the presence of either 4 or 5 distinct exons (Buresi et al. 1989) (Fig. 2B). This allelic polymorphism can be distinguished by restriction fragment length polymorphism (RFLP) analysis and has been studied in five different populations (Buresi et al. 1989 ; Ghanem et al. 1990). 68 % of the alleles in the Black African populations show a TRGC2 gene with triplication of the exon 2 against only 16 % in the French population (Buresi et al. 1989), and 13 % in the Chinese population (Ghanem et al. 1990)
Comparison of the TRGC1 sequence with that of the mouse shows that there has been conservation of the exon 2 cysteine residue, involved in the interchain disulfide bridge whereas this residue is not conserved in exon 2 of the human TRGC2 gene, as shown by analysis of genomic clones (Lefranc et al. 1986b ; Buresi et al. 1989) and complementary DNA clones (Dialynas et al. 1986 ; Littman et al. 1987 ; Krangel et al. 1987). These differences correspond to different types of γ chains at the cell surface of the human T lymphocytes expressing the γ/δ receptor (see references in Lefranc and Rabbitts, 1989) : the 40 kDa γ1 chain, linked by a disulfide bridge to the δ chain, the 40 kDa and the 44 kDa non-disulfide γ2 chains (which represent two different degrees of glycosylation) encoded by a C2 gene with duplication of the exon 2, and for that reason, designated as γ2(2x), and the 55 kDa non-disulfide linked γ2 chain encoded by a C2 gene with triplication of the exon 2, and therefore designated as γ2(3x) (Lefranc and Rabbitts, 1989) (Fig. 3).

Size of the Human TRG Locus

A series of overlapping phage clones spanning 130 kb of genomic DNA has previously been isolated. These clones encompass on the one hand, the 14 known Vγ genes (Lefranc et al. 1986c ; Forster et al. 1987 ; Huck et al. 1988) and, on the other hand, the totality of the C region genes and associated J segments (Lefranc et al. 1986b ; Huck and Lefranc, 1987). Due to the polymorphism of the TRGC2 gene, the distance between JP1 and the exon 3 of TRGC2 is 37 kb for the allele with duplication of the exon 2 (designated as C2(2x)) (Lefranc et al. 1986b ; Huck and Lefranc, 1987) and 39.5 kb for the allele with triplication (or C2(3x)) (Buresi et al. 1989) (Fig. 1). All the V γ genes are contained in a unique 120 kb *Xho*I fragment detected by PFGE which links the V and C regions and the size of the TRG locus can be estimated to be 160 kb (Lefranc et al. 1989). Moreover, we showed that the V and C regions are remarkably close to each other since the distance between V11, the most 3′ V gene and JP1, the most 5′ J segment is only 16 kb (Lefranc et al. 1989). With its 14 Vγ genes spanning 100 kb, the 2 Cγ genes and 5 J segments covering <40 kb and only 16 kb separating the most 3′V gene from the most 5′ J segment, the human T cell receptor gamma locus represents a particularly densely populated region when compared with the other rearranging loci (Lefranc et al.1989)

Rearrangements of the TRG Genes

Interestingly, using a unique probe pH60 (Lefranc and Rabbitts 1985 ; Lefranc et al. 1986a), all the T cell receptor gamma gene rearrangements in normal T cells, T cell leukemias and lymphomas can be assigned to known V and J segments indicating that most, if not all, genes of the human TRG locus have been identified. Since the Jγ1 and Jγ2 segments are highly homologous (Lefranc et al. 1986a), it is possible with the J1 probe pH60, first to detect the V rearrangements to J1 and J2, and second, to identify the rearranged V genes by the sizes of the rearranged *Bam*HI, *Eco*RI and *Hind*III restriction fragments (Forster et al. 1987). Moreover rearrangements to the additional J segments, JP, JP1 and JP2, can be identified by hybridization of the *KpnI* digests to the J1 probe pH60 (Huck and Lefranc 1987) (see Table I in Lefranc and Rabbitts, 1989). This unique probe can, therefore, detect all the γ. rearrangements whatever the J segment involved in the rearrangements. Thus it is a very useful tool to establish the clonality of $\alpha\beta^+$ T cell clones (Moisan et al. 1989), leukemic cells (Chen et al. 1988 ; Migone et al. 1988) and T lymphocytes expressing the γδ receptor (Triebel et al. 1988a, b ; Sturm et al. 1989).

DIVERSITY OF THE HUMAN T CELL RECEPTOR GAMMA CHAINS

The diversity of the T cell receptor gamma chains depends on several mechanims :
a) the combinatorial diversity which is a consequence of the number of V genes and J segments. This diversity is relatively low for the γ locus (40 combinations). This diversity is, in fact, restricted by the preferential usage of some V genes or J segments. Indeed, Vγ9 and JγP are preferentially expressed in peripheral blood T $\gamma\delta^+$ lymphocytes (Triebel et al. 1988a, b) and the corresponding chains

Vγ9-JP-Cγ1 are frequently associated to Vδ2-DJ-Cδ (Sturm et al. 1989). It is noteworthy that the promoter regions of the Vγ9 and Vδ2 genes have several common characteristics : absence of TATA and CAAT boxes, presence of short repeated sequences and a characteristic octanucleotide which could represent a binding site for nuclear factors intervening in the coordinated expression of these genes (Dariavach and Lefranc 1989).
b) the N-diversity which corresponds to the existence of a N region at the V-J junction (Lefranc et al. 1986c). This N-diversity results from the deletion of nucleotides at the extremities of the coding V and J regions by action of an exonuclease and the addition, at random, of nucleotides by the desoxynucleotidyltransferase terminal (dTT) (Alt and Baltimore, 1982). The N regions of the rearranged γ genes can be delimited precisely since the sequences of the germline V genes and J segments are known (reviewed in Huck et al. 1988).

The mechanism of somatic mutations does not seem to exist in the T cell receptor loci and this is an important difference with the immunoglobulin loci. This absence of somatic mutation has clearly been demonstrated in the human γ locus, by the complete identity, at the exception of the N region, of several sequences of rearranged variable genes with those of their germline counterparts (Lefranc et al. 1986c).

REFERENCES

Alt FW, Baltimore D (1982) Joining of immunoglobulin heavy chain gene segments : implication from a chromosome with evidence of three D-JH fusions. Proc Natl Acad Sci USA 79:4118-4122

Bensmana M, Mattei MG, Lefranc MP Localisation of the human T -cell receptor gamma locus at 7p14-p15 by in situ hybridization of the TRG1.4 probe. Cytogenet Cell Genet (in press)

Buresi C, Ghanem N, Huck S, Lefranc, G Lefranc MP (1989) Exon duplication and triplication in the human T-cell receptor gamma (TRG) constant region genes and RFLP in French, Lebanese, Tunisian and Black African populations. Immunogenetics 29: 161-172

Chen Z, Font MP, Bories JC, Degos L, Loiseau P, Lefranc MP, Sigaux F (1988) The human T-cell V gamma gene locus : cloning of new segments and study of V gamma rearrangements in neoplastic T and B cells. Blood 72:776-783

Dariavach P, Lefranc MP (1989) The promoter regions of the T cell Receptor V9 gamma (TRGV9) and V2 delta (TRDV2) genes display short direct repeats but no TATA box. FEBS Lett 256:185-191

Dialynas DP, Murre C, Quertermous T, Boss JM, Leiden JM, Seidman JG, Strominger JL (1986) Cloning and sequence analysis of complementary DNA encoding an aberrantly rearranged human T-cell, chain. Proc. Natl. Acad. Sci. USA 83:2619-2623

Font MP, Chen Z, Bories JC, Duparc N, Loiseau P, Degos L, Cann H, Cohen D, Dausset J, Sigaux F (1988) The Vγ locus of the human T-cell receptor γ gene. Repertoire polymorphism of the first variable gene segment subgroup. J Exp Med 168:1383-1394

Forster A, Huck S, Ghanem N, Lefranc MP, Rabbitts TH (1987) New subgroups in the human T-cell rearranging Vγ gene locus. EMBO J 6:1945-1950

Ghanem N, Buresi C, Moisan JP, Bensmana M, Chuchana P, Huck S, Lefranc G, Lefranc MP (1989) Deletion, insertion, and restriction site polymorphism of the T-cell receptor gamma variable locus in

French, Lebanese, Tunisian, and Black African populations. Immunogenetics 30:350-360
Ghanem N, Soua Z, Zhang XG, Zijun M, Zhiwei Y, Lefranc G, Lefranc MP (1990) Polymorphism of the T-cell receptor gamma variable and constant region genes in a Chinese population. Hum Genet (in press)
Huck S, Lefranc MP (1987) Rearrangements to the JP1, JP and JP2 segments in the human T-cell rearranging gamma gene (TRG) locus. FEBS Lett. 224:291-296
Huck S, Dariavach P, Lefranc MP (1988) Variable region genes in the human T-cell rearranging gamma (TRG) locus : V-J junction and homology with the mouse genes. EMBO J. 7:719-726
Krangel MS, Band H, Hata S, McLean J, Brenner MB (1987) Structurally divergent human T cell receptor γ proteins encoded by distinct Cc genes. Science, 237:64-67
Lefranc MP, Rabbitts TH (1985) Two tandemly organized human genes encoding the T-cell γ constant-region sequences show multiple rearrangements in different T-cell types. Nature 316:464-466
Lefranc MP, Forster A, Rabbitts TH (1986a) Rearrangement of two distinct T-cell γ-chain variable-region genes in human DNA. Nature 319:420-422
Lefranc MP, Forster A, Rabbitts (1986b) Genetic polymorphism and exon changes of the constant regions of the human T-cell rearranging gene γ. Proc Natl Acad Sci USA 83:9596-9600
Lefranc MP, Forster A, Baer R, Stinson MA, Rabbitts TH (1986c) Diversity and rearrangement of the human T-cell rearranging genes : nine germ-line variable genes belonging to two subgroups. Cell 45:237-246
Lefranc MP, Forster A, Rabbitts TH (1987) Organization of the human T-cell rearranging gamma genes (TRG). In : Kappler J, Davis M (eds) The T-cell receptor, vol. 73. AR Liss, New York, pp 25-29
Lefranc MP (1988) The human T-cell rearranging gamma (TRG) genes and the gamma T-cell receptor. Biochimie 70, 901-908
Lefranc MP, Rabbitts TH (1989) The human T-cell receptor gamma (TRG) genes. Trends Bioch Sci 14:214-218
Lefranc MP, Chuchana P, Dariavach P, Nguyen C, Huck S, Brockly F, Jordan B, Lefranc G (1989) Molecular mapping of the human T cell receptor gamma (TRG) genes and linkage of the variable and constant regions. Eur J Immunol 19:989-994
Littman DR, Newton M, Crommie D, Ang SL, Seidman JG, Gettner SN, Weiss A. (1987) Characterization of an expressed CD3-associated Ti γ-chain reveals C domain polymorphism. Nature 326:85-88
Migone N, Casorati G, Francia P, Lusso P, Foa R, Lefranc MP (1988) Non-random TRG gamma variable gene rearrangements in normal human T-cells and T-cell leukaemias. Eur J Immunol 18:173-178
Moisan JP, Bonneville M, Bouyge I, Moreau JF, Soulillou JP, Lefranc MP (1989) Characterization of the T-cell receptor gamma (TRG) gene rearrangements in alloreactive T-cell clones. Human Immunol 24:95-110
Murre C, Waldmann RA, Morton CC, Bongiovanni KF, Waldmann JA, Shows T.B, Seidman JG (1985) Human γ-chain genes are rearranged in leukaemic T-cells and map to the short arm of chromosome 7. Nature 316:549-552
Quertermous T, Strauss WM, van Dongen JJM, Seidman JG (1987), Human T-cell γ-chain joining regions and T-cell development. J Immunol 138:2687-2690

Rabbitts TH, Lefranc MP, Stinson MA, Sims JE, Shroder J, Steinmetz M, Spurr NL, Solomon E, Goodfellow PN (1985) The chromosomal location of T-cell receptor gene and a T-cell rearranging gene : possible correlation with specific translocations in human T-cell leukemia. EMBO J 4:1461-1465
Sturm E, Braakman E, Bontrop R, Chuchana P, van de Griend R, Koning F, Lefranc MP, Bolhuis R (1989) Coordinated V gamma and V delta gene segment rearrangements in human gamma -delta$^+$ lymphocytes. Eur J Immunol 19:1261-1265
Tighe L, Forster A, Clark D, Boylston A, Lavenir I, Rabbitts TH (1987) Unusual forms of T cell γ mRNA in a human T cell leukemia cell line : implications for the γ gene expression. Eur J Immunol 17:1729-1736
Triebel F, Faure F, Graziani M, Jitsukawa S, Lefranc MP, Hercend T (1988a), A unique V-J-C rearranged gene encodes a protein expressed on the majority of CD3$^+$ T-cell receptor α/β-circulating lymphocytes. J Exp Med 167:694-699
Triebel F, Lefranc MP, Hercend T (1988b), Further evidence for a sequentially ordered activation of T-cell rearranging, gamma genes during T lymphocyte differentiation. Eur.J.Immunol. 18: 789-794

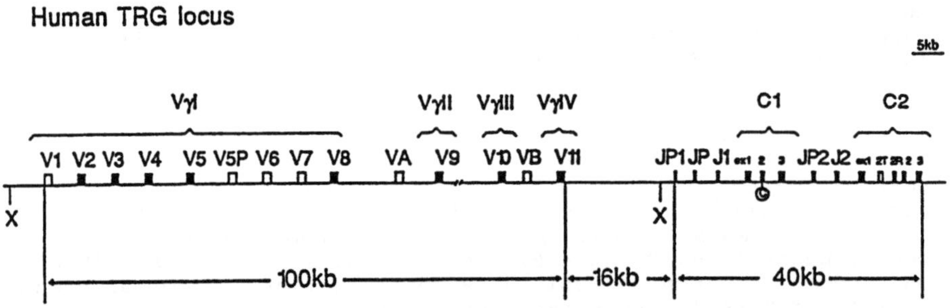

Fig. 1. Organization of the human T cell receptor γ (TRG) locus (from Lefranc et al. 1989) (for review of the human γ locus, see Lefranc and Rabbitts, 1989)

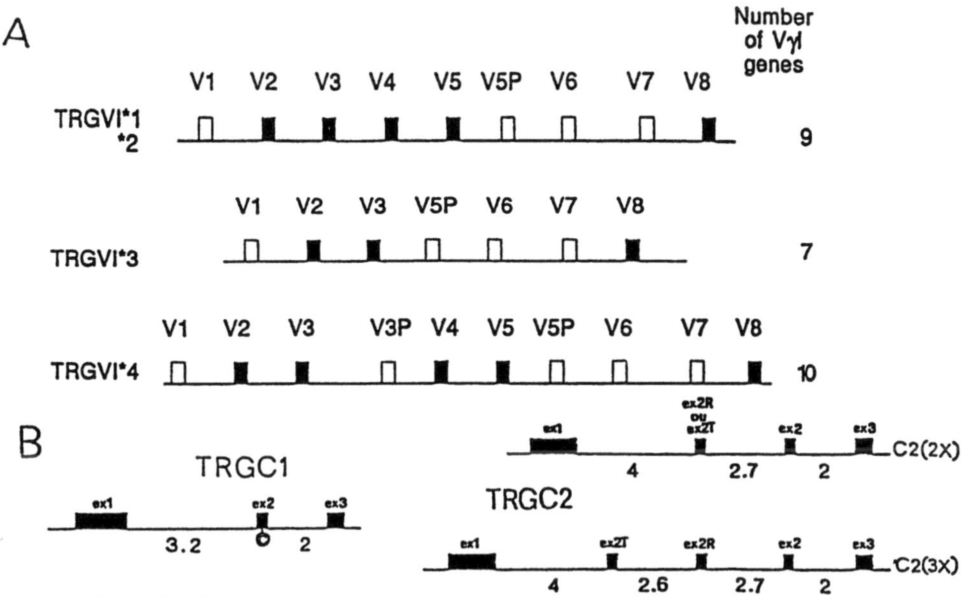

Fig. 2. Allelic polymorphism of the human T cell receptor gamma genes
A - schematic representation of the VγI haplotypes (from Ghanem et al. 1989, 1990). For a detailed map, see Lefranc et al. 1986c and Ghanem et al. 1989.
B - schematic representation of the TRGC genes (from Buresi et al. 1989). Sizes of the introns are indicated in kb. For a detailed map, see Lefranc et al. 1986b and Huck and Lefranc, 1987.

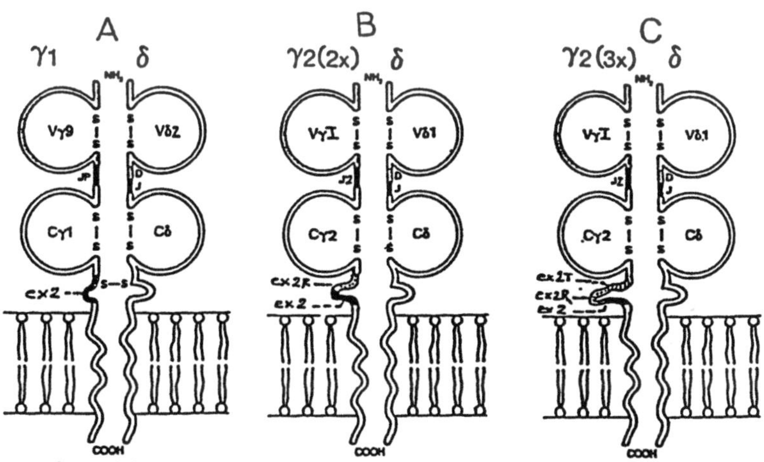

Fig. 3. Schematic representation of the three types of T cell receptors γδ (from Lefranc and Rabbitts, 1989). Depending from the γ chain, there are three types of T cell receptors (A) the γ1-δ receptor, in which the 40 kDa γ1 chain is disulfide-linked to the δ chain (B) the γ2(2x)-δ and γ2(3x)-δ receptors, in which the 40 or 44 kDa γ2(2x) chain and the 55 kDa γ2(3x) chain are characterized by a duplication or triplication of the exon 2, respectively (Buresi et al. 1989) and are non-disulfide linked to the δ chain.

Differentiation *In Vitro*

Development of Human T Cells Expressing TCR γδ *In Vitro*

V. Groh, M. Fabbi[1] and J. L. Strominger

Department of Biochemistry and Molecular Biology, University, Cambridge, MA 02138
[1] Present address: Laboratorio di Immunobiologia, Istituto Nazionale per la Ricerca sul Cancro, Viale Benedetto XV, 16132 Genova, Italy

Human T cell differentiation is believed to occur largely in the thymus. (However, the occurrence of some extrathymic differentiation sites(s), particularly for cells expressing TCRγδ, has been suggested. For example, in adults where the thymus has involuted, T cell development nevertheless continues; one evidence is the reconstitution of the T cell repertoire after bone marrow transplantation.) The following studies represent our initial effort to investigate the role of the human thymus in T cell differentiation *in vitro*, especially in the development of TCRγδ cells.

I. Groh, V., Fabbi, M. and Strominger J.L. Maturation or differentiation of human thymocyte precursors *in vitro*? Proc. Natl. Acad. Sci. USA, 87: 5973-5977 (1990).

Abstract: The differentiation or maturation potential of human thymocyte precursors has been studied by using a population of CD3/TCR$^-$, CD4$^-$, CD8$^-$ ("triple negative") thymocytes isolated by negative selection. This cell population, however, also contained 30-50% previously undescribed cells expressing very low levels of CD3/TCRγδ (CD3/TCRγδ^{low} ; ~60% of which expressed the variable region gene Vδ1). Correspondingly, TCR γ and TCR δ gene rearrangements (predominantly Vδ1/joining region Jδ1) and full-length TCR γ and TCR δ transcripts (but only immature TCR β and no TCR α mRNAs) were found. These cells mobilized Ca^{2+} in response to ligation of CD3 but not following ligation of TCRγδ . When cultured in the presence of interleukin 7 or interleukin 2, these thymocytes gave rise to 30-60% CD3/TCRγδ$^{medium\ and\ high}$ cells (60-70% expressing Vδ1) seen as discrete populations. Thus, the proportion and Vδ phenotype of in vitro generated CD3/TCRγδ cells closely resembled

Current Topics in Microbiology and Immunology, Vol. 173

those of CD3/TCR$\gamma\delta^{low}$ cells in freshly isolated "thymocyte precursor" preparations. Small numbers of TCR$\alpha\beta^+$ cells also appeared. It is thus uncertain whether maturation, differentiation, or both account for the appearance of mature CD3/TCR$^+$ thymocytes, although the former appears most likely.

Thus, this initial study was designed to ask whether previous reports of T cell differentiation, including TCR gene rearrangements, from "triple negative" thymocytes in vitro in the absence of thymic epithelial cells could be reproduced. The conclusion was reached that little or no true differentiation could be observed under these circumstances and that the development of TCR$\gamma\delta$ cells was due to maturation of previously differentiated and incompletely developed precursors.

Several important questions are unanswered:

1. What happened to the similar precursors of TCR$\alpha\beta$ cells which are present in the periphery in at least 10 times the number of TCR$\gamma\delta$ cells? One possibility is that the expression of CD8 or CD4 in T cell development precedes TCR$\alpha\beta$ expression and thus that these putative CD3$^-$, CD4$^-$, CD8$^+$ or CD3$^-$, CD4$^+$, CD8$^-$ precursors of TCR$\alpha\beta$ cells were eliminated by the selection process. The negative selection utilized anti-CD3, anti-CD4 and anti-CD8 mAbs to eliminate cells expressing these molecules.

2. The novel CD3/TCR^{low} cells observed here appear to represent a third subset of cells defined by level of TCR expression, the others being TCR^{medium} (previously termed TCR^{low}) and TCR^{high} cells. Are the different levels of TCR expression correlated with any important physiological event in T cell maturation or differentiation? It is striking that the TCR^{low} cells flux Ca+ in response to anti-CD3 mAb but not to anti-TCR mAb and thus correspond to a thymocyte phenotype which had been observed in the mouse. How are the different levels of TCR expression regulated?

II. Fabbi, M., Groh, V., and Strominger, J. L. IL-7 induces proliferation of CD3$^{-/low}$, CD4$^-$, CD8$^-$ human thymocyte precursors by an IL-2 Independent Pathway, Submitted for publication

Abstract: The proliferation potential of highly purified human $CD3^-CD4^-CD8^-$ (triple negative) and $CD3^{low}CD4^-CD8^-$ thymocyte precursors in response to various cytokines was investigated. High *in vitro* growth ability was observe in response to recombinant interleukin-2 (rIL-2) and rIL-7, both in the absence of any co-mitogen and in combination with phorbol 12-myristate 13 acetate (PMA) . Furthermore, the proliferation of thymocyte precursors in presence of rIL-7, although accompanied by significant increase of IL-2 receptor (IL-2R) p55 expression, appeared independent of that mediated by the autocrine IL-2 pathway, since monoclonal antibodies (mAbs) to IL-2 and IL-2R p55 did not eliminate responsiveness to rIL-7. Synergism of rIL-7 wit rIL-2 was also observed, while no cooperation was detectable with rIL-4 or rIL-6. Analysis of surface phenotype and cell cycle status of cells cultured in presence of rIL-7, both plus and minus PMA, showed that $CD3^-$ as well as $CD3^{low}$ cells readily proliferated to rIL-7. Up regulation of the levels of expression of CD3 antigen was also observed in these cultures. These results, together with the previous characterization of IL-7 as a pre-B cell and matu T cell growth factor, identify IL-7 as a cytokine with biologic activities on a variety of target cells. They also suggest that IL-7, in analogy with the mous system, might play a role in human T cell ontogeny.

This study thus addresses the question of signals which may be mediated either by contact with thymic epithelium through "adhesion" molecules or by lymphokines, such as IL-7, secreted by thymic epithelium. Current efforts in the laboratory are directed toward culturing human thymic epithelial cell lines and attempting to clone the multiple distinct epithelial cells in these lines in order to examine their roles in the proliferation and differentiation of true "triple negative" human thymocytes and in various other maturational steps in T cell development

This research was supported by NIH Research Grant CA 47554.

From the Developmental Expression of γδ T Cell Receptors to the Implications in the Acquisition of Tolerance

C. MARINEZ-A.[1], J. M. ALONSO[1], ALICIA BARCENA[1], P. APARICIO[2], and MARIA L. TORIBIO[1]

[1] Centro de Biologia Molecular, CSIC
[2] Universidad Autonoma de Madrid, Campus de Cantoblanco, 28049 Madrid, Departamento de Bioquimica, Facultad de Medicina, Murcia

Correspondence : C. Martinez-A., Centro de Biologia Molecular, CSIC, Universidad Autonoma de Madrid, Campus de Cantoblanco, 28049 Madrid. Phone: 34.1.3974252 FAX:34.1.397-4799

T cell precursors arising from hematopoietic stem cells colonize the thymus during ontogeny, where they undergo differentiation and maturation events involving genotypic changes that result in the expression of distinct surface molecules. During this process, an ordered rearrangement and expression of the T cell receptor (TCR) genes leads to the acquisition of two distinct types of CD3 associated TCR structures, α,β and γ,δ, which are independently expressed on the surface of distinct intrathymic subpopulations. The first TCR characterized α,β is expressed on the majority of mature thymocytes and peripheral T lymphocytes, and mediates the specific recognition of antigen (peptides) in association with Major Histocompatibility Complex (MHC)-encoded class I and class II molecules. The second TCR, γ,δ is expressed early in thymic ontogeny, and in adulthood its expression is restricted to a small fraction of peripheral blood and thymic T cells. A major difference between the α,β and γ,δ TCR bearing cell subsets resides in the differential distribution of CD4 and CD8 molecules, whose expression appears to be dictated by the TCR specificity.

Developmental studies in mice, support the view that the TCR γ and δ chains are expressed at day 14 of gestation, suggesting that it has a functional role very early in T cell ontogeny, well before the TCR α,β is acquired. Thus, before day 16 of fetal development, when most thymocytes express the CD4-CD8-double negative phenotype, only TCR γ,δ are detected intrathymically. Further maturation events leads to a gradual decrease in TCR γ,δ bearing cells which correlates with the acquisition of CD4 and/or CD8 molecules and with the expression of the TCR α,β genes, resulting in the generation of TCR a,b bearing cells. In humans, the developmetal expression of the two TCR types has been approached by trying to correlate different intrathymic subpopulations with distinct ontogenic stages. Results from these studies indicated that, as described in mice, TCR γ and δ RNA messages are expressed at early ontogenic stages(Davis et al ; Toribio et al 1988) even before the acquisition of the CD2 molecule, while α and β TCR genes remain in germ line configuration. Thereafter, rearrangement and expression of α and β genes lead to the shutdown of the γ and

δ genes resulting in the expansion of the TCR α,β T cells (von Boehmer et al 1988).
The sequence of ontogenic events leading to the acquisition of an MHC-restricted T-cell repertoire remains to be established. It appears clear however, that for the TCR α,β bearing cells, it may be related to the expression of functional TCR, through which interaction with the appropriate MHC restricting elements takes place and results in both positive and negative selection. Interestingly enough, these interactions appear to occur at the double positive (CD4+CD8+) intrathymic stage (Blackman et al 1990: von Boehmer 1990). Much less is known for the γ,δ T cells which appear to differentiate independently of the double positive intermediate stage, throughout different stages of development not yet characterized.

Impinge into the analysis of the TCR γ,δ repertoire.

In mice, various γ,δ T cell subsets are generated within the thymus and in the periphery at different times during ontogeny. Thus, by day 13 of gestation, two homogeneous subsets expressing Vγ5 and Vγ6 genes and associated with Vδ1 are consecutively generated in the thymus, defining two waves of development of γ,δ T cells (Itohara et al 1989). Both subsets have been characterized by the homogeneity of the junctional sequences (canonical sequences) in the productively rearranged genes, while the corresponding nonproductive rearrangements are far more diverse. Because of the homogeneity of their TCR, the two fetal γ,δ T cell subsets are likely to recognize self antigens which are expressed in the fetal thymus and, therefore, it has been suggested that the accumulation of the canonical sequences results from positive thymic selection. In agreement, the Vγ5 gene which is also expressed in the T cells present in the epidermal layer of the skin (S-IEL) shows low diversity and appears to recognize self antigens expressed by keratinocytes or by fibroblast which have been treated with tryptic digest of keratynocytes (Haas et al 1990).
Two futher murinė γ,δ T cell subsets which are generated extrathymically in some, but not all all mouse strains, are characterized by the expression of Vδ5 and Vγ4 genes that are highly homogeneous and demostrate characteristics junctional N-region sequences, respectively. These peripheral γ,δ T cells most likely result as a consequence of the positive selection imposed upon interaction with a still unknown self antigen.
In adult mice, γ,δ T cells represent a very small percentage of the total T lymphocyte subset. In contrast to embryonic γ,δ T cells, they are highly diversified, although Vγ4 and Vγ5 genes are used more frequently than other V gene segments.
In humans, two major γ,δ T cell subsets have been identified expressing either Vδ1(recognized by the mAb dTCS1) or Vδ2 (recognized by the mAb BB3), the latter being frequently associated with the Vγ9Cγ1 chain. Interestingly enough, cells expressing the Vδ1 gene comprise the predominant γ,δ T cell subset in the thymus, as well as during the fetal life, while the Vδ2 subset is predominant in the periphery within the first month of life and thereafter (Parker et al 1990; Krangel et al 1990). The predominant Vδ2 expression in the periphery is thought

to result from the selective expansion of the Vδ2 bearing cells in response to antigenic stimuli in the periphery, which include self antigens (see below).

Mechanisms involved in the acquisition of self tolerance by the g,d T cells.

It is commonly believed that CD4-CD8- lymphoblasts which have entered the thymus develop into CD4+CD8+ that express the α,β TCR. The binding of a certain TCR expressed on this cell population to thymic MHC class I molecules should result in the generation of CD4-CD8+ T cells, while, alternatively the binding to MHC class II antigens would yield CD4+CD8- lymphocytes. In addition to this process of positive selection, mechanisms ensuring the acqusition of self tolerance appear to take place also at the double positive stage.

There are basically three mechanisms for α,β T cells to avoid autoimmunity associated with self recognition. First, T cells bearing receptors that recognize self antigens could be physically eliminated as they develop in the thymus (clonal deletion). Second, self reactive T cells that do mature and enter the periphery could be functionally inactivated by other T cells (immunosuppression). Third, when self reactive T cells encounter the antigen they recognize, they could be functionally inactivated to develop an immune response (anergy) (Mueller et al 1989). While the first mechanism applied clearly for the thymus, there is evidence for the existence of functionally inactivated T cells in vivo (Rammensee et al. 1989), which account for peripheral antigen-specific tolerance. TCR α,β T cells specific for an antigen recognized in the context of I-E and expressing a given Vβ gene segment have been found within both immature CD4+CD8+ double positive and mature CD4+/CD8+ single positive thymocyte subsets in I-E-negative mouse strains; while they are only found in the immature CD4+CD8+ subpopulation in the I-E bearing strains. These data have led to the notion that acquisition of self tolerance occurs as a consequence of thymic elimination (clonal deletion) of T cells bearing particular Vβ genes specific for a particular self-MHC antigenic combination. Using transgenic mice for TCR α.β specific for male H-Y antigen, in association with the b allele of the class I MHC molecules consequently only expressed by CD8+ mature T cells, was found that more than 50% of peripheral T cells from mice carrying the antigen and the appropriate MHC class I molecule expressed high levels of the transgenic TCR, while they lacked expression of CD4 and/or CD8 molecules; therefore displaying the CD4-CD8-double negative phenotype. Interestingly enough, CD8+ cells expressing the transgenic TCR were also identified in those animals, although the level of expression corresponded to one-tenth of the level of CD8 found in normal peripheral T cells (von Boehmer 1990). These results, in agreement with the clonal deletion model, extend the role of the TCR α,β to the CD4 and CD8 molecules in the acquisition of self tolerance.

Developmental analyses of transgenic TCR α,β expression also revealed that double positive thymocytes, the target cells where clonal deletion takes places in the thymus, developed in both female and male embryos even though the presence of the male antigen presented by H-2D^{b} MHC molecules could already be detected by day 16 of gestation. Because the earliest T cell precursor expressing CD8 co-receptors are deleted in adult male mice, the possibility was suggested that the deletion process required expression of the ligand on a cell which develops relatively late in the thymus. Simultaneously, with the decrease in double positive thymocytes in male mice, there is a selective increase in either CD8+ cells from day 17 to birth or double negative T cells, which thereafter account for the majority of T cells expressing the transgenic TCR in the adult male. One can imagine that at a very immature stage of development (before birth) the interaction of the TCR with the antigen-MHC ligand expressed on thymic epithelium results in positive selection of self reactive T cells, whereas the phenotypic changes (clonal deletion and decreased of CD8 level expression) occur relatively soon after this inducing event.

This can also occur with the γ,δ T cells. Thus, the analysis of the expression of CD4 and/or CD8 on γ,δ T cells has revealed drastic differences between fetal and adult γ,δ T cells. At the time of birth, over 60% of splenic γ,δ T cells are CD8+, and only 30% and 5% are double negative and CD4+, respectively. In adults however,γ,δ T cells represent 15% and 80% of CD8 and double negative cells respectively (Bonneville et al 1990). Also of interest is the finding that CD4 and CD8 molecules are expressed in adult γ,δ T cells at relatively low levels, in contrast to the high level of expression detected on fetal γ,δ and α,β T cells. Importantly, very few, if any double positive γ,δ T cells have been identified to date.

The evidence for the existence of mechanisms implicated in the acquisition of clonal anergy comes from experiments conducted by R. Schwartz et al (Mueller et al 1989). These experiments demostrated that when mature α,β T cells encounter the antigen-MHC complex on inappropriate antigen-presenting cells, a profound state of antigen-specific unresponsiveness is induced. Studies performed in transgenic mice for γ and δ genes encoding a TCR specific for a gene of the TL-region of the TL^b haplotype demostrated that γ,δ thymocytes from TL^b strains differ from thymocytes from TL^d mice in cell size, TCR density, and capacity to respond to stimulator cells. Particularly, γ,δ T cells from TL^b transgenic mice did not produce IL 2 and did not proliferate in response to appropriate stimulator cells, but they did proliferate in the presence of exogenous IL 2 (Bonneville et al. 1990).

The characterization of human γ,δ T cell clones derived from adult healthy voluntaries revealed basically the same conclusions: 1) Human γ,δ T cell clones displaying the CD4-CD8- phenotype specifically respond to EBV transformed syngeneic B cells. The mitogenic and/or antigenic stimulation of these cells required the presence of exogenous IL 2, although they are perfectly able to express the IL 2R; 2) Maximal proliferation of γ,δ T cells after TCR crosslinking induced by anti-CD3 antibodies required higher ligand concentrations than the α,β T cell clones with the same specificity; and 3) These γ,δ T cells are unable to produce IL 2 upon antigenic stimulation.

Given the lack of intermediate stages of development where clonal deletion takes place in the pathway of differentiation of γ,δ T cells, we can postulate that one of the ways they achieve self tolerance involves the decrease or down regulation of the CD4 and/or CD8 co-receptor molecules. Furthermore, since the majority of γ,δ T cells in the adult are either CD8+or double negative, we speculate that the repertoire of the γ,δ T cells is biased toward the recognition of self MHC class I or class I related antigens. Alternatively, superimposing with such a mechanism anergy could act as a safety-valve, playing a critical role in the acquisition of the functional γ,δ T cell repertoire (Aparicio et al 1989).

Characterization of fetal and adult human γ,δ T cells.

Recently, we have identified a T cell subpopulation defined as CD3+TCR+ which lack the cell surface CD2 antigen and comprise the majority of γ,δ T cells in the fetal liver and fetal spleen; while γ,δ T cells expressing CD2 molecules were detected in the adult thymus (Aparicio et al 1989).

Striking differences were also found.between fetal and adult human γ,δ T cells regarding their capacity to produce lymphokines. Thus, fetal clones, in contrast to clones derived from adults, produced IL 2 and IL 4 upon activation with either lectins, anti-CD3 monoclonal antibodies or mitogenic mixtures of anti-CD2 mAbs, while both types of clones indeed produced IFN-γ. Analysis of the helper activity displayed by γ,δ T cell clones in a conventional antibody production assay also revealed the differential behaviour of fetal versus adult γ,δ T cell clones. Such activity appears to be associated with the expression of the CD4 co-receptor by the corresponding clones.

Role of interleukins in g,d T cell development.

Accumulating evidence indicates that IL 4 is involved in T cell development (Barcena et al. 1990a; Martinez-A, 1990a; Martinez-A. et al 1990b). Thus, IL 4 appears to induce both growth and differentiation of fetal intrathymic Thy 1-CD4-CD8- T cell precursors into CD8+ cytolytic T cells. Also, adult double negative mouse thymocytes proliferate in response to IL 4. Similarly, human thymic pro T cells, known to display functional properties similar to those of the cells that first colonize the thymus during fetal development, differentiate in the presence of IL 4 into mature functional γ,δ T cells (Barcena et al. 1990b; Plum et al 1990). Moreover, in agreement with the results obtained with fetal γ,δ thymocytes, neonatal pro T cells also produce constitutively IL 4, suggesting the involvement of this lymphokine in the development of γ,δ T cells. As shown by Carding et al,1989, as in mice our results in humans indicate that the production of lymphokines and the lymphokine receptor expression is tightly controlled during the process of development. In particular maximal IL 4 production in both mice and humans correlates with the first wave of γ,δ T cell differentiation. These differences may define the different capacities of fetal and adult γ,δ T cells in the ability to both produce and respond to IL 4. Therefore, in contrast to what was observed in postnatal γ,δ thymocytes, fetal thymocytes produce and respond to IL 4.(see above).

The possible implication of an overexpression of IL 4 during the T cell development process has been recently tested by producing transgenic mice for IL 4 (Tepper et al. 1990). The conclusions obtained can be summarized as follows: 1) The fully active form of the transgene expressed under the control of immunoglobulin regulatory sequences was lethal in newborn mice; 2) By reducing the activity of the promoter/enhancer, the overexpression of the transgene in the T cell compartment results in a severe reduction in the number of CD4+CD8+ thymocytes, a marked expansion of mature thymocytes bearing the CD8 marker, a marked expansion of double negative (CD4-CD8-) thymocytes, and a reduction of peripheral T cells; 3) The T cell defect results from the expression of IL 4 by a bone marrow derived cells, likely comprising the T cell precursors themselves. Interestingly, the phenotype of the thymus subpopulations characterized in IL 4 transgenic mice closely resemble the phenotype of the mature T cells generated " in vitro "from human pro T cells after culture with IL4.

Concluding remarks

Fetal hematopoietic precursors are capable to constitutively produce IL 4 simultaneously expressing IL 4R. before migration to the embryonic thymus. Upon migration to the thymus, T cell precursors interact with epithelial cells promoting γ and δ gene rearrangement and expression of γ,δ TCR very early in development. Based on the major existence of CD8+ γ,δ T cells, we assume class I MHC related antigens as their natural ligands. Antigenic recognition promotes positive selection of this cell population, suffering thereafter the consequences of the "learning process" that occurs in the thymus, where negative selection takes place. This results in the decreased expression or loss of the CD8 co-receptor and therefore most cells will remain double negative. Also, T cell precursors, upon interaction with thymic epithelial cells, promote the activation of the IL 2 pathway implicated in the generation of the α,β T cell population. The production of IL 2 in the thymus dominates the phenotype of the IL4, and therefore the generation of the a,b T cells is assured.

Acknowledgements

We like to thank Drs. J.E. Alex Martinez and G. Kroemer for critical reading of the manuscript, E. Leonardo and J.A. Gonzalo for their excellent technical assistance,. and K. Sweeting for secretarial assistance. This work was partially supported by grants from CICyT, EC and FIS.

REFERENCES

Aparicio, P., J. M. Alonso, M. L. Toribio, J. C. Gutierrez, L. Pezzi and C. Martinez-A. (1989). "Differential growth requirements and effector function of α,β and γ,δ human T cells." Immunol. Rev. **111**: 5-33.

Aparicio, P., J. M. Alonso, M. L. Toribio, M. A. R. Marcos and C. and Martinez-A. (1989). "Isolation and characterization of γ,δ CD4+ T cell clones derived from human fetal liver cells." J. Exp. Med. :

Barcena, A., M. J. Sanchez, J. L. de la Pompa, M. L. Toribio and C. and Martinez-A. (1990). "Involvement of the Interleukin 4 pathway in the generation of functional γ,δ T cells from human Pro-T cells." EMBO J. **in press.**:

Barcena, A., M. L. Toribio, L. Pezzi and C. and Martinez-A. (1990). "A role for interleukin 4 in the differentiation of TCR-γ,δ+ cells from human intrathymic T cell precursors." J. Exp. Med. :

Blackman, M., J. Kappler and P. Marrack. (1990). "The role of the T cell receptor in positive and negative selection of developping T cells." Science. **248**: 1335.

Bonneville, M., I. Ishida, S. Ithoara, S. Verbeeck, A. Berns, O. Kanagawa, W. Haas and S. Tonegawa. (1990). "Self tolerance to transgenic γ,δ T cells by intrathymic inactivation." Nature. **344**: 163-165.

Carding, S. R., E. J. Jenkinson, R. Kingston, A. C. Hayday, K. Bottomly and J. J. T. and Owen. (1989). "Developmental control of lymphokine gene expression in fetal thymocytes during T cell ontogeny." Proc. Natl. Acad. Sci. USA. **86**: 3642.

Haas, W., S. Kaufman and C. and Martinez-A. (1990). "Development and function of γ,δ T cells." Immunol. Today. :

Itohara, S., N. Nakanishi, O. Kanagawa, R. Kubo and S. Tonegawa. (1989). "Monoclonal antibodies specific to native murine T cell receptor g,d: Analysis of γ,δ T cells during ontogeny and in peripheral lymphoid organs." Procc. Natl. Acad. Sci. USA. **86.**: 5094-5098.

Krangel, M. S., H. Yssel, B. C. and H. and Spits. (1990). "A distinct wave of human γ,δ lymphocytes in the early fetal thymus: Evidence for controlled gene rearrangement and cytokine production." J.Exp. Med. :

Martinez-A., C. (1990). "Interleukins and T cell development." Research. in Immunol. **141**: 263-312.

Martinez-A., C., J. C. Gutierrez-Ramos, J. L. de la Pompa, M. J. Sanchez, J. M. Alonso and M. L. and Toribio. (1990). "The thousand and one ways of being a T cell." Thymus. **in press**:

Mueller, D. L., M. K. Jenjins and R. J. Schwarzrtz. (1989). "Clonal expansion vs functional clonal inactivation." Ann. Rev. Immunol. **7**: 445.

Plum, J., M. De Smedt, G. Leclerq and B. and Tison. (1990). "Inhibitory effect of murine recombinant IL 4 on thymocyte development in fetal organ cultures." J. Immunol. **145**: 1066.

Tepper, R. I., D. A. Levinson, B. Z. Stanger, J. Campos-Torres, A. K. Abbas and P. and Leder. (1990). "IL 4 induces allergic like inflammatory disease and alters T cell development in transgenic mice." Cell. **62**: 457.

Toribio, M. L., J. M. Alonso, A. Barcena, J. C. Gutierrez, A. de la Hera, M. A. R. Marcos, C. Marquez and C. Martinez-A. (1988). "Human T cell precursors: Involvement of the IL 2 pathway in the generation of mature T cells." Immunol. Rev. **104**: 29-55.

von Boehmer, H. (1990). "Developmental biology of T cell receptor in T cell receptor transgenic mice." Ann. Rev. of Immunol. **8**: 531-556.

Culture Conditions Dictate Whether Mouse Fetal Thymus Lobes Generate *Predominantly* γ/δ or α/β T Cells

CATHRIN SCHLEUSSNER, AMANDA FISHER and RHODRI CEREDIG

Institut de Chimie Biologique, Faculté de Medicine, 67085 Strasbourg, Cedex, France

SUMMARY

Thymus lobes from 14 day-old mouse embryos cultured submerged in r-IL-2 generated a mixture of CD8α+/CD4- and CD8-/CD4- γδTcR expressing cells (Ceredig et. al. 1989). Based upon Northern analysis with TcR constant region probes, no αβ T cells could be identified in these cultures. Submerged lobes also showed responsiveness to IL-7. In contrast, when cultured at an air liquid interface as organ cultures (OC), most cells appeared to express αβTcR (Ceredig 1988). Thus depending on the mode of culture, fetal thymus lobes generate predominantly γδ or αβ T cells; it is unclear how this difference is regulated. Previous phenotypic and functional experiments suggested that γδ T cells may be present in OC. In order to study γδ T cells in both submerged lobe and OC, we have carried out three colour flow microfluorimetric analysis of γδTcR, abTcR, CD3, J11d and CD8β expression by subpopulations of CD8α and CD4 defined thymocytes. In addition, using Vγ-specific oligonucleotides and the polymerase chain reaction, we have begun identifying and sequencing the Vγ repertoire of γδT cells in these mouse fetal thymus cultures.

RESULTS AND DISCUSSION

Recently, monoclonal antibodies to mouse γδTcR have become available and using such reagents, we have begun to analyze by three colour flow microfluorimetry (FMF), the distribution of γδ T cells among the four CD4/CD8 defined subpopulations of thymocytes. These experiments have been carried out using freshly-isolated fetal thymocytes, OC of 14-day fetal thymus lobes and fetal thymocytes grown in liquid culture as submerged lobes (so-called FT cultures).

In the developing thymus *in vivo*, three colour FMF showed that at day 15, γδ T cells were found in all four CD4/CD8 defined thymocyte subpopulations. The values of γδ-expressing cells in each subset were 5% double negative (DN), 8% double positive (DP), 10% CD8 single positive (SP), and 27% CD4 SP. From day 15 onwards, expression of γδ among DP cells decreased progressively, reaching 1-2% of DP cells at birth. Thus in contrast to a recent report (Itohara et. al. 1989), a major subpopulation of DP cells expressing γδTcR was not detected. Similarly, γδTcR expression by CD4 SP cells also decreased with time, reaching values similar to that of DP cells around birth. It should be noted that the CD4 fluorescence intensity of these CD4 "SP" cells was lower than the equivalent αβTcR-expressing CD4 SP cells

Current Topics in Microbiology and Immunology, Vol. 173

seen at birth. In contrast, $\gamma\delta$TcR expression by CD8 SP cells initially increased from day 15, reaching 22% at day 18 but then decreased to 3-4% in the newborn and adult thymus. Expression of $\gamma\delta$TcR continued to increase among DN thymocytes, reaching 15% at birth and then continued to increase to 20% in the adult thymus. The above results were obtained with the pan-$\gamma\delta$TcR mAb GL3 (Goodman and Lefrancois 1989) and similar results, although with correspondingly decreased percentages were obtained with the Vγ3 mAb 536 (Havran and Allison 1988).

The above three colour FMF analysis was also carried out with mAb to $\alpha\beta$TcR, CD2, CD3, J11d and CD8β. In contrast to $\gamma\delta$TcR staining, a large proportion (40% in the adult) of DP thymocytes expressed $\alpha\beta$TcR; expression of $\alpha\beta$TcR among DN cells increased gradually throughout development, reaching 45% in the adult. At all times tested, DN thymocytes stained in a biphasic manner with anti-CD2 mAb indicating as described elsewhere (Ceredig 1990) that immature (J11d+, IL-2R+) DN thymocytes are $CD2^-$. For J11d, late fetal and early neonatal SP thymocytes were essentially all $J11d^+$; the $J11d^-$ subsets of these cells appeared postnatally (Ceredig 1990). Interestingly, in view of the distinct subset of $\gamma\delta$ cells staining positively with the anti-CD8α mAb 53.6.7, a distinct subpopulation of CD8α+/CD8β- thymocytes were seen in the fetal and neonatal thymus. Such cells have been described previously in the fetal mouse thymus (Habu and Okumura 1984) and suggests that, as in the periphery, $\gamma\delta$T cells that express CD8 can have the $CD8\alpha^+/CD8\beta^-$ phenotype.

We have carried out similar three colour FMF analysis of 14 day fetal thymus lobes maintained in OC. As far as $\gamma\delta$TcR expression is concerned, results were essentially as described above for the thymus *in vivo*. However, expression of $\gamma\delta$TcR by CD4 weak SP thymocytes was not seen, presumably because the earliest OC analysed were three days after setting up, or equivalent to the day 17 thymus *in vivo*. Importantly, no major subpopulation of $\gamma\delta$-expressing DP thymocytes was seen in OC but as in the normal thymus, both DN and CD8 SP subpopulations expressed $\gamma\delta$TcR with % values slightly above those seen *in vivo*. Interestingly, the proportion of $\gamma\delta$T cells expressing Vγ3 decreased with time in culture from 40-50% of $\gamma\delta$ T cells at day 7 to undetectable at day 20. The reason for the disappearance of Vγ3-expressing cells in OC is unclear. In addition to $\gamma\delta$TcR expression, $\alpha\beta$TcR expression by OC thymocytes is also being carried out by three colour FMF and a panel of anti-Vβ TcR mAb.

When 14-day mouse fetal thymus lobes are cultured as FT in IL-2, DN and $CD8\alpha^+/CD8\beta^-$ SP cells are generated both of which express CD3 by FMF and as judged by Northern analysis of total RNA, $\gamma\delta$TcR. Recently, we have been able to show that IL-7 added to FT cultures also results in the growth of $\gamma\delta$T cells. Three colour FMF analysis of both IL-2 and IL-7 supplemented FT cultures shows that both IL-2 and IL-7 promote the outgrowth of $\gamma\delta$T cells from the unmanipulated fetal thymus lobe. IL-2 supplemented cultures contain 49% $\gamma\delta$TcR expressing cells, whereas the value for those containing IL-7 is 16%. In both FT cultures, 75% of $\gamma\delta$-expressing cells are $V\gamma3^+$. Such $\gamma\delta^+$ and $V\gamma3^+$ cells are found among both $CD8\alpha^+$ and DN subpopulations. In IL-2 supplemented cultures, 82% are Thy-1^+ and 62% $CD3^+$ whereas in IL-7 cultures, the values are 94% and 23% respectively. Thus, in IL-7-supplemented cultures, a greater proportion of cells have the Thy-1^+, $CD3^-$

phenotype. Interestingly, 76% of cells grown in IL-7 express the 55kD chain of the IL-2R compared with only 38% in IL-2-supplemented medium. Essentially no cells in these cultures expressed either CD4 or αβTcR. Taken together, these results imply that IL-7 may be maintaining the survival and promoting the growth of more immature fetal thymocytes.

In order to characterise further the γδT cells in these various culture systems, we have begun to clone and sequence the γδTcR expressed by fetal mouse thymocytes. To this end, we have used the polymerase chain reaction (PCR) and pairs of Vγ and Cγ-specific oligonucleotides containing additional 3' and 5' cloning restriction sites to amplify cDNA from populations of freshly isolated, OC and FT-cultured thymus lobes. These PCR-amplified fragments are then cloned and sequenced. These experiments are being carried out firstly to identify which particular γδT cell subsets grow in the different cultures and then to determine their TcR sequence. As these subpopulations of γδT cells are isolated directly from the thymus, it will be of interest to determine their TcR sequence repertoire. Initial experiments have already confirmed the recently-reported data of Carding et al 1990, namely that using PCR, essentially all Vγ genes can be shown to be transcribed in the fresh, 14-day fetal thymus.

Recently, we have perfected a method whereby the cell cycle status of lymphocyte subpopulations defined by two colour (FITC and PE) immunofluorescence can be evaluated. In this system, 7-aminoactinomycin-D (7AAD), a fluorochrome excited at 488nm is used to quantitate the DNA content of FITC and PE-stained cells. Using this single laser, three colour FMF system, we have shown that $\gamma\delta^+$ and $V\gamma3^+$ T cells in the day 16 fetal thymus contain 30-45% of cells in S, G2 + M phases of the cell cycle. Thus, the fetal thymus is producing γδT cells which are actively dividing at the time of their initial generation. In the neonatal thymus, both γδ and Vβ expressing CD4 SP cells were also found to be in cycle (Ceredig 1990).

Thus, in the fetal and early postnatal thymus, both γδ and αβ expressing cells are produced as cycling cells. These results have several implications. Firstly, they may explain the ease with which fetal and neonatal thymocytes form hybridomas when fused with thymoma fusion partners; since it is known that fusion occurs more efficiently when both normal and thymoma cells are actively dividing. Secondly, if CD8α expression by γδ bearing T cells is an indication of proliferation, it is not surprising that CD8α is expressed by developing fetal γδ thymocytes _in situ_. Thirdly, intrathymic proliferation could partly explain the rapid increase in the frequency of immunocompetent cells seen in the perinatal thymus (Ceredig et al 1983). Fourthly, cell proliferation might suggest some degree of oligoclonality in the receptor repertoire of developing T cells. Sequence analysis of fetal γδ TcR has already suggested some degree of oligoclonality in both V region usage and in TcR sequence (Reviwed by Haas et al 1990); similar predictions could be made concerning neonatal αβ sequences. Finally, cell culture experiments (Ceredig and Waltzinger 1990) have shown that unstimulated neonatal, but not adult αβ-expressing CD4 SP thymocytes can continue proliferating in the presence of IL-2 and IL-7. This result means that differences must exist between the fetal and adult thymus in the generation of mature thymocytes. This could reflect either differences between the thymic microenvironment in fetal and adult mice, or to differences in the phenotypic properties of the mature thymocytes themselves.

REFERENCES

Carding SR, Kyes S, Jenkinson EJ, Kingston R, Bottomly K, Owen JJT, Hayday AC (1990) Developmentally regulated fetal thymic and extrathymic T-cell receptor γδ gene expression. Genes and Development 4: 1304-1315.

Ceredig Rh, Dialynas DP, Fitch FW, MacDonald HR (1983) Precursors of T cell growth factor producing cells in the thymus. J Exp Med 158:1654-1671.

Ceredig Rh (1988) Differentiation potential of 14-day fetal mouse thymocytes in organ culture. J Immunol 141: 355-362).

Ceredig Rh, Medveczky J, Skulimowski A (1989) Mouse fetal thymus lobes cultured in IL-2 generate $CD3^+$, TcR γδ expressing $CD4^-/CD8^+$ and $CD4^-/CD8^-$ cells. J Immunol 142: 3353-3360.

Ceredig Rh (1990) Intrathymic proliferation of perinatal mouse αβ and γδ T-cell receptor-expressing mature T-cells. Int Immunol (in press).

Ceredig Rh, Waltzinger C (1990) Neonatal mouse $CD4^+$ mature thymocytes show responsiveness to interleukin 2 and interleukin 7. Int Immunol (in press).

Goodman T, Lefrancois L (1989) Intraepithelial lymphocytes: Anatomical site, not T cell receptor form, dictates phenotype and function. J Exp Med 170: 1569-1581.

Haas W, Kaufman S, Martinez-A C (1990) The development and function of γδ T cells. Immunol Today 11: 340-343.

Habu S, Okumura K (1984) Cell surface antigen marking the stages of murine T cell ontogeny and its functional subsets. Immunol Rev 82:120-142.

Havran WL, Allison JP (1988) Developmentally ordered appearance of thymocytes expressing different T-cell receptors. Nature 335: 443-445

Itohara S, Nakanishi S, Kanagawa O, Kubo R, Tonegawa S (1989) Monoclonal antibodies specific to native murine T cell receptor gamma/delta: analysis of gamma/delta-T cells in thymic ontogeny and peripheral lymphatic organs. Proc Nat Acad Sci 86: 5094-5099.

Preferential Development of γ/δ- Bearing T Cells in Athymic Nude Bone Marrow

R. G. Miller[1], Patricia Benveniste[1] and J. Reimann[2]

[1] Ontario Cancer Institute and Dept. of Immunology, University of Toronto, 500 Sherbourne Street, Toronto, M4X 1K9, Canada
[2] Dept. of Medical Microbiology and Immunology, Ulm University, Ulm, FRG

Pluripotent stem cells present in bone marrow are capable of migrating to the thymus where they differentiate into progeny that populate the peripheral lymphoid tissues with T cells (Abramson et al., 1977; Keller et al, 1985; Dick et al., 1985). Whether and to what extent stem cells become committed to T cells outside the thymus remains controversial. The strongest evidence that this can occur is provided by the fact that functional T cells can develop (albeit very slowly) in the complete absence of a thymus in nude mice (Hünig and Bevan, 1980; Lake et al., 1980; Maleckar and Sherman, 1987). In vivo reconstitution experiments have shown that nude bone marrow contains pluripotent stem cells (Kindred, 1979). Peripheral lymphoid tissues of old nude mice (4 to 5 mo) have been reported to contain T cells by cell-surface phenotype (MacDonald et al., 1986), T cell receptor gene mRNA expression (Yoshikai et al., 1986; MacDonald et al., 1987), and functional criteria (Hünig and Bevan, 1980; Miller et al., 1983; Maleckar and Sherman, 1987). However, peripheral lymphoid tissues of young nude mice (< 11 wk) would appear deficient in T cells (MacDonald et al., 1981; Miller et al., 1983). Whether the environment of the athymic nude mouse is more efficient in allowing the development of α/β or γ/δ T cell receptor bearing cells is unknown although one study (Yoshikai et al., 1986) suggests the latter.

In the study summarized below (Benveniste et al., 1990) we have searched for evidence of T cell differentiation occurring in the bone marrow of young (5-8 wk old) athymic nude mice. Age matched normal mice have been included for controls. Particular attention has been paid to relative numbers of α/β vs γ/δ T cell receptor bearing cells.

Since a cell-surface expressed T cell receptor is always associated with CD3, we first screened BM from young (5-8 wk old) C57BL/6 (B6) nude and normal mice for $CD3^+$ cells. A population

Current Topics in Microbiology and Immunology, Vol. 173

of CD3 dim ($CD3^d$) cells was present in both normal and nude BM and absent in the periphery of nudes. It was also absent in scid BM. It could be greatly enriched (to 8-9% of cells present) by depleting the BM of nylon wool adherent cells and of high density cells (using a BSA gradient), this latter procedure getting rid of any mature T cells which might be recirculating through normal BM. These $CD3^d$ cells were $CD4^-$ $CD8^-$ in both normal and nude BM. In normals, they were almost all (> 95%) α/β but in nudes they were at least 40% γ/δ-bearing. They were 40% and at least 75% respectively $J11d^+$ in normal and nude BM. Note that the J11d marker is present on 85-90% of normal BM cells and 95% of thymocytes but absent on peripheral T cells (Bruce et al., 1981; Crispe and Bevan, 1987).

We consider it unlikely that $CD3^d$ cells are part of the recirculating pool for the following reasons:

(i) $CD3^d$ cells had comparable frequency in nude and normal BM. Nude mice are markedly deficient in mature T cells with none being detectable in mice of the age used here. However, stem cells are normal and there is no reason to expect abnormalities in processes not requiring the thymus or thymus hormones.

(ii) A large fraction of $CD3^d$ cells express J11d, absent on mature recirculating T cells.

(iii) $CD3^d$ cells were not detected in the periphery of the young (5-8 wk old) nude mice analyzed here.

$CD3^d$ cells could be grown by culturing them with mitomycin-treated anti-CD3 hybridoma cells (10^3/well) in the presence of IL-2 (10 units/well). IL-4 enhanced growth but was not required. Growth frequency was 1/650 to 1/3500 BM cells cultured. After 10-12 days of growth, cells in growing cultures were essentially all $CD3^+$ $CD4^-$ $CD8^+$ and displayed a spontaneous LAK-like cytotoxic activity.

Growing wells were tested for the presence of TcR-α or TcR-δ mRNA transcripts using a sensitive dot blot procedure (Benveniste et al., 1988). These two mRNA's were chosen as most accurately reflecting the presence or absence of a TcR-α/β or TcR-γ/δ; β and γ messages were both often found in cell lines expressing only an α/β or a γ/δ TcR. In nude cultures, 76% of growing wells expressed an α message and 54% a δ message. In normals, 85% expressed an α message and only 3% a δ message, figures which reflect fairly accurately the relative numbers of α/β and γ/δ-TcR seen by flow cytometry.

An interesting question is whether a single growing clone can develop both α/β and γ/δ-TcR bearing cells. To address this, we analyzed cultures set up at extreme limiting dilution. The results were consistent with all precursors being already committed to either α/β or γ/δ.

We conclude that both α/β and γ/δ-TcR bearing cells can develop directly in BM. The frequency of α/β-bearing cells is comparable in nude and normal but the frequency of γ/δ-bearing cells that develop in nude BM is markedly higher than in normal BM. Why is this? In normal mice, γ/δ cells develop before α/β cells. Thus the higher frequency of γ/δ in nude could reflect an early stage in normal T cell development. Alternatively, the athymic nude environment could be more favorable to the development of γ/δ cells via an extrathymic pathway.

References

Abramson S, Miller RG, Phillips RA (1977) The identification in adult bone marrow of pluripotent and restricted stem cells of the myeloid and lymphoid systems. J Exp Med 145:1567

Benveniste P et al. (1988) A sensitive dot blot procedure for detecting mRNA in lymphoid cells grown in liquid culture. J Immunol Meth 107:165

Benveniste P et al. (1990) Characterization of cells with T cell markers in athymic nude bone marrow and of their vitro-derived clonal progeny. J Immunol 144:411

Bruce J et al. (1981) A monoclonal antibody discriminating between subsets of T and B cells. J Immunol 127:2496

Crispe N, Bevan MJ (1987) Expression and functional significance of the J11d marker on mouse thymocytes. J Immunol 138:2013

Dick JE et al. (1985) Introduction of a selectable gene into primitive stem cells capable of long term reconstitution of the hemopoietic system of W/W^v mice. Cell 42:71

Hünig T, Bevan MJ (1980) Specificity of cytotoxic T cells from athymic mice. J Exp Med 152:688

Keller G et al. (1985) Expression of a foreign gene in myeloid and lymphoid cells derived from multipotent haematopoietic precursors. Nature 318:149

Kindred B (1979) Nude mice in immunology. Prog Allergy 26:137

Lake JP et al. (1980) Sendai-virus specific, H-2-restricted cytotoxic T lymphocyte response of nude mice grafted with allogenic or semi-allogenic thymus glands. J Exp Med 152:1805

MacDonald HR et al. (1987) T cell antigen receptor expression in athymic (nu/nu) mice. Evidence for an oligoclonal β chain repertoire. J Exp Med 166:195

MacDonald HR et al. (1986) Abnormal distribution of T cell subsets in athymic mice. J Immunol 136:4337

Maleckar R, Sherman A (1987) The composition of the T cell repertoire in nude mice. J Immunol 138:3873

Miller RG et al. (1983) The extent of self MHC restriction of cytotoxic T cells in nude mice varies from mouse to mouse. J Immunol 130:63

Yoshikai Y et al. (1986) Athymic mice express a high level of functional α-chain but greatly reduced levels of α-and β-chain T-cell receptor messages. Nature 324:482

Developmentally Ordered Expression of γδ TCR $V_{\delta}7$ Subfamily Genes

S. Wehr and B. Arden

Max-Planck-Institut für Immunbiologie, D-7800 Freiburg, FRG

INTRODUCTION

A repertoire study of the T-cell receptor α-chain Variable (V) Region genes expressed in thymocytes from adult C57BL/Ka mice comprised V_{α} genes that we divided into 10 different subfamilies (Arden et al. 1985). V gene segments specific for each subfamily were subcloned and used as probes to screen the same cDNA library. The organization of the immunoglobulin heavy chain locus in multiple isotypes led us to test the possibility that the α-chain Variable Region gene pool might be utilized by novel Constant (C) Region genes other than C_{α}. Therefore, we isolated only the cDNA clones that hybridized to the V_{α} probes, but not to a C_{α} probe. Sequence analysis revealed a $V_{\alpha}2$ to $J_{\alpha}TT11$ rearrangement that was not spliced to C_{α}, and one $V_{\alpha}4$ gene segment germline transcript. Suspecting that our remaining $V_{\alpha}^{+}C_{\alpha}^{-}$ cDNA clones were similar transcripts, they were not further characterized. Two years later, a frequent rearrangement in the α-locus was cloned (Chien et al. 1987a) and proven to encode the T-cell receptor δ-chain (Born et al. 1987). We then went back and sequenced our remaining $V_{\alpha}^{+}C_{\alpha}^{-}$ cDNA clones. Two clones contained C_{δ} combined with V gene segments from the $V_{\alpha}7$ and $V_{\alpha}10$ subfamilies (clones A29 and A13 respectively, Table 1).

The V gene segment of cDNA clone A29 shows highest homology to that of cDNA clone Z53 that shares 78% of its nucleotide sequence with $V_{\alpha}7.1$. Z53 was numbered $V_{\delta}6$ by Elliott et al. (1988). We have named it $V_{\delta}7.3$ in keeping with the nomenclature already established for the α-chain cDNA clones of the $V_{\alpha/\delta}7$ subfamily, $V_{\alpha}7.1$ (Arden et al. 1985) and $V_{\alpha}7.2$ (Yague et al. 1988). It had deleted a major portion of the V gene segment through an aberrant splicing event. cDNA clone A13 was homologous to the three V_{α} gene segments known from the $V_{\alpha}10$ subfamily and therefore was named $V_{\delta}10.4$. It differs by only one nucleotide from V_{δ} KN25-D4 (Takagaki et al. 1989). Using a C_{δ}-specific cDNA probe we then isolated three additional δ-chain cDNA clones. The V gene segment of clone A38 was completely identical to $V_{\alpha}7.2$, and therefore named $V_{\delta}7.2$ (Table 1). A1 was identical to V_{δ}DN4 (Chien et al. 1987a), A28 was homologous to V_{δ}M11 (Chien et al. 1987b) with nine nucleotide exchanges resulting in six amino acid replacements (Arden, Zaller, Wang and Hood, in preparation).

Current Topics in Microbiology and Immunology, Vol. 173

RESULTS AND DISCUSSION

Fetal Ontogeny of T-Cell Receptor δ-Chain Expression

During fetal ontogeny, thymocytes exhibit a programmed pattern of Variable (V_γ and V_δ) gene utilization. At different stages of gestation, waves of cells appear that express distinct V gene segments. First, $V_\delta 1$ is predominantly expressed at day 14 to 16 of gestation. Moreover, the $V_\delta 1$-$J_\delta 2$ junctions almost exclusively consist of one canonical sequence. These thymic $V_\delta 1/V_\gamma 3$ clones are the precursors of the dendritic epidermal cells. In Southern analyses a $J_\delta 1$ probe detects several rearranged bands at day 16 (Chien et al. 1987a). We have recently been able to establish the anchored polymerase chain reaction (A-PCR) technique using 5 µg of total RNA from day 16 fetal thymocytes. The homopolymer (dG)-tailed cDNA is amplified in two rounds of PCR using nested C_δ-specific primers at the 3'end and an anchor sequence attached to the oligo(dC) primer at the variable 5'end. We also looked at neonatal thymus with A-PCR, where the V_δ repertoire is more diverse, to distinguish the onset or time intervall of expression of different V_δ genes during ontogeny.

Surprisingly, at day 16, in contrast to newborn thymus, we have not yet identified any transcripts containing V gene segments. Instead, we found transcripts initiated 5' of $J_\delta 1$ and $J_\delta 2$ or from partial $D_\delta 2$-$J_\delta 1$ rearrangements. Conceivably, these could be substrates for subsequent V-DJ rearrangements. Of course, we would not detect V-$D_\delta 2$ rearrangements with the A-PCR technique, which could account for some of the rearranged bands using the $J_\delta 1$ probe. These preliminary data indicate that at day 16 of gestation partial DJ rearrangements occur in large excess over complete V-DJ rearrangements. We are currently investigating whether there is a marked down-regulation of $V_\delta 1$ expression at day 16. In newborn thymus we identified $V_\delta 4.5$ (Table 1) identical to the V_δ gene segment of DN2.3 (Korman et al. 1988). We believe that day 16 represents the time point of a trough between the early fetal wave of $V_\delta 1$ expression and a subsequent wave of more diverse V_δ expression around birth.

Wave of $V_\delta 7$ Expression in Neonatal Thymocytes

RNA dot blot experiments indicate that the crosshybridizing $V_{\alpha/\delta}7$ subfamily (currently designated $V_\delta 6/V_\alpha 7$, Table 1) is expressed in over 60% of a panel of γδ- and not in any of the αβ hybridomas from newborn thymus (Happ et al. 1989a). Many of these unselected hybridomas are reactive with mycobacterial PPD and require a V_δ identical to the $V_\alpha 7.1$ gene segment for recognition of PPD (called $V_\delta 6.3$ by Happ et al. 1989b). We have amplified by PCR RNA from the whole population of neonatal thymocytes using primers specific for C_δ and the leader sequence that appeared to be conserved within the $V_{\alpha/\delta}7$ subfamily. We found dominant usage of the $V_\alpha 7.1$ gene segment in productive V_δ rearrangements, and therefore name it $V_\delta 7.1$ when it occurs in δ-chain message. Surprisingly, we obtained another V_δ identical to $V_\alpha 7.2$ (Table 1) as frequently in frame as $V_\delta 7.1$.

Although this $V_\delta 7.2$ gene segment shares 86% of its amino acid sequence with $V_\delta 7.1$ it was not utilized in any of the PPD-reactive hybridomas. Neither V gene segment was detected in δ-chain message from day 16 fetal and adult thymus. Therefore, we interpret this as yet another wave of δ-chain expression around birth, utilizing predominantly $V_\delta 7.1$ and $V_\delta 7.2$.

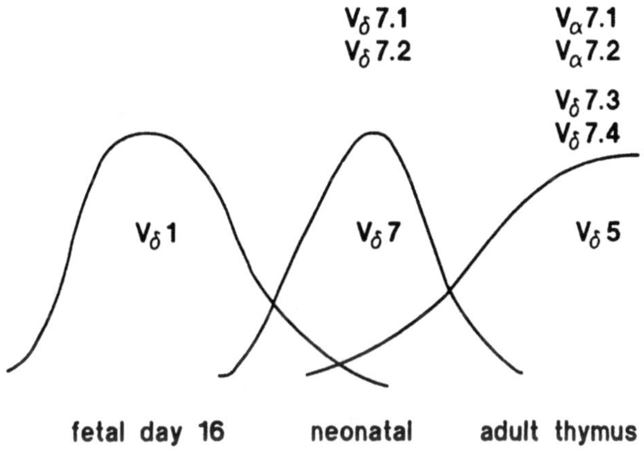

Fig. 1. Wave of thymic $V_\delta 7$ expression at birth

$V_\delta 7.4$ is another member of the $V_{\alpha/\delta} 7$ subfamily with only about 65% similarity at the protein level to $V_\delta 7.1$ or $V_\delta 7.2$. Whenever we found $V_\delta 7.4$ expressed in frame, its message was inactivated by an aberrant splice that resulted in the deletion of the major portion of the V gene segment. In adult thymocytes we did not observe this aberrant splice event in functionally rearranged $V_\delta 7.4$ or $V_\delta 7.3$ that share 92% of their amino acids (Table 1). These results suggest that $V_\delta 7.1$ and $V_\delta 7.2$ are exclusively utilized in neonatal thymocytes, whereas $V_\delta 7.3$ along with $V_\delta 7.4$ were functionally expressed only in adult thymocytes (Fig. 1)

Table 1. Proposed common nomenclature for V_α and V_δ gene segments

Current V_δ Designation[a]	Homologous V_α Subfamily[b]	Family Size	Clone Name[c]	Proposed $V_{\alpha/\delta}$ Nomenclature
$V_\delta 1$	-	1 member	M21	
$V_\delta 2$	-	1	M11 A28	
$V_\delta 3$	$V_\alpha 6$	3	TA1 Z68	V_α 6.1 V_δ 6.2
$V_\delta 4$	-	2	DN4, A1	
$V_\delta 5$	-	1	Z44	
$V_\delta 6$	$V_\alpha 7$	4	TA27 RL6.14 RL6.8 A38 A29, Z53 AT9.4, Z49	$V\alpha$ 7.1 V_δ 7.1 V_α 7.2 V_δ 7.2 V_δ 7.3 V_δ 7.4
$V_\delta 7$	$V_\alpha 10$	4	TA57, FN1-18 D6 1F8 A13, KN25	V_α10.1 V_α10.2 V_α10.3 V_δ10.4
$V_\delta 8$	$V_\alpha 2$	5	TA39 TA19 41BNT-117	V_α 2.1 V_α 2.2 V_δ 2.3
	$V_\alpha 4$	8	TA28 TA65 112-2 BDFL2 DN2.3	V_α 4.1 V_α 4.2 V_α 4.3 V_α 4.4 V_δ 4.5
	$V_\alpha 11$	5	C.F6 2B4 G8	V_α11.1 V_α11.2 V_δ11.3

[a] from Elliott et al. (1988).
[b] from Arden et al. (1985), subfamily members having greater than 75% similarity to one another at the DNA level.
[c] clones TA1, TA19, TA27, TA28, TA39, TA57, TA65 are from Arden et al. (1985); 2B4, FN1-18 (Becker et al. 1985); BDFL2 (Dembic et al. 1986); C.F6 (Fink et al. 1986); 112-2 (Hochgeschwender et al. 1987); DN4 (Chien et al. 1987a); M11, M21 (Chien et al. 1987b); Z44, Z49, Z53, Z68 (Elliott et al. 1988); KN25 (Takagaki et al. 1988); D6, 1F8 (McElligott et al. 1988); DN2.3 (Korman et al. 1988); G8 (Bluestone et al. 1988); 41BNT-117 (Happ et al. 1989b).

Selective Expression of $V_\delta 7.1$ by $B2A2^- CD4^- CD8^-$ $\gamma\delta$ Thymocytes

A minor subset of adult $CD4^- CD8^-$ thymocytes lacks the heat stable antigen (recognized by monoclonal antibodies B2A2 or J11d). 90% of these $B2A2^-$ double negative thymocytes express $\alpha\beta$ T-cell receptors utilizing predominantly $V_\beta 8.2$. The remaining 10% are $\gamma\delta$ cells which are phenotypically and developmentally distinct from the majority of adult $\gamma\delta$ thymocytes which express the B2A2 antigen and preferentially use $V_\delta 5$ and $V_\gamma 2$. The latter may give rise to $\gamma\delta$ emigrants that lose the B2A2 determinant upon maturation. RNase protection experiments indicated that $V_\delta 7.1$ was dominantly expressed in $B2A2^-$ $\gamma\delta$ thymocytes (Miescher et al. 1990). With the polymerase chain reaction we found $V_\delta 7.1$ productively rearranged (RL6.14). In one hybridoma (RL6.8) from the $\alpha\beta$ TCR expressing $B2A2^-$ population we found an out-of-frame α-chain transcript of $V_\alpha 7.2$. We are trying to amplify RNA from the $B2A2^-$, $\alpha\beta$ TCR^+ subset by anchored PCR, to determine whether a particular V_α is used along with the dominant $V_\beta 8.2$. (Arden, Miescher and MacDonald, in preparation).

Overlapping V_α and V_δ Usage

We failed to amplify α-chain message from neonatal thymocytes utilizing any gene segments from the $V_\alpha 7$ subfamily. Only in adult thymocytes $V_\alpha 7.1$ and $V_\alpha 7.2$ occurred in α-chain message, however, in all cases rearranged out of frame (Fig. 1). These results suggest that these V gene segments can be rearranged both to J_α and J_δ gene segments, however, at a post-transcriptional level they are selected to be expressed exclusively as δ chains. Due to false priming of the $V_{\alpha/\delta}7$ specific leader primer within the $V_\delta 2$ gene segment we amplified α-chain message that contained $V_\delta 2$ joined in frame to J_αpHDS58. However, when we then used a combination of $V_\delta 2$- and C_α specific primers we obtained three clones that were aberrantly spliced from a cryptic splice site within $V_\delta 2$ to C_α. This differential splice most likely occured subsequent to a $V_\delta 2$ to J_α rearrangement, again suggesting that there is no strict distinction between α-ness or δ-ness at the level of rearrangement. It is possible that the observed differential splicing events occur in thymocytes that are selected against expression of a particular antigen receptor.
RNA dot blot experiments indicate that V gene segments from almost every V_α subfamily are expressed in a panel of $\gamma\delta$ hybridomas from newborn thymus (Happ et al. 1989a). There is increasing evidence that only very few V_δ gene segments are unique, lacking homology to any of the known V_α subfamilies. Therefore, we propose to use a common nomenclature (Table 1) for both V_α and V_δ gene segments.

REFERENCES

Arden B, Klotz JL, Siu G, Hood LE (1985) Diversity and structure of genes of the α family of mouse T-cell antigen receptor. Nature 316:783-787

Becker DM, Patten P, Chien Y, Yokota T, Eshhar Z, Giedlin M, Gascoigne NRJ, Goodnow C, Wolf R, Arai K, Davis MM (1985) Variability and repertoire size of T-cell receptor V_{α} gene segments. Nature 317:430-434

Bluestone JA, Cron RQ, Cotterman M, Houlden BA, Matis LA (1988) Structure and specificity of T-cell receptor γ/δ on major histocompatibility complex antigen-specific $CD3^{+}$, $CD4^{-}$, $CD8^{-}$ T Lymphocytes. J Exp Med 168:1899-1916

Born W, Miles C, White J, O'Brien R, Freed JH, Marrack P, Kappler J, Kubo RT (1987) Peptide sequences of T-cell receptor δ and γ chains are identical to predicted X and γ proteins. Nature 330:572-574

Chien Y, Iwashima M, Kaplan KB, Elliott JF, Davis MM (1987a) A new T-cell receptor gene located within the α-locus and expressed early in T-cell differentiation. Nature 327:677-682

Chien Y, Iwashima M, Wettstein DA, Kaplan KB, Elliott JF, Born W, Davis MM (1987b) T-cell receptor δ gene rearrangements in early thymocytes. Nature 330:722-727

Dembic Z, Haas W, Weiss S, McCubrey J, Kiefer H, von Boehmer H, Steinmetz M (1986) Transfer of specificity by murine α and β T-cell receptor genes. Nature 320:232-238

Elliott JF, Rock EP, Patten PA, Davis MM, Chien Y (1988) The adult T-cell receptor δ-chain is diverse and distinct from that of fetal thymocytes. Nature 331:627-631

Fink P, Matis LA, McElligott DL, Bookman M, Hedrick SM (1986) Correlations between T-cell specificity and the structure of the antigen receptor. Nature 321:219-226

Happ MP, Palmer E (1989a) Thymocyte development: an analysis of T-cell receptor gene expression in 519 newborn thymocyte hybridomas. Eur J Immunol 19:1317-1325

Happ MP, Kubo RT, Palmer E, Born WK, O'Brien RL (1989b) Limited receptor repertoire in a mycobacteria-reactive subset of γδ T lymphocytes. Nature 342:696-698

Hochgeschwender U, Simon H, Weltzien HU, Bartels F, Becker A, Epplen JT (1987) Dominance of one T-cell receptor in the $H\text{-}2K^{b}$/TNP response. Nature 326:307-309

Korman AJ, Marusic-Galesic S, Spencer D, Kruisbeek AM, Raulet DH (1988) Predominant variable region gene usage by γ/δ T-cell receptor-bearing cells in the adult thymus. J Exp Med 168:1021-1040

McElligott DL, Sorger SB, Matis LA, Hedrick SM (1988) Two distinct mechanisms account for the immune response (Ir) gene control of the T-cell response to pigeon cytochrom c. J Immunol 140: 4123-4131

Miescher GC, Liao NS, Lees RK, MacDonald HR, Raulet DH (1990) Selective expression of $V_{\delta}6$ genes by $B2A2^{-}CD4^{-}CD8^{-}$ T-cell receptor γ/δ thymocytes. Eur J Immunol 20:41-45

Takagaki Y, Nakanishi N, Ishida I, Kanagawa O, Tonegawa S (1989) T-cell receptor-γ and -δ genes preferentially utilized by adult thymocytes for the surface expression. J Immunol 142:2112-2121

Yague J, Blackman M, Born W., Marrack P, Kappler J, Palmer E (1988) The structure of V_{α} and J_{α} segments in the mouse. Nucleic Acids Res 16:11355-11363

Maturation/Selection

Tissue Distribution and Repertoire Selection of Human γδ T Cells: Comparison With the Murine System

J. Borst[1], Thea M. Vroom[2], J. D. Bos[3], and J. J. M. van Dongen[4]

[1] Division of Immunology, The Netherlands Cancer Institute, Plesmanlaan 121, 1066 CX Amsterdam
[2] Department of Pathology, Rotterdam Cancer Center, Rotterdam
[3] Department of Dermatology, Academic Medical Center, Amsterdam
[4] Department of Immunology, Erasmus University, Rotterdam, The Netherlands

INTRODUCTION

With the discovery of T cell receptor (TCR) γδ, the question has arisen what contribution T cells expressing this receptor type make to the immune system. This question can also be formulated as follows: Do γδ T cells complement or overlap with αβ T cells with respect to 1) antigenic specificities (repertoire), 2) functions, and/or 3) sites of action within the body? In the murine and the human systems a great deal of data have been gathered regarding repertoire and tissue distribution of γδ T cells. In the mouse, there appears to be a connection between repertoire and tissue distribution, in that cells expressing TCR γδ characterized by certain Vγ and Vδ gene segments are preferentially localized in certain tissues. We have made an inventory of the tissue distribution of human γδ T cells and looked into the repertoire bias of peripheral human γδ T cells. Here, we would like to present a simplified picture of the species differences between man and mouse with respect to these two issues.

TISSUE DISTRIBUTION OF HUMAN AND MURINE γδ T CELLS.

We have analysed a great variety of normal human lymphatic and non-lymphatic tissues, in which $CD3^+$, TCR $\alpha\beta^+$ and TCR $\gamma\delta^+$ cells were identified in serial sections of freshly frozen, acetone fixed material by means of a sensitive immunoalkaline-phosphatase method. Special attention was paid to the various epithelial layers within these tissues in order to shed light on the possible specialized role of γδ T cells in the immunosurveillance of epithelia. Samples were derived from autopsies or from normal parts of operative specimina. This inventory has included the lymphoid organs, skin, tongue, salivary glands, oesophagus, duodenum, jejunum, ileum, coecum, appendix, lung, trachea, ureter, testis, epididymis, uterus, cervix, and vagina (Vroom et al., submitted).

In the mouse, >90% of T cells in the epidermis express TCR γδ (Koning et al., 1987; Kuziel et al., 1987), while γδ T cells have also been reported to constitute a major proportion of total T cells in the epithelia of reproductive organs and tongue (Itohara et al., 1990). The epidermal γδ T cell population expresses V5/J1/Cγ1 and V1/D2/J2/Cδ encoded receptor chains (Asarnow et al., 1989), while the γδ T cell population in

Current Topics in Microbiology and Immunology, Vol 173

reproductive organs and tongue, expresses V6/J1/Cγ1 and V1/D2/J2/Cδ encoded receptor chains (Lafaille et al., 1989). Strikingly, these chains do not display junctional diversity so that this population can be considered clonal. In contrast, in the human epidermis (Groh et al., 1989; Bos et al., 1990) and epithelia of reproductive organs and tongue (Vroom et al., submitted), γδ T cells are a minority of total T cells. Also, the absolute number of epidermal T cells per unit of surface area is significantly lower than in mice, while the morphology of the few detectable γδ T cells is not comparable to that of the highly dendritic murine epidermal γδ T cells (Bos et al., 1990).

It has been reported that the majority of T cells present in the epithelium of the intestinal tract of the mouse express TCR γδ (Goodman and Lefrancois, 1988). These murine γδ T cells preferentially but not exclusively express the Vγ7 gene segment, with significant diversity at the V-J junction (Bonneville et al., 1988; Asarnow et al., 1989). Moreover, this γ chain is found in combination with δ chains that use V4, V5, V6 or V7 gene segments and two D segments, and also display a great deal of junctional diversity. In other words, the repertoire of γδ T cells in the murine intestinal epithelium is large. In our studies, the human intestinal epithelium was found to contain significant amounts of γδ T cells, but αβ T cells still formed the majority of total T cells (80-90%). This ratio is different from that found in other epithelia, where γδ T cells did not exceed 5% of the total T cell population. A striking observation was that γδ T cells localized preferentially within the epithelium of the crypts. Within the lamina propria significant amounts of αβ T cells were present, while γδ T cells were only rarely detected. In our hands, the situation in mouse and man with respect to the localization and relative abundance of αβ and γδ T cells in the intestine was quite comparable.

We conclude that γδ T cells are not the predominant T cell population in all human epithelia. Perhaps the differences in tissue distribution of γδ T cells between man and mouse can be summarized by saying that apparently in man the equivalents of the two murine subpopulations of γδ T cells expressing homogeneous receptors of either the Vγ5/Vδ1 type or the Vγ6/Vδ1 type that are found in epidermis and epithelium of the reproductive tract and tongue are lacking. In the mouse, these two populations are generated during fetal ontogeny in two distinct waves, where the Vγ5/Vδ1 population is the first to be observed (Havran and Allison 1988, 1990). At day 15 of fetal development the only TCR/CD3 expressing thymocytes use Vγ5/Vδ1. The second population is observed within the thymus around birth (Ito et al., 1989). Studies with TCR γδ transgenic mice have clearly demonstrated that the homing characteristics of the two murine γδ T cell subpopulations are not determined by the TCR itself, since cells expressing other receptor types also end up at these locations in the transgenic animals (Ferrick et al., 1989; Bonneville et al., 1990). Therefore, it might be hypothesized that concomitant with γδ T cell differentiation certain homing receptors are expressed, that are different for the Vγ5/Vδ1 and the Vγ6/Vδ1 populations and determine their selective migration. Sequential gene rearrangement patterns within the γ and the δ locus may at least in part determine the sequential development of murine $\gamma\delta$ T cell subpopulations and the γ/δ chain combinations.

It has recently been demonstrated that also in human T cell development, TCR γ and δ loci may be rearranged and expressed in an ordered fashion, where as in the murine locus, the more C-gene segment proximal V segments would recombine first (Krangel et al., 1990). The repertoire in early human fetal thymocytes was found to be dominated by Vγ9/Jγ1/Cγ1-V$\delta 2^+$ cells. This is in contrast to the situation at later times, where more upstream Vγ's of the VγI subfamily are used, joined to Jγ2/Cγ1, in combination with the upstream located Vδ1 gene segment. The early receptor types display significantly less junctional diversity than the later ones. Apparently there is a significant amount of similarity between the events occurring during human and murine $\gamma\delta$ T cell development. Yet, the homing properties if these early $\gamma\delta$ T cells seem to differ between species. In this respect it is also striking that at no time during human T cell development the thymus can been found to harbor only TCR $\gamma\delta^+$ cells and no TCR $\alpha\beta^+$ cells (Campana et al., 1989).

REPERTOIRE BIAS OF $\gamma\delta$ T CELLS IN HUMAN PERIPHERAL BLOOD.

Various investigators have described that the great majority of $\gamma\delta$ T cells in peripheral blood of most healthy individuals express receptors using the Vγ9-Jγ1/Cγ1 and Vδ2/Cδ gene segments (Triebel et al., 1988a,1988b; Borst et al., 1989; Casorati et al., 1989). Such a distribution is also found in peripheral lymphoid organs of adult donors (Falini et al., 1989). Apparently the repertoire of $\gamma\delta$ T cells in the peripheral lymphoid compartment of healthy donors is more limited than the germline diversity would allow. However, the receptors do show junctional diversity (Loh et al., 1988; Takihara et al., 1989).

What could be the explanation for the preferred use of Vγ9/Vδ2 encoded receptors? Possibilities that may play a role either alone or in combination are: 1) Preferential or highly effective rearrangement of Vγ9 and Vδ2 gene segments. 2) Preferential chain pairing. 3) Negative selection of $\gamma\delta$ T cells expressing receptors other than Vγ9/Vδ2. 4) Positive "selection" of Vγ9/Vδ2 expressing cells. Here, we would like to propose that Vγ9/V$\delta 2^+$ $\gamma\delta$ T cells undergo a selective expansion in the peripheral compartment, on the basis of two lines of experimental evidence. One is the predominance of Vγ9/V$\delta 2^+$ cells in a patient with thymic aplasia, the other is the change in repertoire and absolute number of $\gamma\delta$ T cells in peripheral blood during the first year of life (Van Dongen et al., submitted).

DiGeorge anomaly (DGA) is characterized by facial, cardiac, parathyroid and thymic defects, due to malformations or disruptions of the third and fourth pharyngeal arches and pouches from which these organs develop. The thymic defect involves epithelial mass, i.e. it is a quantitative defect. DGA is associated with T cell deficiency of variable degree, that is assumed to be directly proportional to the degree of thymic hypoplasia.

Complete DGA, characterized by total thymic aplasia and virtually complete absence of T cells has an incidence of, roughly estimated, one in 10^6-10^7 newborns.

Patient, A.L., who was clinically diagnosed as having complete DGA, total $CD3^+$ peripheral T cells constituted only 1.1% of mononuclear cells. Absolute numbers of $CD3^+$ cells were 0.03 x 10^9/liter peripheral blood, which is 100-fold lower than in healthy age group controls. The absolute number of αβ T cells was about 1000-fold reduced compared to control values. Strikingly, the absolute number of γδ T cells was only about 4-fold lower than normal. This resulted in a complete inversion of the γδ/αβ T cell ratio, with γδ T cells constituting over 85% of all T cells in PB. The repertoire of the γδ T cells of patient A.L. was analysed by immunofluorescence with V region specific mAbs TiγA (Triebel et al., 1988a), BB3 (Ciccone et al., 1988) and δTCS-1 (Wu et al., 1988). The great majority of γδ T cells in peripheral blood employed the Vγ9 and/or Vδ2 gene segments (80 and 79% respectively), while Vδ1-Jδ1 expression was found on 1% of γδ T cells. This repertoire does not differ from that of healthy donors. We conclude that the predominance of $V\gamma9/V\delta2^+$ peripheral T cells can arise in the absence of detectable thymic epithelial mass. This is an argument in favour of positive, peripheral selection or peripheral expansion of cells expressing this receptor type.

A second argument in favour of positive selection/expansion in the periphery of $V\gamma9/V\delta2^+$ γδ T cells comes from the observation that the peripheral repertoire of γδ T cells changes greatly during the first year of life. The absolute number of γδ T cells in neonatal cord blood is about 0.04 x 10^9/liter, but 4 to 5-fold higher (0.15-0.20 x 10^9/liter) in peripheral blood of children of different age groups. This increase in absolute numbers of γδ T cells takes place during the first months of life and subsequently numbers remain stable. The question arose which subpopulation of γδ T cells might be responsible for the observed increase. Immunofluorescence staining with V-region specific mAbs pointed out that in neonatal cord blood approximately half of the γδ T cells expressed Vδ1-Jδ1, while 30-40% expressed Vγ9 and/or Vδ2. This repertoire is comparable to that of infant γδ thymocytes. However, in peripheral blood of children up to 6 months of age this distribution has changed to about 20% $V\delta1\text{-}J\delta1^+$ and 75% $V\gamma9^+$ and/or $V\delta2^+$. In most older children and adults this change is even more prominent: 10-15% $V\delta1\text{-}J\delta1^+$ and 80-85% Vγ9 and/or $V\delta2^+$ cells (Van Dongen et al., submitted).

Therefore, the substantial increase in absolute numbers of γδ T cells during the first half year after birth can best be explained by assuming that the $V\gamma9/V\delta2^+$ population expands in the periphery. An observation that is in line with this idea is that $V\gamma9/V\delta2^+$ cells express the CD45RO isoform (Parker et al., 1990; Miyawaki et al., 1990), that is a marker of "memory" cells, suggesting that the γδ T cells have previously been activated. What is the molecule these γδ TCR interact with? Studies from Parker et al. (1990) have indicated that it is most likely not a self-protein, being either MHC or another genetically determined factor, since great repertoire differences may exist between identical twins. The increase in peripheral $V\gamma9/V\delta2^+$ cells occurs in the first months after birth, which would be in line with the idea that it is exposure to foreign antigens that drives the expansion. The fact that TCR γδ recognizes such hypothetical antigen(s) independent of its junctional diversity, but dependent on V gene usage, suggests a role for a superantigen. Recognition of the superantigen Staphylococcal enterotoxin A by γδ T cells has been shown to occur and to be dependent on Vγ9 expression (Rust et al., 1990). Obviously the nature of the antigen(s) playing a role in vivo remains to be elucidated.

It is interesting that in human as well as in mouse the γδ T cell repertoire in peripheral lymphoid organs and blood differs significantly from that in the intestine. In human intestinal epithelium γδ T cells express predominantly the Vδ1 gene segment (Halstensen et al., 1989). Possibly in both species these repertoire differences are related to differences in exposure to foreign antigens at either site.

REFERENCES

Asarnow D, Goodman T, Lefrancois L, Allison JP (1989) Distinct antigen receptor repertoires of two classes of murine epithelium-associated T cells. Nature 341: 60-62.

Bonneville M, Janeway CA, Ito K, Haser W, Ishida I, Nakanishi N, Tonegawa S (1988) Intestinal intraepithelial lymphocytes are a distinct set of γδ T cells. Nature 336: 479-481.

Bonneville M, Itohara S, Krecko EG, Mombaerts P, Ishida I, Katsuki M, Berns A, Farr AG, Janeway CA, Tonegawa S (1990) Transgenic mice demonstrate that epithelial homing of γ/δ T cells is determined by cell lineages independent of T cell receptor specificity. J Exp Med 171: 1015-1026.

Borst J, Wicherink A, Van Dongen JJM, De Vries E, Comans-Bitter WM, Wassenaar F, Van den Elsen P (1989) Non-random expression of T cell receptor γ and δ variable gene segments in functional T lymphocyte clones from human peripheral blood. Eur J Immunol 19: 1559-1568.

Bos JD, Teunissen MBM, van der Kraan J, Cairo I, Krieg SR, Kapsenberg ML, Das PK, Borst J (1990) T-cell receptor γδ bearing cells in normal human skin. J Invest Dermatol 94: 37-42.

Campana D, Janossy G, Coustan-Smith E, Amlot PL, Tian W-T, Ip S, Wong L (1989) The expression of T cell receptor-associated protein during T cell ontogeny in man. J Immunol 142: 57-66.

Casorati G, De Libero G, Lanzavecchia A, Migone N (1989) Molecular analysis of human γ/δ$^+$ clones from thymus and peripheral blood. J Exp Med 170: 1521-1535.

Ciccone E, Ferrini S, Bottino C, Viale O, Progione I, Pantaleo G, Tambussi G, Moretta L (1988) A monoclonal antibody specific for a common determinant on the human T cell receptor γ/δ directly activates CD3$^+$WT31$^-$ lymphocytes to express their functional programs. J Exp Med 168: 1-11.

Falini B, Flenghi L, Pileri S, Pelicci P, Fagioli M, Martelli MF, Moretta L, Ciccone E (1989) Distribution of T cells bearing different forms of the T cell receptor γ/δ in normal and pathological human tissues. J Immunol 143: 2480-2488.

Ferrick DA, Sambhara SR, Ballhausen W, Iwamoto A, Pircher H, Walker CL, Yokoyama WM, Miller RG, Mak TW (1989) T cell function and expression are dramatically altered in T cell receptor Vγ1.1Jγ4Cγ4 transgenic mice. Cell 57: 483-492.

Goodman T, Lefrancois L (1988) Expression of the γ-δ T-cell receptor on intestinal CD8$^+$ intraepithelial lymphocytes. Nature 333: 855-858.

Groh V, Porcelli S, Fabbi M, Lanier LL, Picker LJ, Anderson T, Warnke RA, Bhan AK, Strominger JL, Brenner MB (1989) Human lymphocytes bearing T cell receptor γδ are phenotypically diverse and evenly distributed throughout the lymphoid system. J Exp Med 169: 1277-1294.

Halstensen TS, Scott H, Brandtzaeg P (1989) Intraepithelial T cells of the TCR γ/δ$^+$CD8$^-$ and Vδ1/Jδ1$^+$ phenotypes are increased in coeliac disease. Scand J Immunol 30: 665-672.

Havran WL, Allison JP (1988) Developmentally ordered appearance of thymocytes expressing different T-cell antigen receptors. Nature 335: 443-445.

Havran WL, Allison JP (1990) Origin of Thy-1$^+$ dendritic epidermal cells of adult mice from fetal thymic precursors. Nature 344: 68-70.

Ito, K., Bonneville, M., Takagaki, Y., Nakanishi, N., Kanagawa, O., Krecko, E.G., and Tonegawa, S. 1989. Different γδ T-cell receptors are expressed on thymocytes at different stages of development. Proc Natl Acad Sci USA 86: 631-635.

Itohara S, Farr AG, Lafaille JJ, Bonneville M, Takagaki Y, Haas W, Tonegawa S (1990) Homing of a γδ thymocyte subset with homogeneous T-cell receptors to mucosal epithelia. Nature, 343: 754-757.

Koning F, Stingl G, Yokoyama WM, Yamada H, Maloy WL, Tschachler E, Shevach EM, Coligan JE (1987) Identification of a T3-associated γ/δ T cell receptor on Thy-1+ dendritic epidermal cell lines. Science 236: 834-837.

Krangel MS, Yssel H, Brocklehurst C, Spits H (1990) A distinct wave of human T cell receptor γδ lymphocytes in the early fetal thymus: Evidence for controlled gene rearrangement and cytokine production. J Exp Med 172: 847-859.

Kuziel WA, Takashima A, Bonyhadi M, Bergstresser PR, Allison JP, Tigelaar RE, Tucker PW (1987) Regulation of T-cell receptor γ-chain RNA expression in murine Thy-1+ dendritic epidermal cells. Nature 328: 263-266.

Lafaille JJ, DeCloux A, Bonneville M, Takagaki Y, Tonegawa S (1989) Junctional sequences of T cell receptor γδ genes: implications for γδ T cell lineages and for a novel intermediate of V-(D)-J joining. Cell 59: 859-870.

Loh EY, Cwirla S, Serafini AT, Phillips JH, Lanier LL (1988) Human T-cell receptor δ chain: genomic organization, diversity, and expression in populations of cells. Proc Natl Acad Sci USA 85: 9714-9718.

Miyawaki T, Kasahara Y, Taga K, Yachie A, Taniguchi N (1990) Differential expression of CD45R0 (UCHL1) and its functional relevance in two subpopulations of circulating TCR-γ/δ+ lymphocytes. J Exp Med 171: 1833-1838.

Parker CM, Groh V, Band H, Porcelli SA, Morita C, Fabbi M, Glass D, Strominger JL, Brenner MB (1990) Evidence for extrathymic changes in the T cell receptor γ/δ repertoire. J Exp Med 171: 1579-1612.

Rust CJJ, Verreck F, Vietor H, Koning F (1990) Specific recognition of Staphylococcal enterotoxin A by human T cells bearing receptors with the Vγ9 region. Nature 346: 572-574.

Takagaki Y, DeCloux A, Bonneville M, Tonegawa S (1989) Diversity of γδ T cell receptors on murine intestinal intra-epithelial lymphocytes. Nature 339: 712-714.

Takihara Y, Reimann J, Michalopoulos E, Ciccone E, Moretta L, Mak TW (1989) Diversity and structure of human T cell receptor δ chain genes in peripheral blood γ/δ-bearing T lymphocytes. J Exp Med 169: 393-405.

Triebel F, Faure F, Graziani M, Jitsukawa S, Lefranc M-P, Hercend T (1988a) A unique V-J-C rearranged gene encodes a γ protein expressed on the majority of CD3+ T cell receptor α/β- circulating lymphocytes. J Exp Med 167, 694-699.

Triebel F, Faure F, Mami-Chouaib F, Jitsukawa S, Giscelli A, Genevée C, Roman-Roman S, Hercend T (1988b) A novel human Vδ gene expressed predominantly in the TiγA fraction of γ/δ+ peripheral lymphocytes. Eur J Immunol 18: 2021-2027.

Wu Y-J, Tian W-T, Snider RM, Rittershaus C, Rogers P, LaManna L, Ip SH (1988) Signal transduction of γ/δ T cell antigen receptor with a novel mitogenic anti-δ antibody. J Immunol 141, 1476-1479.

Evidence For Controlled Gene Rearrangements and Cytokine Production During Development of Human TCRγδ+ Lymphocytes

H. Spits[1], H. Yssel[1], Cathy Brockelhurst[2], and M. Krangel[2]

[1] DNAX Research Institue for Molecular and Cellular Biology, 901 California Avenue, Palo Alto, CA 94304-1104, USA
[2] Dana Farber Cancer Institute, 44 Binney Street, Boston MA 02115, USA

INTRODUCTION

Molecular analysis of the TCRγ and δ loci has established that the number of V gene segments at these loci is limited. In man, nine functional Vγ gene segments have been described (Huck *et al.* 1988) and five Vδ gene segments (Takihara *et al.* 1989, Krangel *et al.* 1990). Potentially TCRγδ cells can use 54 different combinations of Vγ and δ gene segments. However, only a few combinations have been observed so far Phenotypical analysis of TCRγδ+ T cell from peripheral blood has revealed that almost all of these cells use one of only two Vδ gene segments, Vδ1 and Vδ2. The majority of the TCRγδ+ T cells in most adults uses Vδ2 that pairs exclusively with Vγ2 (nomenclature of Strauss *et al.* 1987, Vγ2 = Vγ9 in the nomenclature of Huck *et al.* 1988). In contrast, in postnatal thymus TCRγδ+ T cells using Vδ1 predominate, whereas the Vδ2 using cells form a minority (Casorati *et al.* 1989). Parker *et al.* (1990) recently found that Vδ2+ cells undergo a postnatal extrathymic expansion which may explain why these cells constitute a majority of the TCRγδ+ T cell pool in most individuals. Those results, however, did not explain why there is a preferential pairing of γ and δ chains expressing Vγ2 and Vδ2. Transfection experiments have excluded that there is a physical constraint that prevents Vδ2 encoded δ chains to pair with other γ chains (Solomon *et al.* 1989). In this paper we discuss recent evidence that in man like in mouse, rearrangements at the γ and δ loci occur in a coordinated, ordered fashion during thymic development. This ordered rearrangements may partly explain why Vδ2 pairs predominantly with Vγ2. In addition, we discuss the findings that early fetal TCRγδ+ T cells and postnatal TCRγδ+ T cells produce different cytokines following activation.

Current Topics in Microbiology and Immunology, Vol 173

ORDERED REARRANGEMENTS OF γ AND δ GENES DURING DEVELOPMENT

Studies in mouse systems have indicated that rearrangements at the TCRγ and δ loci occur in different waves. Each wave of rearrangements involves different V gene segments. It has been suggested previously that rearrangements at the γ locus join the downstream Vγ gene segments (Vγ2, Vγ3, Vγ4) to upstream J segments (Jγ1.1, Jγ1.2, Jγ1.3) associated with Cγ1 and subsequent rearrangements would join upstream Vγ gene segments of the Vγ1 family to the downstream J segments associated with Cγ2 (Treibel *et al.* 1988a). Indeed, analysis of rearrangements and expression of γ and δ genes in a series of cultures derived from early fetal thymocytes (8.5 week, 12 and 15 week of gestational age) support the notion that human γ and δ genes rearrange in an orderly fashion (Krangel *et al* .1990). Most rearrangements detected in three TCRγδ+ clones isolated from thymocytes of a 8.5 week old fetus involved Vγ2 and the Jγ1 cluster. In addition, Vγ1 rearrangements were detected in early fetal thymocyte clones as well, but these are not random, since they only involve the most downstream Vγ1 segment, Vγ1.8 that is most proximal to Vγ2. In agreement with these results, one other group has recently described two fetal TCRγδ clones that also use Vγ1.8 (Carding *et al.* 1990). In three TCRγδ+ clones obtained from the 8.5 week old thymus, Vδ2 was the only Vδ rearrangement detected. In addition, virtually all cultured fetal TCRγδ thymocytes obtained from 8.5, 12 and 15 weeks fetuses expressed the Vδ2 gene segment as assessed with the monoclonal antibody BB3 that specifically reacts with Vδ2 encoded determinants (Ciccone *et al.* 1988, Treibel *et al.* 1988b, Sturm *et al.* 1989). These data indicate that δ rearrangements in most, if not all, TCRγδ+ T cells in early fetuses involve Vδ2. Sequence analysis of V-D-J junctions of the fetal thymocyte cultures revealed limited junctional diversity as compared to postnatal TCRγδ+ clones. Moreover, the fetal TCRγδ+ cells exclusively use the Dδ3 segment. The patterns of rearrangements in postnatal TCRγδ cells are strikingly different from those in early fetal thymus (Casorati *et al.* 1989, Krangel *et al.* 1990). In postnatal TCRγδ cells, Vδ1 as well as some other Vδ segments are primarily joined to the upstream Dδ segments Dδ1 and 2. In the TCRγ locus, upstream Vγ gene segments in the Vγ1 family are joined to downstream Jγ2 segments. Moreover, diversity in the junctions in postnatal γδ+ thymocytes is extensive (Krangel *et al.* 1990). Thus, the patterns of TCRγδ rearrangements emerging from our studies in man is similar as in mouse. In both systems, the repertoire of the V gene segments that rearrange is restricted and usage of homologous Dδ gene segments (Dδ3 in human, Dδ2 in mouse) and Jδ gene segments (Jδ3 in human, Jδ2 in mouse) predominates. These early rearrangements are furthermore characterized in both systems by minimal insertions of N nucleotides, presumably as a result of low levels of terminal transferase activity (Landau *et al.* 1987). One important difference was noted in that the γδ receptors of early murine fetal thymocytes that use specific VγVδ pairs are highly homogeneous (Asarnow *et al.* 1988, Lafaille *et al.* 1989, Havran *et al.* 1990), while the human Vδ2-Dδ3-Jδ3 junctions we analyzed are heterogeneous. This may suggest that early human TCRγδ+ cells are not subjected to the same types of intrathymic selection as their murine counterparts. On the other hand, with a few exceptions, all cultured fetal TCRγδ+ T cells used Vδ2 in combination with Vγ2, while usage of the more downstream Vγ3 and 4 segments was not observed. There may be a mechanism that prevents rearrangement of Vγ3 and 4, or Vγ3 and Vγ4 positive cells are subject to negative selection. It is also possible that Vγ2Vδ2 positive cells are positively selected. In that case, the V-D-J junctions are not critical in this selection process. Although our studies have provided useful information about rearrangements at the γ and δ loci in early human fetal thymus, more information is required for a comprehensive picture of the sequence and timing of these rearrangements in man. It is, for example, unclear at what gestational age the rearrangement patterns switch from an early fetal type to a postnatal type of rearrangement .

CYTOKINE PRODUCTION BY FETAL AND POSTNATAL THYMIC TCRγδ+ T CELL CLONES AND THE POSSIBLE CONTRIBUTION OF TCRγδ+ T CELLS TO TCRαβ DEVELOPMENT

Striking differences were observed in the production of cytokines by early fetal or postnatal thymic TCRγδ+ T cell clones (Krangel *et al.* 1990). All fetal Vδ2+ clones that we have tested produced significant levels of IL-4 and IL-5, while in contrast none of the postnatal TCR$\gamma\delta^+$ clones produced those cytokines following *in vitro* activation. Both sets of clones produced IL-2, IFN-γ and GM-CSF after activation. Also murine fetal TCR$\gamma\delta^+$ thymocytes can produce IL-4 upon stimulation with anti-CD3 mAbs *in vitro* (Tentori *et al.* 1988). This could reflect a capacity of fetal TCR$\gamma\delta^+$ thymocytes to secrete IL-4 *in vivo*. Carding *et al.* (1989) have demonstrated that high proportion (45%) of cells in the thymus of 15 days old CBA embryo's express IL-4 mRNA *in vivo*. This percentage of IL-4 mRNA expressing cells was reduced dramatically (to 2%) at day 16, when the first TCR$\alpha\beta^+$ cells start to appear. The authors did not determine which cells actually expressed IL-4, but it is possible that TCR$\gamma\delta^+$ cells are, at least partly, responsible for IL-4 production *in vivo*. In addition to IL-4, high levels of IL-2 mRNA in the murine fetal thymus were also detected at day 15 which is strongly diminished at day 16. These findings should be considered in view of the suggestions that TCR$\gamma\delta^+$ cells in the thymus are involved in development of TCR$\alpha\beta^+$ T cells. If this speculation is correct, it is likely that cytokines, produced by the TCR$\gamma\delta^+$ T cells, are involved. No effects of IL-5 on cells other than eosinophils and B cells have been reported, but IL-2, IL-4 and IFN γ can affect T cells, thymic stromal cells or both. IL-2 and IL-4 are growth factors for activated thymocytes (Spits *et al.* 1987). IL-4 can induce CD8α on mature human CD4$^+$ T cells (Paliard *et al.* 1988) and CD4$^-$CD8$^-$ T cells (Paliard and Spits, unpublished), suggesting that IL-4 plays a role in the regulation of CD8 expression in the thymus. Studies with mice expressing an IL-4 encoding transgene have provided evidence that IL-4 influences patterns of intrathymic T cell differentiation (Tepper *et al.* 1990). An increase in CD8$^+$CD4$^-$ thymocytes, as compared to control mice, was observed in these transgenic mice, supporting the notion that IL-4 is implicated in the development of CD8$^+$CD4$^-$ T cells. In addition, a strong decrease in levels of CD4$^+$CD8$^+$ immature cells was observed, and it was suggested by the authors that IL-4 may play a role in the physiological process of cell death that accompanies clonal deletion (Tepper *et al.* 1990). However, the results of the experiments with IL-4 transgenic animals should be considered with care, since

IL-4 is expressed in the transgenics at any time during development, while expression in normal thymus seems to be tightly regulated (Carding *et al.* 1989). Furthermore, IL-4 can have effects not only on T cells, but on thymic epithelial cells and macrophages as well. Notably, low concentrations of IL-4 inhibit strongly the production of IL-6, GM-CSF and Leucocyte Inhibitory Factor (LIF) by human thymic epithelial cells (Galy *et al.*, manuscript in preparation). In the mouse system, it has been shown that IL-4 upregulates class II MHC antigens on thymic macrophages (Ransom *et al.* 1987). This effect was also observed with human monocytes (te Velde *et al.* 1988). Like with epithelial cells, cytokine production by human monocytes is strongly inhibited by IL-4 (te Velde *et al.* 1988). Thus, the effects of IL-4 can exert stimulatory, as well as inhibitory effects. IFN-γ, which is produced in high concentrations by the fetal thymic TCR$\gamma\delta^+$ T cell clones upon activation, induces class II MHC antigens on purified thymic epithelial cells. Finally Ferrick *et al.* (1989) have reported that expression of a functional TCRγ transgene strongly enhanced cellularity of the thymus, while the relative proportions of different subsets of TCR$\alpha\beta^+$ cells remained unaltered. It was furthermore found that the T cells in the peripheral lymphoid tissues of these young transgeneic mice were much more immunoreactive than those in control mice. These results may be consistent with the idea that TCR$\gamma\delta^+$ cells influence development of TCR$\alpha\beta^+$ T cells in the thymus.

In order to produce these cytokines, the TCR$\gamma\delta^+$ T cells need to be activated. It is not clear whether that happens in the developing thymus. It is, however, intriguing that most fetal TCR$\gamma\delta^+$ T cells in early fetus express the Vγ2δ2 pair. Perhaps there is a common ligand for these cells present in the fetal thymus. Recently, Fisch *et al.* (1990) provided evidence that Vγ2δ2$^+$ T cells from peripheral blood are able to specifically kill cells from the Burkitt lymphoma cell line Daudi. We have recently found that, although fetal thymic Vγ2δ2$^+$ T cells fail to lyse Daudi cells, these latter cells induce cytokine production by fetal TCR$\gamma\delta^+$ clones that use Vγ2 and Vδ2 (Spits *et al.*, manuscript in preparation). The nature of the antigen on Daudi cells that is recognized by the TCR$\gamma\delta^+$ is presently unknown, but it is perhaps possible that this antigen is also present in the developing thymus.

ACKNOWLEDGMENTS

I thank J.E. de Vries, and J.A. Waitz for critically reading this manuscript. Mrs. J. Dillon is gratefully acknowledged for her secretarial assistance. DNAX Research Institute of Molecular and Cellular Biology is supported by Schering-Plough Corporation.

REFERENCES

Asarnow DM, Kuziel WA, Bonyhadi M, Tigelaar RE, Tucker PW, Allison JP (1988) Limited diversity of $\gamma\delta$ antigen receptor genes of thy-1+ dendritic epidermal cells. Cell 55:837-847.

Carding SR, Jenkinson EJ, Kingston R, Hayday AC, Bottomly K, Owen JJT (1989) Developmental control of lymphokine gene expression in fetal thymocytes during T cell ontogeny. Proc Natl Acad Sci USA 86:3342-3345.

Carding SR, McNamara JG, Pan M, Bottomly K (1990) Characterization of γ/δ T cell clones isolated from human fetal liver and thymus. Eur J Immunol 20:1327-1335

Casorati G, De Libero G, Lanzavecchia A, Migone N (1989) Molecular analysis of human γ/δ^+ clones from thymus and peripheral blood. J Exp Med 170:1521-1535.

Ciccone E, Ferrini S, Bottino C, Viale O, Prigione I, Panteleo G, Tambussi G, Moretta A, Moretta L (1988) A monoclonal antibody specific for a common determinant of the human T cell receptor γ/δ directly activates CD3+WT31- lymphocytes to express their functional program(s). J Exp Med 168:1-6.

Ferrick DA, Sambhara SR, Ballhausen W, Iwamoto A, Pircher H, Walker CL, Yokoyama WM, Miller RG, Mak TW (1989) T cell function and expression are dramatically altered in T cell receptor Vγ1.1Jγ4Cγ4 transgenic mice. Cell 57:483-492.

Fisch P, Malkovsky M, Braakman E, Sturm E, Bolhuis RLH, Prieve A, Sosman JA, Lam VA, Sondel P (1990) Gamma/delta T cell clones and natural killer cell clones mediate distinct patterns of non-MHC restricted cytolysis. J Exp Med 171:1567-1579.

Havran WL, Allison JP (1990) Origin of Thy-1+ dendritic epidermal cells of adult mice from fetal thymic precursors. Nature 344:68-70.

Huck S, Dariavach P, LeFranc MP (1988) Variable region genes in the human T cell rearranging γ (TRG) locus: V-J junction and homology with the mouse genes. EMBO J 7:719-726.

Krangel M, Yssel H, Brocklehurst C, Spits H (1990) A distinct wave of human T cell receptor γδ lymphocytes in early fetal thymus: Evidence for controlled gene rearrangement and cytokine secretion. J Exp Med 172:847-860.

Lafaille JJ, DeCloux A, Bonneville M, Takagaki Y, Tonegawa S (1989) Junctional sequences of T cell receptor γδ genes: Implications for γδ T cell lineages and for a novel intermediate of V-(D)-J joining. Cell 59:859-870.

Landau NR, Schatz DG, Rosa M, Baltimore D (1987) Increased frequency of N-region insertion in a murine pre-B cell line infected with a terminal deoxynucleotidyl transferase retroviral expression vector. Mol Cell Biol 7:3237-3243.

Paliard X, de Waal Malefyt R, de Vries JE, Spits H (1988) Interleukin-4 mediates CD8 induction on human $CD4^+$ T cell clones. Nature 335:642-644.

Parker CM, Groh V, Band H, Porcelli SA, Morita C, Fabbi M, Glass D, Strominger JL, Brenner MB (1990) Evidence for extrathymic changes in the TCRγδ repertoire. J Exp Med 171:1597-1612.

Ransom J, Fischer M, Mercer L, Zlotnik A (1987) Lymphokine-mediated induction of antigen presenting ability in thymic stromal cells. J Immunol 139:2620-2628.

Solomon KR, Krangel MS, McLean J, Brenner MB, Band H (1990) Human T cell receptor-γ and -δ chain pairing analyzed by transfection of a T cell receptor-δ negative mutant cell line. J Immunol 144:1120-1126.

Spits H, Yssel H, Takebe Y, Arai N, Yokota T, Lee F, Arai K, Banchereau J, de Vries JE (1987) Recombinant interleukin-4 promotes the growth of human T cells. J Immunol 139:1142-1147.

Strauss WM, Quertermous T, Seidman JG (1987) Measuring the human T cell receptor γ chain locus. Science 237:1217-1219.

Sturm E, Braakman E, Bontrop RE, Chuchana P, Van de Griend RJ, Koning F, LeFranc MP, Bolhuis RL (1989) Coordinated V γ and V δ gene segment rearrangements in human T cell receptor γ/δ^+ lymphocytes. Eur J Immunol 19:1261-1265.

Takihara Y, Reimann J, Michalopoulos E, Ciccone E, Moretta L, Mak TW (1989) Diversity and structure of human T cell receptor δ chain genes in peripheral blood γ/δ-bearing T lymphocytes. J Exp Med 169:393-405.

Tentori L, Pardoll DM, Zuñiga JC, Hu-Li J, Paul WE, Bluestone JA, Kruisbeek AM (1988) Proliferation and production of IL-2 and B cell stimulatory factor 1/IL-4 in early fetal thymocytes by activation through Thy-1 and CD3. J Immunol 140:1089-1094.

Tepper RI, Levinson DA, Stanger BZ, Campos-Torres J, Abbas AK, Leder P (1990) IL-4 induces allergic-like inflammatory disease and alters T cell development in transgenic mice. Cell 62:457-467.

te Velde AA, Klomp JPG, Yard BA, de Vries JE, Figdor CG (1988) Modulation of phenotypic and functional properties of human peripheral blood monocytes by IL-4. J Immunol 140:1548-1554.

Treibel F, Faure F, Mami-Chouaib F, Jitsukawa S, Griscelli A, Genevée C, Roman-Roman S, Hercend T (1988) A novel human Vδ gene expressed predominantly in the TiγA fraction of γ/δ^+ peripheral lymphocytes. Eur J Immunol 18:2021-2027.

Thymic and Extrathymic Development of Human γ/δ T Cells

L. D. McVay[1,2], A. C. Hayday[1,2], K. Bottomly[1], and S. R Carding[1]

[1] Section of Immunobiology, Yale University School of Medicine
[2] Department of Biology, Yale University, New Haven, CT 06510, USA

1. Introduction

By immunohistochemical criteria, γ/δ cells represent only a minor population of cells in the developing human thymus (0.02%-0.5%) which remains relatively the same in post-natal thymus (0.02%-0.1%) (Campana et al, 1989). However, it is not clear whether different populations of γ/δ T cells reside in the thymus at different times during fetal gestation. Molecular analysis of human γ/δ leukemic lines suggests that there is a sequential and ordered activation of the human T cell receptor gamma locus during thymic ontogeny (Kimura et al, 1987; VanDongen et al, 1987; Triebel et al, 1988;). In order to achieve a better understanding of the origin and development of human γ/δ T cells, the developmental regulation and expression of all of the known Vγ and Vδ genes during fetal ontogeny needs to be studied.

2. Human Thymic γ/δ Development

Our studies , first described in McVay et al, 1991, have utilized the polymerase chain reaction (PCR) to analyze the expression of all Vγ and Vδ genes in situ in human fetal tissues at different times during ontogeny and in early post-natal life. This analysis was performed immediately after excision of the tissue, in the absence of any in vitro manipulation that may have selected for particular populations. mRNA from fetal tissues was reverse transcribed, and cDNA amplified in the PCR using oligonucleotide primers specific for each of the 4 Vγ gene subfamilies (VγI to VγIV; Lefranc and Rabbitts, 1991) as well as each of the Vδ genes defined to date, in conjunction with appropriate Cγ or Cδ oligos, respectively. PCR products were identified by Southern blotting and probing with a pool of Jγ or Jδ oligos that lie internal to the V and C primers, thus detecting only conventionally rearranged and expressed genes. Furthermore, the structure of the transcripts isolated in the thymus at early and mid-gestation have been compared to post-natal thymic transcripts. These and similar studies, analysing TCR composition of γ/δ$^+$ T cell lines and clones, define the Vγ and Vδ genes and proteins expressed throughout gestation which is summarized in Table 1 and 2, respectively.

2.1 Ontogenetic Expression of Vγ and Vδ genes

We have identified expression of all Vγ and Vδ genes, with the exception of Vδ4, from 11 weeks to 22 weeks of human fetal thymic ontogeny, in neonatal thymus and in 3 year post-natal thymus . The primary data presented in McVay et al, 1991, are summarized in Table 1 and Table 2. V gammma gene expression was detected in fetal thymic samples at all time points examined; the expression of Vγ9 (VγII) was relatively low during thymic ontogeny and was not detectable at all in the 3 year post-natal thymus. Early fetal (8.5 weeks) thymocyte clones expressed the Vγ9 gene and receptor (Krangel et al, 1990) while detection of Vγ9 as a post-natal receptor was rare.

Current Topics in Microbiology and Immunology, Vol 173

Vδ2 was the predominant Vδ gene expressed during fetal thymic ontogeny. However, analysis of post-natal thymus (3 year sample) shows that, relative to the expression of other Vδ genes, there is a marked loss of Vδ2 gene expression. Thus, expression of Vδ2 appears to be strongly developmentally regulated (McVay et al, 1991). In addition, there was increased expression of Vδ1 gene in post-natal thymus (McVay et al, 1991) which is consistent with the majority of γ/δ thymocytes expressing Vδ1 surface receptors at that time (Table 2).

TABLE 1. Ontogeny of TCR-γ expression in human fetal thymocytes.

Fetal Age (w)	Vγ mRNA I(1-8)	9	10	11	Vγ protein expressed
8.5	+	+	-	-	I (8), 9
12	+	+	+	+	9
14	+	+	+	+	I (8), 9
15	+	+	+	+	9
24	+	+	+	+	9
6 mo	+	+	+	+	I (3, 4, 8), 9
3y	+	+	+	+	I, 9

TCR-Vγ gene expression and the encoded cell-surface glycoproteins were detected in either poyclonal or clonal populations of fetal thymocytes (McVay et al, 1991; Carding et al, 1990a; Krangel et al, 1990; Kabelitz and Conradt, 1988; Casorati et al., 1989).

Similar findings were reported by Krangel and coworkers (1990) who detected expression of the Vδ2 gene in the thymus at 8.5 weeks of gestation. This is consistent with our findings and suggests that Vδ2 is the predominant, and perhaps the first, Vδ gene segment used in early fetal thymocytes. In contrast, post-natal thymic clones predominantly expressed Vδ1. (Bottino et al, 1988; Lanier et al, 1988; Casorati et al, 1989; Krangel et al, 1990).

2.2 TCR-γ/δ Protein Expression

Analysis of fetal γ/δ$^+$ T cell populations during thymic development provides insight into which of the Vγ and Vδ genes that are productively rearranged and expressed can be used for the generation of cell-surface TCR proteins (Table 1 and 2). Analysis of early fetal (8.5 weeks) thymus clones revealed that in most cases, Vγ8 gene was most frequently utilized (Krangel et al, 1990). Furthermore, γ/δ$^+$ T cell clones expressing the Vγ8/Vδ2 receptor pair have also been isolated from 14 week thymus (Carding et al, 1990a). The selective expression of VγI(8)/Vδ2 gene-encoded receptor proteins on a majority of early fetal thymocyte clones demonstrates the developmental control or selection of rearranged and expressed Vγ and Vδ gene segments that are utilized for production of surface TCRs. Later in gestation (after 15 weeks) thymic γ/δ cells expressing Vγ9 can be detected, consistent with a sequential activation and rearrangement of the TCR-γ locus during ontogeny (Kimura et al, 1987; VanDongen et al, 1987; Triebel et al, 1988). In 3 year post-natal thymus, the proportion of thymic γ/δ T cells expressing $V\gamma9/V\delta2$ receptors declines, consistent with the lack of $V\gamma9$ and $V\delta2$ gene expression in post-natal thymus. Thus, the fetal thymic receptor composition can be distinguished from the post-natal one by a selective loss of cells bearing $V\gamma9/V\delta2$ receptors and an accumulation of $V\delta1^+$ γ/δ cells in the post-natal thymus, reflected both at the level of gene and surface receptor expression. T cells bearing $V\gamma9$ and $V\delta2$ receptors represent the predominant γ/δ T cell population in the adult peripheral blood suggesting that $V\gamma9/V\delta2^+$ T cells may be exported from the thymus to the periphery between 22 weeks of fetal gestation and 3 years post-natal life.

TABLE 2. Ontogeny of TCR-δ expression in human fetal thymocytes.

Fetal Age (w)	Vδ mRNA 1	2	3	4	5	6	Vδ protein expressed
8.5	-	+	-	-	-	-	2
11	+	+	+	-	+	+	ND
12	+	+	+	-	+	+	2
14	+	+	+	-	+	+	2
15	+	+	+	-	+	+	2
6 mo	+	+	+	-	+	+	1, 2, 3, 5
3 y	+	-	+	-	+	+	1

TCR-Vδ gene expression and the encoded cell-surface glycoproteins were detected in polyclonal or clonal populations of fetal thymocytes (McVay et al, 1991; Carding et al, 1990b; Krangel et al, 1990). ND; not determined.

The data shown in Tables 1 and 2 illustrate that although expression of Vγ genes is not a good predictor of Vγ receptor proteins expressed, Vδ gene expression is a more reliable predictor. This situation, first discussed by McVay et al, 1991, is analogous to the ontogeny of γ/δ cells in the mouse (Carding et al, 1990b). Table 2 shows that up until 15 weeks of fetal gestation, most γ/δ^+ cells express Vδ2 receptor, indicating that γ/δ receptor expression may, at least in part, be determined by the differential regulation of Vδ gene expression. The predominance of γ/δ cells bearing restricted receptor pairs may be due to a capacity of the receptors to successfully interact with thymic stroma (thymic selection).

2.3 Structure of Fetal Thymic γ/δ Transcripts

The relatively small size of the γ and δ gene families limits their capacity to generate receptor diversity by combinatorial use of gene segments. Thus junctional diversity, in which N nucleotides are added at the V-J and V-D-J joins during the recombination process (Tonegawa, 1983; Toyonaga and Mak, 1987 and refs. therein), may provide the major mechanism for diversifying γ and δ receptors, especially in δ receptors, where 3 Dδ gene segments in humans can potentially assemble together.

We (McVay et al, 1991) and others (Krangel et al, 1990) have determined whether there are qualitative changes in the structure of γ and δ transcripts generated during thymic ontogeny. Table 3 is a review of the pattern of changes in the structure of Vδ1 and Vδ5 transcripts during thymic ontogeny, as determined by McVay et al, 1991.

Generally, the junctional sequences of Vδ1 and Vδ5 genes at 11 weeks are simple and result from direct fusions of gene segments. Dδ3 is the earliest gene segment to be utilized in Vδ-Jδ recombinations, with little or no nucleotide insertions detected (McVay et al, 1991; Krangel et al, 1990). At 17 weeks, Vδ1 transcripts exhibit more N1 insertion as well as some recurrent use of D1, N2 and D2. By contrast, the structure of post-natal transcripts are quite different. There is a marked increase in junctional diversity primarily due to extensive N nucleotide insertion as well as some P (palindromic) syntheses (Lafaille et al, 1990). The overall increase in N1, N2, N3 and N4 in post-natal transcripts paralleled the increase in usage of Dδ1 and/or Dδ2 gene segments later in ontogeny. These observations suggest a regulated utilization of different gene segments in different populations of cells present in the thymus at any one time. Analysis of murine fetal thymic γ and δ transcripts also reveals an increase in junctional diversity of receptors as ontogeny progresses (Lafaille et al, 1989). Compared to fetal thymic transcripts, post-natal transcripts exhibited increased nucleolytic activity that was particularly evident between Dδ3-Jδ sequences; there was little or no evidence of this enzymatic activity in the V-gene segments of the same transcripts at any time points examined (McVay et al, 1991). Thus, as ontogeny proceeds, the structure of the transcripts change, due to several genetic mechanisms that operate on the genes during the process of recombination (McVay et al, 1991).

TABLE 3. Structural diversity of fetal thymic Vδ- chains during ontogeny.

		5'-3' Gene Segments							
Age	Vδ	N1	D1	N2	D2	N3	D3	N4	Jδ
11w	1						+	+/-	1, 2 or 3
17w	1	+/-	+/-	+/-			+	+/-	1,2 or 3
3y	1	+	+	+	+	+	+	+	1
11w	5						+		1, 2 or 3
17w	5	+	+	+	+		+		1, 2 or 3
3y	5	+	+	+	+	+	+	+	1

Expression of junctional sequences of Vδ1 and Vδ5 fetal and post-natal cDNA clones was determined as described by McVay et al (1991). +/-: Indicates variation in expression of sequences between indivdual clones.

3. Extrathymic Development of Human γ/δ T Cells

The differential expression of different Vγ and Vδ genes and surface receptors in populations of T cells in the adult thymus and blood was originally proposed as evidence for the extrathymic development of peripheral blood γ/δ T cells (Borst et al, 1988). Peripheral blood may also contain T cell precursor populations which, in vitro, have been shown to give rise to γ/δ+ T cells (Preffer et al 1989).

An age-related increase in the number of γ/δ T cells expressed in the periphery, and a shift from a Vδ1 predominance to a Vδ2 predominance was seen in the absence of any equivalent population in the thymus, suggesting a thymic-independent age-related expansion of the Vδ2+ population (Parker et al,

1990). Analysis of patients with DiGeorge anomaly (DGA) provides additional in vivo evidence for the extrathymic development of γ/δ^+ T cells in man (VanDongen et al, 1990). The absolute number of α/β^+ T cells in DGA patients was found to be reduced in proportion to the extent of thymic hypoplasia or aplasia. In contrast, the absolute numbers of γ/δ^+ T cells in all patients were found to be comparable to those in normal age-matched individuals, indicating that γ/δ^+ T cells have a capacity to develop and expand in peripheral compartments.

Despite these studies, however, the potential for extrathymic activation of TCR γ and δ genes outside of the thymus has not been studied during normal embryonic development. Our studies of T cell ontogeny have identified cells expressing Vγl and Vδ2 genes in the fetal liver as early as 6weeks of gestation (Table 4), prior to thymic colonization.

TABLE 4. Profile of T cell receptor γ/δ gene expression in the fetal liver prior to, and after, thymic colonization.

	% Positive Cells							
Fetal Age (w)	Cγ	Vγ1-8	Vγ9	Vγ10	Vγ11	Cδ	Vδ1	Vδ2
6	0.1	0.1	0	0	0	0.2	0	0.1
7	2.2	2.4	0.4	0	0	3.5	0.3	2.9
12	1.6	0.6	0.3	0.2	0.4	1.9	0.4	1.4
14	2.3	0.7	1.5	0.4	0.2	2.7	0.5	1.8

Frequency of fetal liver-mononuclear cells expressing TCR-γ/δ genes was determined by hybridization in situ with ^{35}S-labelled variable (V) and constant (C) region-specific probes (Carding et al, 1990a).

This observation has been independently corroborated by Southern blotting analysis of TCR gene rearrangement in genomic fetal liver DNA samples identifying rearrangement of members of the Vγl gene family, including Vγ8 (McVay, Hayday, Bottomly and Carding, manuscript in preparation). Interestingly, the first γ/δ^+ T cells identified in the fetal thymus, at 8.5 weeks , have also been shown to utilize the Vγ8 and Vδ2 gene segments on their surface TCR (Krangel et al, 1990) suggesting that these cells may not be generated intra-thymically, but are derived from the fetal liver, which then seeds the thymus. Later in gestation, other Vγ genes are expressed in the fetal liver (Table 4).

PCR analysis of fetal liver and gut indicate that the pattern of γ and δ gene expression is similar to the pattern of γ and δ gene expression in the fetal thymus with all γ and δ genes, except Vδ4, expressed (data not shown. McVay, Hayday, Bottomly and Carding manuscript in preparation) Vδ2 is the predominant Vδ gene expressed in the fetal liver and gut during fetal ontogeny. While the expression of Vγ and Vδ genes is present in the fetal gut at all time points examined, expression of most Vδ genes in the liver cannot be detected in several donor-matched 15-16 week samples, suggesting that populations of cells expressing different Vγ and Vδ genes are present in the liver at different times .

Taken together, these data suggest that there appears to be programmed rearrangement and expression of γ and δ genes in fetal liver, which occurs prior to thymic colonization. The expression of Vγ and Vδ genes in fetal liver and gut is similar to the pattern seen in the thymus, suggesting the possibility of an extrathymic origin of TCR γ/δ-bearing thymic progenitors.

4. Conclusions

The results from our studies (McVay et al, 1991, and manuscript in prep.) and those of others of human γ/δ T cell ontogeny have shown: 1) Activation of the TCR-γ and δ gene loci in intrathymic T cell progenitors is developmentally regulated. 2) All Vγ and Vδ genes, with the possible exception of Vδ4, are expressed as rearranged genes in the thymus, indicating that there is no apparent preferrential rearrangement and expression of Vγ and Vδ genes in the thymus. 3) The expression of a single Vδ gene (Vδ2) is predominant in the human fetal thymus. 4) The pattern of expression of Vδ genes in 3 year post-natal thymus is distinguished from the that in the fetal thymus by the loss of Vδ2 and Vγ9 expression. 5) There is a selective utilization of Vγ and Vδ gene products for the generation of cell-surface γ/δ TCRs during intrathymic T cell ontogeny: Vγl(8)/Vδ2 succeeded by Vγ9/Vδ2+ cells . 6) Vδ gene expression is a more reliable predictor of the TCR γ/δ protein composition than Vγ gene expression. 7) Junctional diversity of γ- and δ-chains is limited in fetal thymic transcripts, but is extensive in post-natal transcripts. As ontogeny proceeds, there is an increase in structural diversity of γ and δ genes. 8) At extrathymic sites (liver and gut), unlike the thymus, there appears to be a programmed rearrangement and expression of Vγ and Vδ genes, which in the fetal liver, is initiated prior to thymic colonization. 9) Vδ2 is the predominant Vδ gene expressed in the fetal liver and gut during ontogeny. 10) The γ/δ receptor positive cells first detected in the fetal thymus may represent cells of extrathymic origin.

References

Borst J, VanDongen JJM, Bolhuis RLH, Peters PJ, Hafler D, De Vries R, Van de Griend RJ (1988) Distinct molecular forms of human T cell receptor γ/δ detected on viable T cells by a monoclonal antibody. J Exp Med 167: 1625-1644

Bottino C, Tambussi G, Ferrini S, Ciccone E, Varese P, Mingari MC, Moretta L, Moretta A (1988) Two ubsets of human T lymphocytes expressing γ/δ antigen receptor are identifiable by monoclonal antibodies directed to two distinct molecular forms of the receptor. J Exp Med 168: 491-505

Campana D, Janossy G, Coustan-Smith E, Amlot PL, Tian W-T, Ip S, Wong L (1989) The expression of T cell receptor-associated proteins during T cell ontogeny in man. J Immunol 142: 57-66.

Carding SR, McNamara JG, Pan M, Bottomly K (1990a) Characterization of γ/δ T cell clones isolated from human fetal liver and thymus. Eur J Immunol 20: 1327-1335

Carding SR, Kyes S, Jenkinson EJ, Kingston R, Bottomly K, Hayday AC, (1990b) Developmentally regulated fetal thymic and extrahymic T cell receptor γ/δ gene expression. Genes and Develop 4: 1304-1312

Casorati G, De Libero G, Lanzevecchia A, Migone N (1989) Molecular analysis of human γ/δ+ clones from thymus and peripheral blood. J Exp Med 170: 1521-1530

Kabelitz D, Conradt P (1988) Identification of $CD2^-/CD3^+$ T cells in human fetal tissue. J Exp Med 168: 1941-1946

Kimura N, Du R-P, Mak TW (1987) Rearrangement and organization of T cell receptor gamma chain genes in human leukemic T cell lines. Eur. J. Immunol 17: 1653-1656

Krangel MS, Yssel H, Brockelhurst C, Spits H (1990) A distinct wave of human T cell receptor γδ lymphocytes in the early fetal thymus: Evidence for controlled gene rearrangement and cytokine production. J Exp Med . 172; 847-859

Lafaille JJ, DeCloux A, Bonneville,M, Takagak Y, Tonegawa S (1989) Junctional sequences of T cell receptor γ/δ genes: implications for γ/δ T cell lineages and for a novel intermediate of V-(D)-J joining. Cell 59: 859-870

Lanier LL, Ruitenberg J, Bolhuis RLH, Borst J, Phillips JH, Testi R (1988) Structural and serological heterogeneity of γ/δ T cell antigen receptor expression in thymus and peripheral blood. Eur J Immunol 18: 1985-1990

Lefranc M-P, Rabbitts TH (1991) A nomenclature to fit the organization of the human T cell receptor γ and δ genes Research in Immunology (in press)

McVay LD, Carding SR, Bottomly K, Hayday, AC (1991) Regulated expression and structure of T cell receptor γ/δ transcripts during human thymic ontogeny. EMBO J,. (in press)

Preffer FI, Kim CW, Fisher KH, Sabga EM, Kradin RL, Colvin RB (1989) Identification of pre-T cells in human peripheral blood. J Exp Med 170: 177-190

Parker CM, Groh V, Band H, Porcelli SA, Morita C, Fabbi M, Glass D, Strominger JL , Brenner MB (1990) Evidence for extrathymic changes in the T cell γ/δ repertoire. J Exp Med 171: 1597-1612

Tonegawa S (1983) Somatic generation of antibody diversity. Nature 302: 575-581

Toyonaga B, Mak TW (1987) Genes of the T cell antigen receptor in normal and malignant T cells. Ann Rev Immunol 5: 585-620

Triebel F, Lefranc M-P, Hercend T (1988) Further evidnce for a sequentially-ordered activation of T cell rearranging gamma genes during T lymphocyte differentiation. Eur. J. Immunol. 18: 789-794

VanDongen JJM, Wolvers-Tettero ILM, Seidman JG, Ang S-L, Van de Griend RJ, DeVries EFR, Borst J (1987) Two types of gamma T cell receptors expressed by T cell acute lymphoblastic leukemias. Eur. J. Immunol. 17: 1719-1726

VanDongen JJM, Comans-Bitter WM, Friedrich W, Neijens HJ, Belohradsky BH, Kohn T. Hagemeijer A, Borst J (1990) Analysis of patients with DiGeorge anomaly provides evidence for extrathymic development of T cell receptor $\gamma\delta^+$ lymphocytes in man. PhD. Thesis, JJM VanDongen, Erasmus University, Rotterdam, The Netherlands.

Differential Effects of Anti-CD3 Antibodies In Vivo and In Vitro on α β and γ δ T Cell Differentiation

B. A. KYEWSKI

Institute of Immunology and Genetics, German Cancer Research Centre, Im Neuenheimer Feld 280, D-6900 Heidelberg, FRG

The generation of the α β and γ δ T cell repertoire occurs within the thymus. T cells of the α β lineage are submitted to a dual selection process: thymocytes are positively selected towards recognition of self-MHC determinants and negatively selected for recognition of self-MHC complexes which present self peptides. These selection events operate at the level of immature CD4/8 positive thymocytes and are mediated by direct cell-cell interactions between T cells and thymic stromal cells (v. Boehmer 1990). Positive selection is thought to be strictly dependent on cortical epithelial cells (Benoist and Mathis 1990), while negative selection is obviously less restricted in its accessory cell requirements. Bone marrow-derived thymic stromal cells (macrophages and dendritic cells) efficiently mediate negative selection by deletion (Speiser et al. 1989), whereas thymic epithelial cells have been implicated in causing T cell deletion and antigen-specific T cell anergy (Houssaint and Flajnik 1990). Negative selection by deletion is assumed to occur via induction of programmed cell death ("apoptosis"), a process which can be mimicked *in vitro* by anti-T cell receptor (TCR)/CD3 antibodies (Smith et al 1989). Induction of T cell tolerance by anergy is assumed to occur by inappropriate signalling ("lack of a second signal") as demonstrated and analyzed *in vitro* (Schwartz 1989). In contrast, little is known about the molecular mechanisms responsible for positive selection given the lack of appropriate *in vitro* models.

While mature γ δ T cells may display a TCR repertoire as diverse as α β T cells the necessities and rules for selecting the immature γ δ T cell repertoire are hardly understood. γ δ T cells are not restricted by polymorphic MHC determinants and thus would not be selected according to the same criteria as α β T cells (Born et al 1990). Nevertheless, evidence for positive and negative selection of γ δ T cells has been derived from detailed molecular analysis of γ δ TCRs and from studies in transgenic mice (Lafaille et al 1990; Dent et al 1990). Understanding selection of the γ δ T cell repertoire, however still has to await the definition of the restriction (presentation) molecule(s) for and the range of foreign and self antigens recognized by γ δ T cells.

Since the TCR confers antigen-specificity on T cells it plays a key role in T cell repertoire selection. In order to define more precisely the stage and site at which the TCR determines the fate of developing thymocytes, we applied anti-CD3 antibodies to thymic organ cultures (TOC) and to newborn C3H/He mice and assessed their effect on intra-thymic and post-thymic

Current Topics in Microbiology and Immunology, Vol 173

development of α β and γ δ T cells. Mice were injected every second day for 2-3 weeks with increasing amounts of purified intact IgG antibodies or F(ab')2 fragments. Thymocytes and peripheral lymphoid cells were analyzed *ex vivo* or after 5 days *in vitro* culture in the presence of Concanavalin A and interleukin-2. In addition, the frequencies of distinct intrathymic lymphostromal-cell interactions were monitored.

Table 1

	Control	α–CD3
α β TCR negative	26,1 #	94,5
α β TCR low	62,7	3,5
α β TCR high	9,8	0,0
γδ TCR	0,2	0,7

Surface antigen phenotype of freshly isolated thymocytes from C3H/He mice treated for 17 days postnatally with anti-CD3 Abs (α CD3) or PBS (for details see Kyewski et al 1989). A similar phenotype is expressed after 22 hr *in vitro* culture in the absence of anti-CD3 Abs.
% positive cells, mean values of three animals.

Anti-CD3 Abs arrest T cell differentiation intra-thymically and thus prevent the maturation of single-positive CD4 or CD8 T cells (Kyewski et al 1989). In addition to the loss of mature $CD3^{hi}$ T cells (~10 % of all thymocytes) the immature $CD3^{lo}$ T cells (~60 %) are depleted (Table 1). CD3 expression is not merely modulated, since it does not recover within 22 hr of *in vitro* culture in the absence of Abs. The ablation of the CD3-positive compartment is, however, compensated for by an expansion of CD3 negative, CD4/8 double-positive immature thymocytes thus preserving the pool size of CD4/8 thymocytes. Thymi of Ab-treated mice display a normal sized cortex and a highly condensed, hypocellular medulla. In addition, the arrest of T cell development correlates with a significant reduction of interaction complexes between immature thymocytes and cortical epithelial cells (thymic nurse cells, TNC) and medullary dendritic cells (DC-Rosettes, DC-ROS) and the preservation of interactions between thymocytes and cortical macrophages (M0-Rosettes, M0-ROS, Fig. 1).
Thymocyte-M0 interactions have been mapped to an early cortical, cortical epithelial cell interactions to a late cortical and dendritic cell interactions to an early medullary stage of T cell development (Kyewski 1987) Anti-CD3 Abs thus arrest T cell development at an intermediate stage of the cortical phase which precedes T cell interactions with MHC class II positive stromal cells (epithelial and dendritic cells). The loss of CD3-positive thymocytes indicates that the arrest of differentiation is due to the continuous deletion

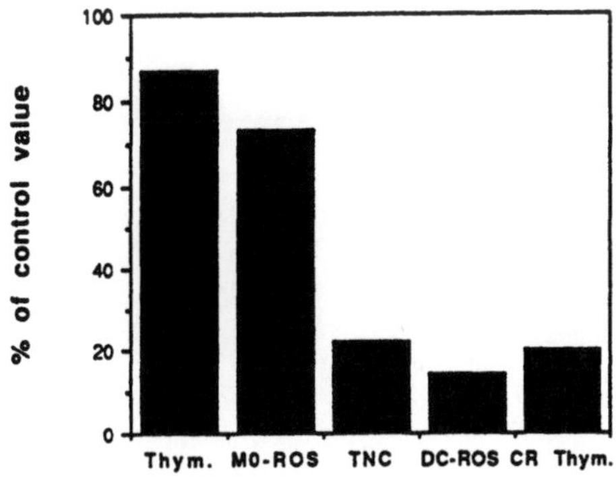

Fig. 1 Relative frequencies of unselected thymocytes (Thym.), MO-ROS, TNC, DC-ROS and cortisone-resistant thymocytes (CR Thym.) of anti-CD3 Ab-treated mice as compared to control animals.

of immature thymocytes as soon as they express surface TCR/CD3 complexes, a process which mimics negative selection of autoreactive T cell clones. As a consequence all downstream events depending on TCR expression are obliterated including interactions with MHC class II positive stromal cells and development of mature T cells. This mechanism is thus different from blockade of T cell differentiation by Abs against MHC class I and II molecules or Abs against CD4 and CD8 which presumably interfere with positive selection, i. e. recognition of epithelial cells without depleting immature $CD3^{lo}$ thymocytes (Zuniga-Pfücker et al 1990).

The postnatal arrest of intrathymic T cell development by anti-CD3 Abs prevents the seeding of peripheral lymphoid organs with mature ("Con A responsive") T cells as reflected by a tight absence of mature T cells in peripheral lymphoid organs even after *in vitro* culture for 4-7 days leading to the loss of T cell-dependent immune functions (Kyewski et al 1989) In contrast, thymic suspension cultures invariably showed the stimulation and growth of CD3 positive T cells. While thymic suspension cultures of PBS-injected control mice yielded typical $\alpha\beta$ TCR positive, mature T cells, cells of the anti-CD3 treated group displayed the phenotype of $\gamma\delta$ T cells.; CD4, CD8, CD2-negative, $\gamma\delta$ TCR positive. These $\gamma\delta$ T cells became gradually detectable *in vitro* (due to differentiation or expansion ?) (see Fig. 2) and did not accumulate *in vivo* during anti-CD3 treatment at the expanse of $\alpha\beta$ T cells. $\gamma\delta$ T cells could not be recovered from spleen or lymph node cultures of anti-CD3 treated mice. Similarly, in TOC anti-CD3 Abs selectively suppressed the development of $\alpha\beta$ but not $\gamma\delta$ TCR positive cells (data not shown).

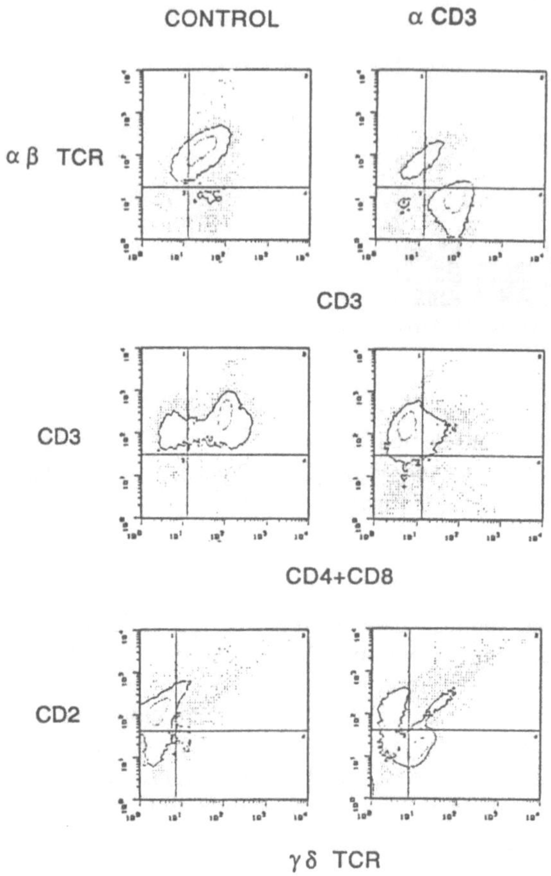

Fig. 2 Surface antigen phenotype of thymocytes after *in vitro* culture for 5 days in the presence of Con A and Interleukin-2. Mice treated with anti-CD3 Abs (α CD3) are compared to control animals.

Two explanations may account for these observations:
Firstly, anti-CD3 Abs deplete both CD3 positive α β and γ δ T cells *in vivo* and spare their CD3 negative precursors. γ δ T cells develop *in vitro* from CD3 negative precursor cells whereas the suspension culture conditions are insufficient to promote the differentiation of α β T cells from their precursors. Alternatively, γ δ T cells expand from a minor subset of CD3 positive γ δ T cells due to a relative resistance of γ δ T cells (in contrast to α β T cells) to anti-CD3 Abs *in vivo* or in TOC. Anti-CD3 Abs would neither delete nor favor γ δ T cell development, but rather reveal γ δ T cells due to

deletion of α β T cells. Available experimental data are insufficient to distinguish between these alternatives.

Acknowledgement: The author wishes to acknowledge the generous gift of Abs by Dr. O. Kanagawa, expert technical help by S. Höflinger and support of these studies by the Deutsche Forschungsgemeinschaft.

REFERENCES

Benoist C, Mathis D (1990) Positive selection of the T cell repertoire: where and when does it occur? Cell 58:1027-1033

Born W, Happ MP, Dallas A, Reardon C, Kubo R, Shinnick T, Brennan P, O'Brien R (1990) Recognition of heat shock proteins and γ δ cell function. Immunol Today 11: 40-43

Dent AL, Matis LA, Hooshmand F, Widacki SM, Bluestone JA, Hedrick, SM (1990) Self-reactive γ δ T cells are eliminated in the thymus. Nature 343: 714-719

Houssaint E, Flajnik M (1990) The role of thymic epithelium in the acquisition of tolerance. Immunol Today 11: 357-360

Kyewski BA (1987) Seeding of thymic microenvironments defined by distinct thymocyte-stromal cell interactions is developmentally controlled. J Exp Med 166: 520-538

Kyewski BA, Schirrmacher V, Allison JP (1989) Antibodies against the T cell receptor/CD3 complex interfere with distinct intra-thymic cell-cell interactions *in vivo* : correlation with arrest of T cell differentiation. E J Immunol 19: 857-863

Lafaille JJ, Haas W, Cotinho A, Tonegawa S (1990) Positive selection of γ δ T cells. Immunol Today 11: 75-78

Schwartz RH (1989) A cell culture model for T lymphocyte clonal anergy. Science 248: 1349-1356

Smith CA, Williams GT, Kingston R, Jenkinson EJ, Owen JJT (1989) Antibodies to CD3/T cell receptor complex cause controlled cell death in immature T cells in thymic cultures. Nature 337: 181-187

Speiser DE, Lees RK, Hengartner H, Zinkernagel RM, MacDonald HR (1989) Positive and negative selection ot T cell receptor V β domains controlled by distinct populations in the thymus. J Exp Med 170: 2165-2170

v. Boehmer H (1990) Developmental biology of T cells in T cell receptor transgenic mice. Annu Rev Immunol 8: 551-556

Zuniga-Pflücker JC, Jones LA, Longo AL, Kruisbeek AM (1990) CD8 is required during positive selection of $CD4^-/CD8^+$ T cells. J Exp Med 171: 427-437

The Beginning and the End of the Development of TCRγδ Cells in the Thymus

K. SHORTMAN, LI WU, KATHERINE A. KELLY, and R. SCOLLAY

The Walter and Eliza Hall Institute of Medical Research, Melbourne, Victoria 3050, Australia

1. Introduction

The thymus is primarily a site for the expansion, differentiation and selection of developing cells of the TCR-αβ lineage. However, TCR-γδ cells also develop in the thymus, and are found primarily within the $CD4^{-}8^{-}$ subpopulation (reviewed by Raulet, 1989). In our past studies (Pearse et al, 1988; Pearse et al, 1989) where we aimed to isolate subpopulations of adult mouse $CD4^{-}8^{-}$ thymocytes representing sequential stages of development along the TCR-αβ pathway, we have noted thymocytes with various degrees of rearrangement of TCR-γ genes, of expression of γ-mRNA, and of surface expression of TCR-γδ. Although the balance between TCR-αβ and TCR-γδ expression differed in different subpopulations, we have never found a combination of surface markers, apart form the TCR itself, which would separate the developing cells of the two lineages. The developmental pathways are intertwined, and it is not clear if there are two quite separate pathways, or if there are common streams at some stages.

In this presentation we will not attempt to untangle these developmental strands, but will ask some simple questions at the level of the earliest and the latest stages of intrathymic development. Firstly, does the earliest intrathymic precursor population for the TCR-αβ lineage also serve as a precursor of the γδ lineage? Secondly, are

Current Topics in Microbiology and Immunology, Vol. 173

the TCR-$\gamma\delta$ cells produced within the thymus continually exported to the periphery, in the same manner as mature cells of the TCR-$\alpha\beta$ lineage?

2. The earliest intrathymic precursor cells of the TCR-$\alpha\beta$ lineage.

T cell development in the adult thymus requires the continuous input of some form of stem cell or pro-thymocyte from bone marrow (reviewed by Scollay et al, 1986). This input leads to a small population of intrathymic T-precursor cells which, while unable to sustain long-time lymphopoiesis, can give a wave of reconstitution of all the other thymocyte populations. Most of such precursor activity is found within the minor $CD4^-8^-$ population (Fowlkes et al, 1985), which has led to the impression that all T precursors are of this phenotype. We set out to isolate the earliest intrathymic precursor cell by subdividing the $CD4^-8^-$ population (Pearse et al, 1989), searching for a cell with extensive thymic reconstitution activity, unrearranged TCR genes, and a surface antigen phenotype resembling that of bone marrow haemopoietic stem cells or bone marrow prothymocytes. We failed to find such a cell within the $CD4^-8^-$ population. We have recently succeeded in isolating such early precursors after realizing that they express low levels of the mature T-cell marker CD4 (Wu et al, 1990). We have termed this thymus population the "low CD4 precursor".

The strategy for isolating these earliest precursors was to first extensively deplete thymocytes of mature T cells, of most cortical thymocytes, and of the more developed of the $CD4^-8^-$ precursors (using anti-CD3, anti-CD8, anti-CD2 and anti-IL-2R) and of non-T-lineage cells (using antibodies against B cells, erythroid cells, macrophages and polymorphs). This left less than 1% of thymocytes. This small residue was then labeled in three fluorescent colors and the precursor cells sorted according to the expression of any three of a particular set of surface markers, namely: $CD4^+$, Thy 1^+, HSA^{++}, Pgp-1^{+++}, H-$2K^{+++}$, Sca-1^{++}, Sca-2^{++} (where + = low but positive, ++ = intermediate, +++ = very high). A single discrete subpopulation bearing this unique set of surface markers was then apparent, representing about 0.05% of the original thymocytes. Apart from the expression of Sca-2, these markers resemble those of bone-marrow hematopoietic stem cells (Spangrude et al, 1988). It is of particular interest that Fredrickson and Basch (1989) have recently found that such stem cells also express low levels of CD4.

We have determined the status of TCR genes in this low CD4 precursor population. Both TCR-β and TCR-γ genes have been found by Southern gel hybridisation analyses to be in the germline state, an unusual situation for a thymic lymphoid population, since even the earliest $CD4^-8^-$ precursors have already partially arranged TCR genes (Pearse et al, 1989).

On intrathymic transfer these low CD4 precursors reconstitute all major thymocyte populations, with an efficiency that is 50 to 80-fold greater than with $CD4^-8^-$ precursors. However, reconstitution takes around 7 days longer than with $CD4^-8^-$ precursors. Thus 1 week post-transfer the low CD4 precursors lose CD4 to produced $CD4^-8^-$ progeny; 2 weeks post-transfer most of the progeny are $CD4^+8^+$; not until 3 weeks post-transfer do mature $CD4^-8^+3^{++}$ and $CD4^+8^-3^{++}$ begin to accumulate in significant quantities and emigrate to the periphery.

3. Low CD4 precursors also generate TCR-γδ cells

To determine if this early thymocyte population contained the precursors of the γδ, as well as the αβ lineage, the purified low CD4 precursors were transferred intrathymically into irradiated recipients differing at the Thy 1 locus, and the recipient thymuses analysed at various times thereafter by four-color immunofluorescent staining and flow cytometry. Anti-Thy 1 was used to detect cells derived from the donor low CD4 precursors, and anti-CD4 and anti-CD8, together with low-angle light scatter and level of Thy 1 expression, was used to subdivide these donor-derived cells into the various thymic populations. Monoclonal antibodies specific for determinants on TCR-αβ (Kubo et al, 1989) or TCR-γδ (Goodman and Lefrancois, 1989) were used to determine which class of TCR was eventually expressed. The results are summarised in Table 1.

No cells expressing surface TCR were found 7 days post-transfer. At later times cells expressing low levels of TCR-αβ were found within the $CD4^{+}8^{+}$ donor-derived population and, as expected, cells expressing high levels of TCR-αβ were found amongst the mature "single positives". Some TCR-$\alpha\beta^{+}$ cells were also found amongst the $CD4^{-}8^{-}$ population. Cells expressing high levels of TCR-γδ were seen at 2 and 3 weeks post-transfer. Although these were infrequent amongst the progeny of the low CD4 precursors, they were a distinct and readily measured group. In contrast to the TCR-αβ cells, those expressing TCR-γδ were restricted to the $CD4^{-}8^{-}$ population. Since a low incidence of TCR-γδ expressing cells, concentrated in the $CD4^{-}8^{-}$ population, is the situation in a normal adult thymus, we conclude the "low CD4 precursors" include precursors of the TCR-γδ lineage.

4. Lineage restriction of the low CD4 precursor cells

The surface antigenic phenotype of the low CD4 precursors, and their lack of TCR-gene rearrangement, raises the question of their relationship to bone-marrow multipotent stem cells. The expression of Sca-2, absent from purified bone marrow stem cells (Spangrude et al, 1988) but acquired by them soon after injection into an irradiated thymus (Spangrude and Scollay, 1990), suggests the thymic population is a developmental step later than the multipotent stem cell. In contrast to bone marrow stem cell preparations, the low CD4 precursor population lacks CFU-s activity. On intravenous transfer to irradiated recipients these low CD4 precursors have so far been unable to reconstitute the myeloid lineages, but have produced both B-cell and T-cell progeny (unpublished data). Thus it is possible that the low CD4 precursors are lymphoid restricted stem cells. However, we cannot exclude the possibility that this population consists of three separate restricted precursors (for B cells, αβ-T cells and γδ-T cells), all with identical surface antigenic phenotype.

5. Exit of γδ T cells from the thymus

Turning to the other extreme of the intrathymic developmental process, we have asked whether the TCR-γδ bearing cells in the thymus represent a "dead end", with no relationship to the γδ-T cells in the periphery, or whether the thymus actually exports these cells. The murine thymus exports about 10^6 T cells per day, representing only 3-4% of the daily cell production within the organ. These few emigrants may be located, counted and phenotyped using the technique devised by Scollay et al, 1980, namely injection of fluorescein isothiocyanate into the thymus, followed by assay for green fluorescent cells in spleen and lymph nodes 3-24 hr later. Although the recent emigrants amount to less than 1% of peripheral T cells, they may with sufficient patience and care be immunofluorescent stained (using fluorochromes other than FITC) for the expression of CD4, CD8, CD3, TCR-αβ and TCR-γδ, and analysed by flow cytometry. A summary of the surface antigenic phenotype of recent thymic emigrants is given in Table 2, extending our recently published data (Kelly and Scollay, 1990). The dominant population exported from the thymus clearly has the phenotype of mature T cells ($CD4^-8^+$ or $CD4^+8$,

$CD3^{++}$). Virtually all of these mature emigrants express TCR-αβ, as expected. However, a small proportion of $CD4^{-}8^{-}$ cells are found amongst the recent emigrants. Of these about 25% express TCR-αβ and 75% express TCR-γδ. These very few γδ-bearing cells were brightly stained and stood out clearly above the background.

Thus, although the rate of export of γδ-T cells from the thymus is very low, about 10^4 cells per day, it definitely occurs and is roughly in line with the relative incidence of γδ-T cells within the adult thymus itself. We conclude that at least some of the γδ-T cells which develop within the thymus are destined for export to the periphery, and at least some of the γδ-T cells found in the periphery are of thymic origin.

Table 1 Reconstitution of TCR-αβ and TCR-γδ T lineages by the low CD4 precursors

Day of assay	Percent thymocytes of donor origin	Ratio recovered/injected donor cells	Population of donor derived cells	Percent of all donor derived cells	Percent of donor derived population expressing TCR αβ^{+}	αβ^{++}	γδ^{++}
13	17	320	CD4$^-$8$^-$	0.4	12	5	14
			CD4$^+$8$^+$	97.0	32	1	0
			CD4$^+$8$^-$3^{++} and CD4$^-$8$^+$3^{++}	0.7	0	100	0
21	70	740	CD4$^-$8$^-$	0.1	26	26	10
			CD4$^+$8$^+$	91.0	38	5	0
			CD4$^+$8$^-$3^{++} and CD4$^-$8$^+$3^{++}	8.0	0	100	0

The low CD4 subpopulation was isolated from C57BL/Ka Thy 1.2 mouse thymuses by depletion of most other cells followed by immunofluorescent labeling and sorting, using in this instance the markers Thy 1$^+$ HSA$^+$ H-2K^{+++}. Sorted cells (20,000) were injected into one lobe of an irradiated C57BL/Ka Thy 1.1 mouse thymus. At 7, 13 or 21 days after injection the thymocytes present in the lobe were analysed using four-color immunofluorescence and flow cytometry, gating for Thy 1.2$^+$ donor-derived cells. Separate samples were analyzed for TCR-αβ and TCR-γδ; + = low level of surface expression, ++ = high level of surface expression. Four mice were analysed per timepoint. No donor-derived cells surface positive for TCR-αβ or TCR-γδ were seen at day 7.

Table 2

Incidence of cells bearing TCR-$\gamma\delta$ amongst recent thymic emigrants

Subpopulation of migrants	Percent of all migrants	Percent of subpopulation TCR-$\gamma\delta^{++}$
CD4^{-}8^{-}	1	75
CD4^{+}8^{+}	2	2
CD4^{+}8^{-}	76	<1
CD4^{-}8^{+}	21	<1

Recent emigrants from the thymus to lymph nodes were determined as green fluorescent cells 16 hr after intrathymic injection of fluorescein isothiocyanate. The lymph node cells were stained with anti-CD4, anti-CD8 and anti-TCR-$\gamma\delta$, and the surface phenotype of the migrants determined by four-color flow cytometric analysis, gating for the green fluorescent cells. Only 0.3% of all lymph node cells were green fluorescent migrants. Of these 100% were CD3^{+}, but only about 1% were TCR-$\gamma\delta^{+}$. Similar results were obtained by analyzing thymic emigrants in the spleen.

5. References

1 Fowlkes BJ, Edison L, Mathieson BJ and Chused TM (1985) Early T lymphocytes: differentiation in vivo of adult intrathymic precursor cells. J Exp Med 162:802-822

2 Fredrickson GG, Basch RS (1989) L3T4 antigen expression by hemopoietic precursor cells. J Exp Med 169:1473-1478

3 Goodman T and Le Francois L (1988) Expression of the γδ T cell receptor on intestinal $CD8^+$ intraepithelial lymphocytes. Nature 333:855-858.

4 Kelly KA, Scollay R (1990) Analysis of recent thymic emigrants with subset and maturity related markers. Int Immunol 2:419-425

5 Kubo RT, Born W, Kappler JW, Marrack P, Pigeon M (1989) Characterisation of a monoclonal antibody which detects all murine αβ T cell receptors. J Immunol 142:2736-2742

6 Pearse M, Gallagher P, Wilson A, Wu L, Fisicaro N, Miller J, Scollay R, Shortman K (1988) Molecular characterization of T-cell antigen receptor expression by subsets of $CD4^-CD8^-$ murine thymocytes. Proc Natl Acad Sci USA 85:6082-6086

7 Pearse M, Wu L, Egerton M, Wilson A, Shortman K, Scollay R (1989) An early thymocyte development sequence marked by transient expression of the IL-2 receptor. Proc Natl Acad Sci USA 86:1614-1618.

8 Raulet DH (1989) The structure, function and molecular genetics of the γ/δ T cell receptor. Am Rev Immunol 7:175-207

9 Scollay R, Butcher E, Weissman I (1980) Thymus migration: quantitative studies on the rate of migration of cells from the thymus to the periphery in mice. Eur J Immunol 10:210-218

10 Scollay R, Smith J, Stauffer V (1986) Dynamics of early T cells: Prothymocyte migration and proliferation in the adult mouse thymus. Immunol Rev 91:129-157

11 Spangrude GJ, Heimfeld S, Weissman IL (1988) Purification and characterization of mouse hematopoietic stem cells. Science 241:58-62

12 Spangrude GJ, Scollay R (1990) Differentiation of hematopoietic stem cells in irradiated mouse thymic lobes: kinetics and phenotype of progeny. J Immunol, to be published.

13 Wu L, Scollay R, Egerton M, Pearse M, Spangrude GJ, Shortman K (1990) The earliest T-lineage precursor cells in the adult murine thymus express low levels of CD4. Submitted for publication.

Cyclosporin A (CsA) Prevents the Generation of Mature Thymic α/β T Cells but Spares γ/δ T Lymphocytes

PIA BADER, SYLVIA BENDIGS, H. WAGNER, and K. HEEG

Institute of Medical Microbiology and Hygiene, Technical University of Munich, Trögerstraße 4a, D-8000 Munich 80, FRG

The immunosuppressive agent Cyclosporin A (CsA) is a potent pharmakon known to suppress activation of immunocompetent T lymphocytes in vivo and in vitro (Shevach 1985). The mode of its action has focused once on the binding of CsA to calmodulin (Colombani et al. 1985) shown to be important for intracellular Ca^{++} regulation, and secondly, to specific inhibition of peptidyl-prolyl cis-trans isomerases essential for correct protein folding (Takahashi et al. 1989, Fischer et al. 1989). At the level of mature T cells, CsA inhibits lymphokine secretion by interfering with the corresponding m-RNA synthesis (Kroenke et al. 1984), as well as with the transition into cell cycle of resting T cells by a yet undefined mechanism (Heeg et al. 1984).

Recently it has been recognised that CsA not only affects mature peripheral T cells, but in addition has also dramatic effects on intrathymic T cell maturation in vivo (Jenkins et al. 1988, Gao et al. 1988, Kosugi et al. 1989a, Fukuzawa et al 1989, Heeg et al. 1989) as well as on T cell maturation in thymic organ cultures (Kosugi et al. 1989b, Bucy et al. 1990). CsA treatment of adult mice led to a rapid disappearance of single positive $CD4^+CD8^-$ thymocytes, yet the number of single positive $CD8^+CD4^-$ thymocytes was reduced only by 50%. Moreover, using Vβ17 (Jenkins et al. 1988) or Vβ11 (Gao et al. 1988) as marker for thymic negative selection, it has been shown that these T cell receptor (TCR) phenotypes were still generated during CsA treatment. Since control animals readily deleted Vβ17 or Vβ11 TCR bearing thymocytes, it was concluded that CsA prevented negative selection in the thymus.

To circumvent the problem posed by mature T cells already generated in adult mice, or bone marrow-reconstituted mice, when subjected to CsA treatment, we decided to analyse the effect of CsA on thymic T cell maturation in newborn mice. Pregnant mice were injected daily with 20mg CsA/kg body weight starting at day 15 of pregnany, and CsA treatment was continued with newborn mice for one to four weeks (Heeg et al. 1989). As expected CsA treatment prevented almost completely the generation of single

Current Topics in Microbiology and Immunology, Vol. 173

positive CD4+CD8- thymocytes but reduced the number of single positive CD8+CD4- thymic cells only marginally (Fig. 1). However, these CD8+CD4- thymocytes expressed only low levels of CD3-ε and α/β-TCR (Fig. 2) and were nonfunctional in vitro (Heeg et al. 1989). These cells have thus to be scored as immature thymic precursors. Moreover, mature CD8+CD4- or CD8-CD4+ T cells were virtually absent in the peripheral lympoid organs of CsA-treated mice (Fig. 3). In contrast, CsA treatment had no effect on the number of double negative CD8-CD4-CD3+ thymoctes (Fig. 2). The majority of these cells did not bear α/β-TCRs (Fig. 2). Table I shows that after CsA treatment

Figure 1. CD4/CD8 expression in thymocytes from normal and CsA-treated newborn mice.

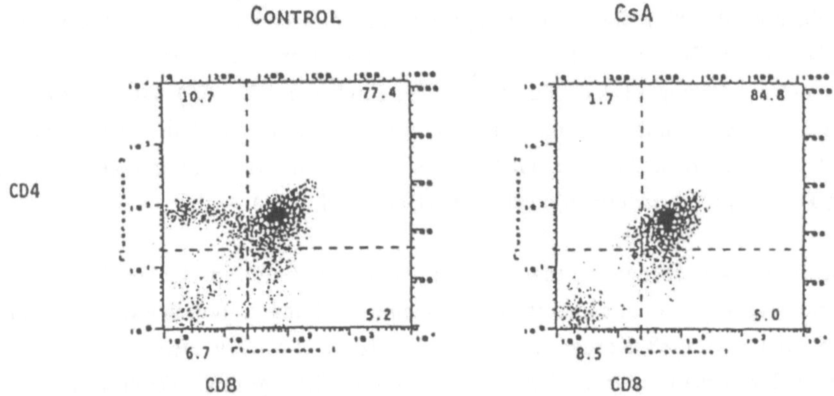

Thymocytes from seven days old normal (left) or CsA-treated (right) mice were stained with FITC-labelled anti-CD8 mAb and PE-conjugated anti-CD4 mAb. 3x10^4 cells were then analysed cytofluometrically.

Table 1. Thymic cell phenotype after CsA-treatment.

THYMIC CELL NUMBER x 10^6		CD3+ x 10^6		CD3+CD4-CD8- % OF CD3+	
CONTROL	CsA	CONTROL	CsA	CONTROL	CsA
44	42	4.2	0.9	6.4	51.1
50	19	8.5	2.4	7.2	25.5
75	38	9.8	2.5	4.9	22.1
61	18	4.5	0.2	5.3	65.3
50	42	14.1	3.0	5.1	48.9

Figure 2. **CD3-ϵ and α/β TCR expression in thymocytes from control and CsA-treated mice.**

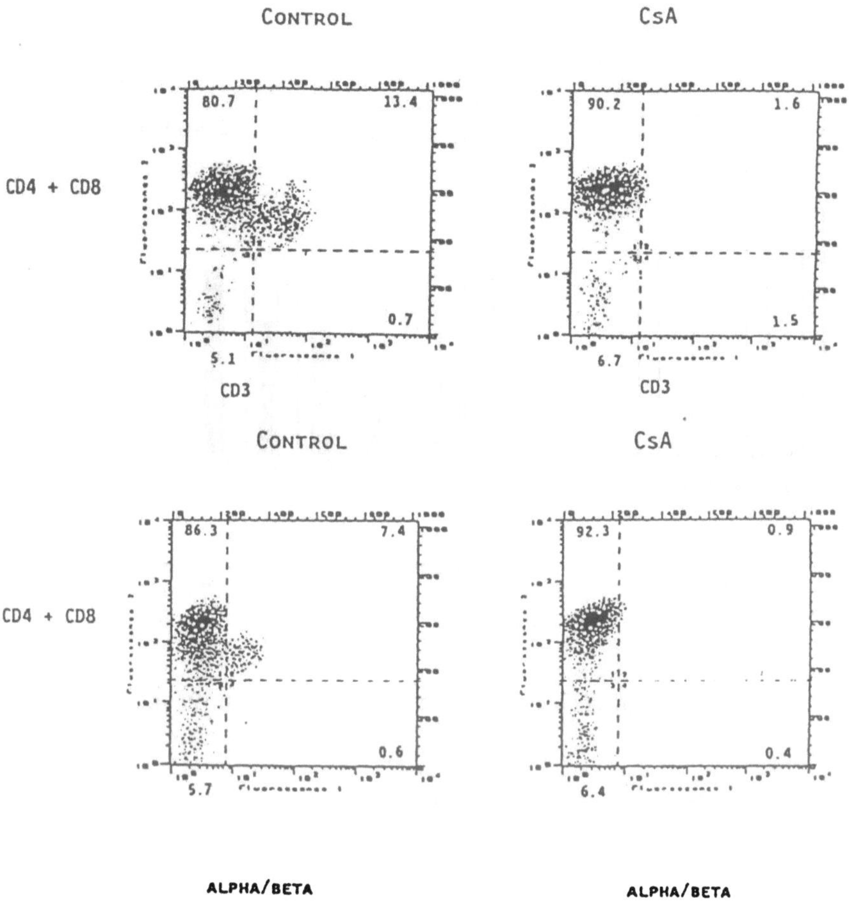

Thymocytes from seven days old normal (left) or CsA-treated (right) mice were stained with biotin-conjugated anti-CD4 and biotin-conjugated anti-CD8 mAb followed by streptavidin-PE. Then the cells were counterstained with FITC-labelled anti-CD3-ϵ mAb or FITC-conjugated anti-α/β framework mAb. $3x10^4$ cells were analysed.

double negative thymocytes represent the majority of the mature CD3-positive thymic T cell pool. This double negative $CD8^-CD4^-$ subpopulation was also detectable in peripheral lymphoid organs of CsA-treated newborn mice. A time-course of their appearance in the spleen revealed that these cells seem to accumulate (Fig. 3). It should be stressed, however, that peripheral lympoid tissues from CsA-treated animals are rather hypoplastic (Heeg et al. 1989).

Figure 3. Mature CD3+ T cells in splenocytes from normal and CsA-treated mice.

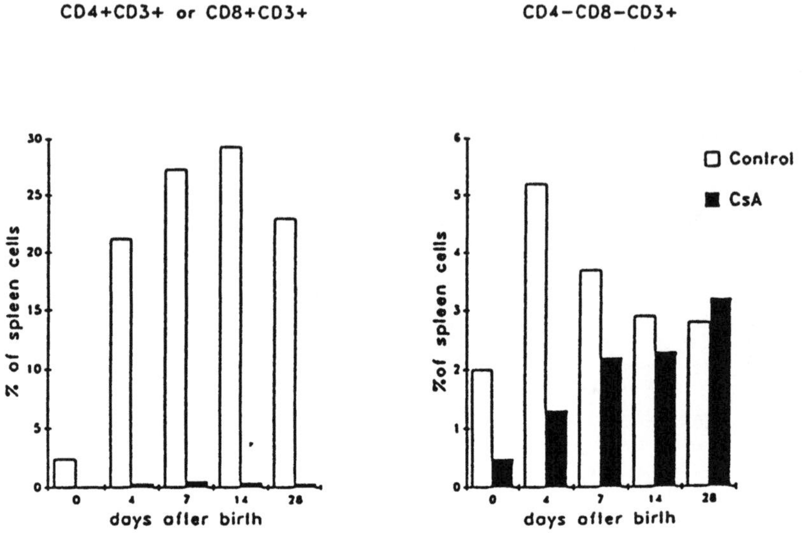

Splenocytes from normal (open column) or CsA-treated (closed column) mice were stained with anti-CD4, anti-CD8, and anti-CD3 mAb and analysed cytofluometrically The percentage of CD4+CD3+ plus CD8+CD3+ (left) and CD4-CD8-CD3+ T cells in the spleens is given.

Figure 4. T cell phenotype of splenocytes after in vitro expansion via anti-CD3 hybridomas.

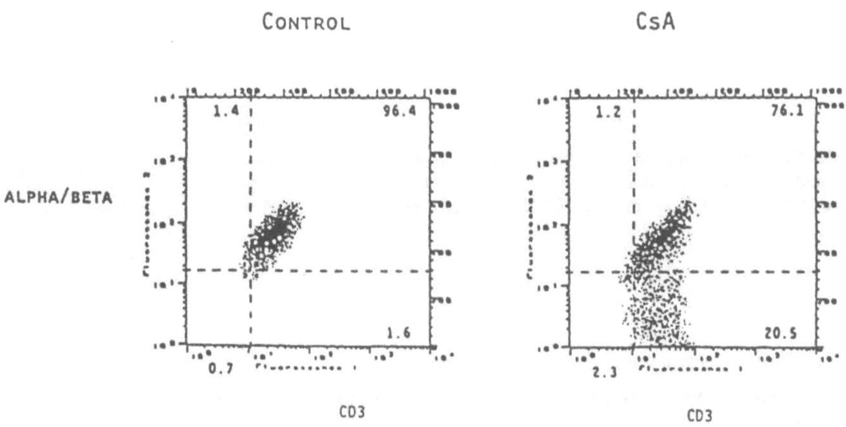

Spleen cells from normal (left) or CsA-treated (right) mice were expanded in vitro with anti-CD3-hybridomas and Il-2. After seven days the cells were harvested and stained with biotin-coupled anti-α/β mAb plus streptavidin PE and FITC-coupled anti-CD3 mAb. $3x10^4$ cells were analysed.

To further analyse the functional phenotype of the CsA-resistant $CD8^-CD4^-CD3^+$ thymocytes and peripheral T cells we expanded in vitro thymic and splenic T cells from CsA-treated (or control) mice by stimulation with anti-CD3-ϵ mAb producing hybridomas. After seven days of culture the growing T cells were analysed phenotypically (Fig. 4, 5) as well as functionally (Fig. 6). A large fraction of growing T cells from the spleen from CsA-treated mice was $CD3^+CD8^-CD4^-\alpha/\beta^-$ (Fig. 4), this T

Figure 5. T cell phenotype of thymocytes after in vitro expansion via anti-CD3 hybridomas.

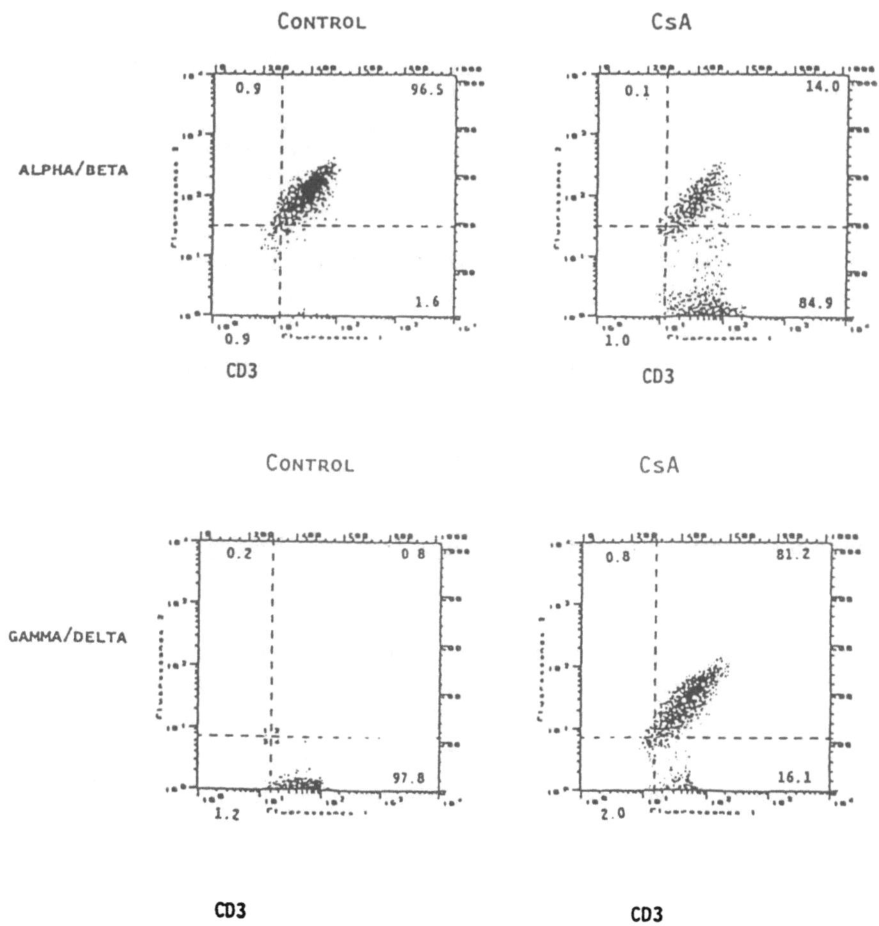

Thymocytes from normal (left) or CsA-treated (right) mice were expanded in vitro with anti-CD3-hybridomas and Il-2. After seven days growing cells were harvested and stained with biotin-coupled anti-α/β mAb plus streptavidin PE or anti-γ/δ mab plus PE-conjugated rat anti-hamster IgG and FITC-coupled anti-CD3 mAb. $3x10^4$ cells were analysed.

Fig. 6. CD3-facilitated cytotoxicity of in vitro expanded thymocytes from CsA-treated newborn mice.

Thymocytes were expanded in vitro with anti-CD3 hybridomas and Il-2. After 7 days the growing blast cells were tested in a ^{51}Cr release assay against CD3-hybridoma target cells.

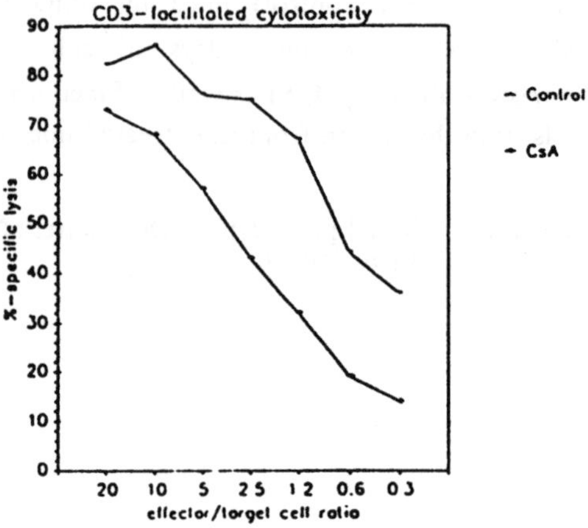

cell phenotype was dominating in T cells growing from thymi of CsA-traeted mice (Fig. 5). Staining with a mAb specific for γ/δ-TCRs (a kind gift from Dr. S.Tonegawa) revealed that these cells were indeed $CD3^{+}\gamma/\delta^{+}$ (Fig. 5) and were negative for CD8 or CD4 expression (data not shown). Next, growing T cell populations from thymi of normal or CsA-treated mice were then tested for cytolytic potential using anti-CD3-ϵ hybridomas as target cells. Both polyclonally activated α/β T cells from normal as well as γ/δ T cells from CsA-treated mice displayed high lytic activity against anti-CD3-expressing target cells (Fig. 6). However, in contrast to α/β T cells, we found no lysis of allogeneic or syngeneic ConA or LPS blast target cells by γ/δ bearing T cells from CsA-treated mice (data not shown).

There is agreement that CsA prevents thymic positive selection of both $CD8^{+}CD4^{-}$ as well as of $CD8^{-}CD4^{+}$ mature α/β T cells (Jenkins et al. 1988, Gao et al. 1988, Kosugi et al. 1989a, Fukuzawa et al 1989, Heeg et al. 1989, Kosugi et al. 1989b, Bucy et al. 1990). Since the mode of action of CsA in mature T cells involves prevention of activation as well as prevention of lymphokine secretion (Bunjes et al. 1981, Heeg et al. 1984, Shevach 1985), it is envisaged that such steps are critical in the process of positve selection of α/β T cells. At first glance this appears not to be the case for γ/δ T cells in vivo (Jenkins et al. 1988) and in thymic organ cultures (Kosugi et al. 1989b, Matsuhashi et al. 1989, Bucy et al. 1990). This T cell subpopulation seems to be CsA-resistant (Figs. 2, 5). Moreover these cells accumulate in the periphery (Fig. 3). However, until now there are only limited data available on the $V\gamma/V\delta$ repertoire expressed in CsA-treated mice. Therefore the question remains whether the expression of certain $V\gamma/V\delta$ TCRs might be prevented and/or facilitated by CsA-treatment.

The paradoxical finding that in the presence of CsA syngeneic or allogeneic bone marrow transplanted animals develop fulminant graft-versus-host diseases provided CsA treatment is terminated (Glazier et al. 1983), has led to the conclusion that CsA in addition to its influence on thymic positive selection also prevents thymic negative selection of selfreactive T cells (Jenkins et al. 1988, Gao et al. 1988). Using $V\beta 17$ and $V\beta 11$ as marker for IE reactive α/β T cells it has been shown that such T cells are still present in the periphery of CsA-treated mice (Jenkins et al. 1988, Gao et al. 1988), although they were detectable only in low numbers, and nonfunctional in vitro (Gao et al. 1988). When we used the superantigen SEB to induce tolerance in newborn CsA-treated mice and analysed the corresponding $V\beta 8$-expression in the remaining small numbers of mature thymic α/β T cells, we found a complete deletion of this $V\beta$-phenotype suggesting an negative selection mechanism for α/β thymocytes is operative in CsA-treated mice (Pia Bader, unpublished observation). What then are the candidates for selfreactive GvHD causing T cells in CsA-treated animals? Since after termination of CsA-treatment thymectomy in such animals results rather in excerabration than in prevention of GvHD (Sakaguchi and Sakaguchi 1989), we are left to speculate that γ/δ T cells, whose generation is not affected by CsA, might be effector cells in the GvHD caused by CsA.

Aknowledgements:

We thank Mrs. Birgit Deeg for her excellent technical assistance.

This work was supported by the SFB 322.

Correspondence to:
Klaus Heeg, Insitute of Medical Microbiology and Immunology, Technical University of Munich, Trogerstr.4a, 8000-Munich 80, Germany

References:

Bunjes D, Hardt C, Roellinghoff M, Wagner H (1981) Cyclosporin A mediates immunosuppression of primary cytotoxic T cell responses by impairing the release of interleukin 1 and interleukin 2. Eur J Immunol 11: 657-661

Bucy RP, Li JM, Xu XY, Char D, Chen CL (1990) Effect of cyclosporin A on the ontogeny of different T cell sublineages in chickens. J Immunol 144:3257-3265

Colombani PM, Robb A, Hess AD (1985) Cyclosporin A binding to calmodulin: a possible site of action on T lymphocytes. Science 228:337-339

Fischer G, Wittmann-Liebold B, Lang K, Kiefhaber T, Schmid FX (1989) Cyclophilin and peptidyl-prolyl cis-trans isomerase are probably identical proteins. Nature 337:476-478

Fukuzawa M, Sharrow SO, Shearer GM (1989) Effect of cyclosporin A on T cell immunity. II. Defective thymic education of CD4 T helper cell function in cyclosporin A-treated mice. Eur J Immunol 19:1147-1152

Gao EK, Lo D, Cheney R, Kanagawa O, Sprent J (1988) Abnormal differentiation of thymocytes in mice treated with cyclosporin A. Nature 336:176-179

Glazier A, Tutschka PJ, Farmer ER, Santos GW (1983) Graft versus host disease in cyclosporine A-treated rats after syngeneic and autologous bone marrow reconstitution. J Exp Med 158:1-8

Heeg K, Deusch K, Solbach W, Bunjes D, Wagner H (1984) Frequency analysis of cyclosporine sensitive cytotoxic T lymphocyte precursors. Transplantation 38:532-536

Heeg K, Bendigs S, Wagner H (1989) Cyclosporine A prevents the generation of single positive ($Lyt2^+L3T4^-$, $Lyt2^-L3T4^+$) mature T cells, but not single positive ($Lyt2^+L3T4^-$) immature thymocytes, in newborn mice. Scand J Immunol 30:703-710

Jenkins , Schwartz RH, Pardoll DM (1988) Effects of Cyclosporine A on T cell development and clonal deletion. Science 2411655-1658

Kosugi A, Sharrow SO, Shearer GM (1989a) Effect of cyclosporin A on lymphopoiesis. I. Absence of mature T cells in thymus and periphery of bone marrow transplanted mice treated with cyclosporin A. J Immunol 142:3026-3032

Kosugi A, Zuniga-Pflucker JC, Sharrow SO, Kruisbeek AM, Shearer GM (1989b) Effect of cyclosporin A on lymphopoisis. II. Developmental defects of immature and mature thymocytes in fetal thymus organ cultures treated with cyclosporin A. J Immunol 143:3134-3140

Kroenke M, Leonard WJ, Depper JM, Arya SK, Wong-Staal F, Gallo RC, Waldmann TA, Greene WC (1984) Cyclosporin A inhibits T-cell growth factor gene expression at the level of mRNA transcription. Proc Natl Acad Sci USA 81:5214-5218

Matsuhashi N, Kawase Y, Suzuki G (1989) Effects of cyclosporine A on thymocyte differentiation in fetal thymus organ culture. Cell Immunol 123:307-315

Sakaguchi S, Sakaguchi N (1989) Organ-specific autoimmune disease induced in mice by elimination of T cell subsets. V. Neonatal administration of cyclosporin A causes autoimmune disease. J Immunol 142:471-480

Shevach EM (1985) The effects of cyclosporine A on the immune system. Ann Rev Immunol 3:397-423

Takahashi N, Hayano T, Suzuki M (1989) Peptidyl-prolyl cis-trans isomerase is the cyclosporin A-binding protein cyclophilin. Nature 337:473-475

Other Species

Lack of Alloantigen-Specific Cytotoxic T Cell Activity in the TCR-γδ T Cell Subpopulation of Alloantigen-Immune Chickens

J. Cihak, H. Merkle, and U. Losch

Institute of Animal Physiology, University of Munich, Veterinärstraße 13, D-8000 München 22, FRG

INTRODUCTION

Recently T cell receptor (TCR) homologues of the mammalian TCR-αß (TCR2) and the TCR-γδ (TCR1) have been defined in the chicken (Sowder et al. 1988; Cihak et al. 1988; Chen et al. 1988), where the TCR1 cells may constitute approximately 20-65 % of the T cell pool (Cihak et al. in preparation). In addition to the TCR1 and TCR2 cells a third subpopulation of T cells has been identified in the chicken (Chen et al. 1989; Cooper et al. 1989). These cells, provisionally named "TCR3" cells, are TCR-αß T cells which use the Vß2 gene product whereas TCR2 cells use the Vß1 gene product (M.D. Cooper, personal communication and in this issue).

Knowledge about the function of the TCR-γδ T cells is relatively limited, being restricted primarily to the study of cultured cell lines. Recently several groups reported on alloreactive TCR-γδ murine and human cytotoxic T lymphocyte (CTL) clones that recognized class I MHC antigens (Matis et al. 1987; Bluestone et al. 1988; Rivas et al. 1989; Ciccone et al. 1989; Spits et al. 1990).

While most of the chicken TCR1 cells in blood are negative for the CD4 and CD8 accessory molecules, most of the splenic TCR1 cells express the CD8 antigen (Chen et al. 1988). This suggests a possible cytotoxic function against target cells expressing class I or class I like molecules. Hence, we asked whether alloimmunization of chickens induces the TCR1 subpopulation in the spleen to express alloantigen-specific CTL activity. As we will show herein, in vivo immunization with allogeneic erythrocytes induces the TCR2 and "TCR3" cells to exert alloantigen-specific CTL activity, while TCR1 cells lack such an activity.

MATERIALS AND METHODS

Animals. Chickens of strain RPRL6 (B2/B2) and CB (B12/B12) were used.

Monoclonal Antibodies. The production of monoclonal antibodies specific for chicken molecules TCR1, TCR2, CT3 (anti-CD3), CT4 (anti-CD4) and CT8 (anti-CD8) has been described previously (Chen et al. 1988; Cihak et al. 1988; Sowder et al. 1988; Chen et al. 1986; Chan et al. 1988). All antibodies are of the IgG1 class.

Immunofluorescence Staining. Immunofluorescence staining and flow cytometry were performed as described (Cihak et al. 1988).

Induction and Assay of Alloantigen-Specific Cytotoxic T Lymphocytes. Induction and assay of allospecific CTL were performed according to a modification (Cihak et al. 1988) of a published method (Chi et al.1981).

Current Topics in Microbiology and Immunology, Vol 173

Negative Selection of TCR1, TCR2 and "TCR3" Cells. TCR1, TCR2 and "TCR3" cells were negatively selected using rosetting with antibody-coated sheep erythrocytes according to Wilhelm et al (1986). In brief, when selecting TCR1 cells chicken spleen cells were mixed with anti-TCR2 and anti-"TCR3" antibody-coated sheep erythrocytes. For negative selection of TCR2 cells anti-TCR1 and anti-"TCR3" antibody coated erythrocytes and for selection of "TCR3" cells anti-TCR1 and anti-TCR2-coated erythrocytes were used. Cell mixtures were incubated for 1 h on ice. Rosetted cells were centrifuged on Ficoll (Pharmacia, Uppsala, Sweden) and the negatively selected nonrosetted cells were collected from the interphase.

Concanavalin A stimulation and assay of lymphocyte DNA synthesis as well as "panning" procedure were performed according to Cihak et al. (1988).

RESULTS

We immunized B2 chickens with allogeneic B12 cells and used the spleen cells from the immunized animals as effector cells in a ^{51}Cr release assay. When analyzing the specificity of the cytotoxicity it was observed that only the B12 targets were lysed by the spleen effector cells while autologous B2 and unrelated B5 and B19 cells were unaffected (Fig. 1). In order to delineate the type of cells involved in this specific cytotoxicity response, we performed inhibition studies with purified monoclonal antibodies directed against CD3, CD4 and CD8 molecules. As might be expected from the mammalian precedent, both the CT3 and CT8 antibodies virtually abolished the cytotoxic activity indicating the $CD8^+$ nature of the cytotoxic T lymphocytes. By contrast, no inhibition was seen with the CT4 antibody and two irrelevant isotype control antibodies F71 and G2. Also, elimination of plastic adherent cells or B cells did not affect the cytotoxicity (Fig. 1).

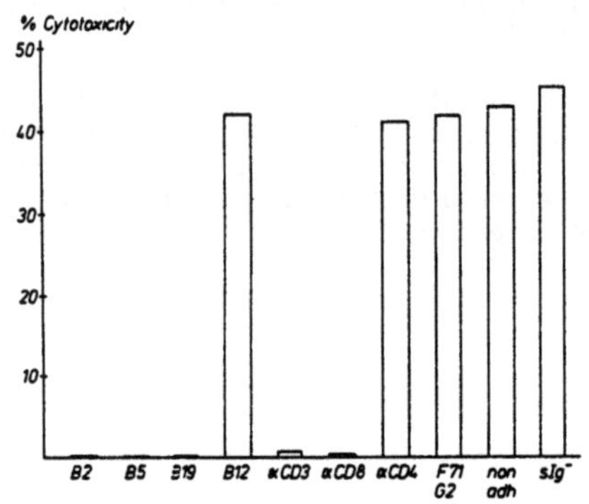

Fig. 1 Specificity and T cell nature of cytotoxic effector cells. Cytotoxic activity of alloantigen-immune B2 spleen cells was assayed against B12, B2, B5 and B19 erythrocyte targets at an E/T ratio of 40:1 in an 14h ^{51}Cr release assay. The ^{51}Cr release assay with B12 target cells was performed in the presence or absence of 10 µg/ml purified anti-CD3, anti-CD4, anti-CD8 or F71 and G2 isotype control antibodies. Adherent cells were eliminated by incubation of effector spleen cells on tissue culture plates for 2h at 37°C and Ig^+ cells were eliminated by "panning".

Spleen cells from B2 chickens were stained with F71 antibody as isotype control or with anti-TCR1, anti-TCR2, anti-"TCR3" or anti-CD3 antibodies in indirect immunofluorescence and the percentage of cells positive for each antibody was evaluated by FACS analysis. The distribution of the different types of T cell receptor-bearing cells in chicken spleen cells is shown in Table 1. While a high percentage of $CD3^+$ spleen cells expressed TCR1 or TCR2 (42±12 and 43±10% respectively), "TCR3" cells accounted only for 14±5% of all $CD3^+$ spleen cells (n=10). Spleen cells from B2 chickens exhibited 64±4% $CD3^+$ cells (Table 2).

In order to examine TCR1, TCR2 and "TCR3" cells for alloantigen-specific CTL activity, we separated these T cell subpopulations from spleen of alloantigen-immune chickens using rosetting with antibody-coated sheep erythrocytes. The negatively selected TCR1 and "TCR3" cells were >95% pure, the TCR2 cells >90% pure (Table 1).

Table 1. Negative selection of TCR1, TCR2 and "TCR3" cells from spleen cells by direct mab rosetting

Before rosetting				After rosetting		
% total $CD3^+$ cells						
TCR1	TCR2	"TCR3"	Selection	TCR1	TCR2	"TCR3"
42.4±11.7*	42.9±10.3	14.3±4.8	TCR1	96.5±3.9	0.4±0.5	2.5±3.3
			"TCR3"	1.0±1.5	1.1±1.5	97.5±2.2
40.0±20.0+	45.2±14.1	14.8±10.2	TCR2	0.2±0.2	90.6±2.3	9.2±2.1

*Mean ± SD of values from 10 negative selections
+Mean ± SD of values from 3 negative selections

Table 2 shows the percentage of T cells in TCR1, TCR2 and "TCR3" cells negatively selected from alloantigen-immune spleen cells and used as effector cells in the ^{51}Cr-release assay. The negatively selected TCR1 cells exhibited 18.3 to 40.1% $CD3^+$ cells, the TCR2 cells 56.6% and the "TCR3" cells 21.3 and 24.4% $CD3^+$ cells.

Table 2. Percentage of T cells in unseparated and in negatively selected effector spleen cells.

Effector cells	% $CD3^+$ cells		
	Exp. 1	Exp.2	Exp.3
Unseparated SPC	66.4	58.8	66.6
Selected TCR1 cells	18.3	40.1	30.1
Selected TCR2 cells	n.d.	n.d.	56.6
Selected "TCR3" cells	21.3	24.4	n.d.

In order to exclude the possibility that the separation procedure adversely affected the viability and functional activity of the negatively selected cells which were examined for alloantigen-specific CTL activity, we measured the proliferative response of the separated cells to concanavalin A. All negatively selected T cell subpopulations could be activated by Con A and no marked difference in Con A response of TCR1, TCR2 and "TCR3" cells could be observed (Table 3).

Table 3. Proliferative response of negatively selected TCR1, TCR2 and "TCR3" cells to concanavalin A

Stimulus	^{3}H-Thymidine incorporation (cpm)			
	Exp. 1		Exp. 3	
	TCR1 cells	"TCR3" cells	TCR1 cells	TCR2 cells
Culture medium	987*	434	820	620
Con A	4467	1310	3448	6072

*Numbers represent ^{3}H-thymidine incorporation by 5 x 10^5 cells from triplicate cultures

The alloantigen-specific CTL activity of TCR1, TCR2 and "TCR3" cells was examined in comparison to unseparated immune spleen cells in three experiments. While negatively selected TCR2 and "TCR3" cells exerted alloantigen specific CTL activity ranging from 10 to 47.4 % specific lysis - an activity comparable to cytotoxic activity of unseparated spleen cells - the TCR1 cells did not exert cytotoxic activity in any experiment (Fig. 2).

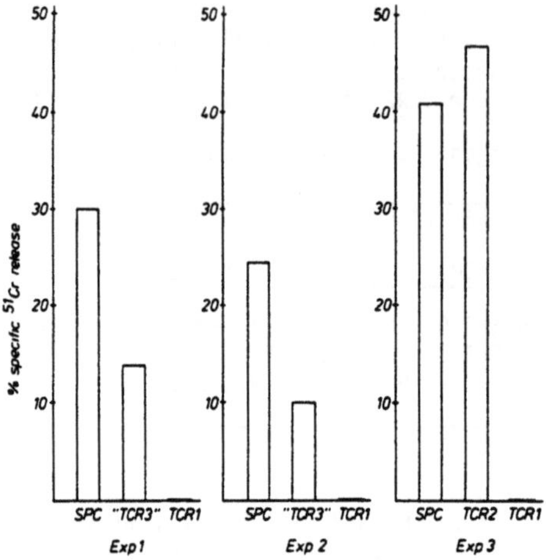

Fig. 2. Alloantigen-specific CTL activity of unseparated immune spleen cells and negatively selected TCR1, TCR2 and "TCR3" cells. The cytotoxic activity of unseparated alloantigen-immune B2 spleen cells (SPC) and of TCR1, TCR2 and "TCR3" cells negatively selected from the immune spleen cells against B12 (nucleated) erythrocyte targets was assayed at an E/T ratio of 40:1 (Exp. 1 and 2) or 80:1 (Exp. 3) in an 14h ^{51}Cr release assay. Mean spontaneous ^{51}Cr release from B12 target cells (2.5×10^4 cells in 200 µl culture medium/well) was 1%.

DISCUSSION

The function of the TCR1 T cells remains elusive. Insight into the nature of the ligands that are recognized by TCR1 is crucial for an understanding of the function of TCR1 cells. Recently, murine and human alloreactive TCR1 cell clones have been described that recognized class I MHC antigens (Matis et al. 1987; Bluestone et al. 1988; Rivas et al. 1989; Ciccone et al. 1989; Spits et al. 1990). These reports prompted us to investigate, whether alloreactivity can be induced in the TCR1 subpopulation in vivo. The high frequency of TCR1 cells makes the chicken an excellent model for this type of study.

Alloantigen-specific CTL response was induced by immunization of B2 chickens with allogeneic B12 erythrocytes. The antigen recognized by the alloantigen-specific CTL in the present study is presumably a class I MHC antigen, because the allogeneic target cells, nucleated erythrocytes in this case, express MHC class I (B-F) but not class II (B-L) antigens. The biochemical structure, adult tissue distribution, and functional attributes of B-F antigens show that they are the chicken equivalents of the mammalian class I MHC molecules (Guillemot et al. 1989).

The cell surface antigen profile of alloantigen-specific cells in the chicken indicates their CD8$^+$ cytotoxic T cell nature. Thus, anti-CD3 and anti-CD8 antibodies completely inhibited the cytotoxic activity, whereas no inhibition was seen with the anti-CD4 antibody, suggesting the absence of the CD4 accessory molecule on alloantigen-specific CTL in the chicken. These findings suggest a functional conservation of the CD3, CD4 and CD8 homologues in birds and mammals.

We separated TCR1, TCR2 and "TCR3" cells from spleens of alloantigen-immune chickens using a negative selection method. The negatively selected TCR1 cells were >95% pure and viable, as shown by their proliferative response to concanavalin A. When examining the negatively selected cells for alloantigen-specific CTL activity it was found that TCR2 and "TCR3" cells expressed specific cytotoxic activity against B12 targets, but the TCR1 cells did not exert any cytotoxic activity. These findings indicate that chicken TCR1 cells either do not recognize the immunizing alloantigen or that the frequency of alloantigen-specific CTL is too low in this subpopulation to be detectable in the ^{51}Cr release assay.

SUMMARY

We examined the in vivo generation of alloantigen-specific cytotoxic T lymphocytes in chicken TCR-α/β and TCR-$\gamma\delta$T cells. The TCR-$\alpha\beta$ and TCR-$\gamma\delta$ subpopulations were separated from spleen of alloantigen-immune chickens using a negative selection method. The separated cells were examined for alloantigen-specific CTL activity in a ^{51}Cr release assay. While negatively selected TCR-α/β cells exerted alloantigen-specific CTL activity, no cytotoxic activity could be detected with TCR-$\gamma\delta$ cells.

Acknowledgements

The authors thank Dr. M.D. Cooper for the gift of the monoclonal antibodies, Dr. W. Hartmann and Dr.K.Hala for providing B2 eggs and B12 chickens and Mrs M. Reuter for typing the manuscript. This work has been supported by DFG grant Lo 279/5-1.

REFERENCES

Bluestone JA, Cron G, Cotterman M, Houlden BA, Matis LA (1988) Structure and specificity of T cell receptor γ/δ on major histocompatibility complex antigen specific CD3$^+$, CD4$^-$, CD8$^-$ T lymphocytes. J Exp Med 168:1899-1916
Chan MM, Chen CH, Ager LL, Cooper MD (1989) Identification of the avian homologues of mammalian CD4 and CD8 antigens. J Immunol 140: 2133-2138
Chen CH, Ager LL, Gartland GL, Cooper MD (1986) Identification of a T3/T cell receptor complex in chickens. J Exp Med 164: 375-380
Chen CH, Cihak J, Lösch U, Cooper MD (1988) Differential expression of two T cell receptors, TCR1 and TCR2, on chicken lymphocytes. Eur J Immunol 18:539-543
Chen CH, Sowder JT, Lahti JM, Cihak J, Lösch U, Cooper MD (1989) TCR3: A third T cell receptor in the chicken. Proc Natl Acad Sci USA 86:2351-2355
Chi DS, Blyznak N, Kimura A, Palladino MA, Thorbecke GJ (1981) Cytotoxicity to allogeneic cells in the chicken. Cell Immunol 64:246-257
Ciccone E, Viale O, Pende D, Malnati M, Ferrara GB, Barocci S, Moretta L (1989) Specificity of human T lymphocytes expressing a γ/δ T cell antigen receptor. Recognition of a polymorphic determinant of HLA class I molecules by a γ/δ clone. Eur J Immunol 19:1267-1271
Cihak J, Ziegler-Heitbrock HWL, Trainer H, Schranner I, Merkenschlager M, Lösch U (1988) Characterization and functional properties of a novel monoclonal antibody which identifies a T cell receptor in chickens. Eur J Immunol 18:533-537
Cooper MD, Sanchez P, Char D, George JF, Lahti JM, Bucy RP, Chen CH, Cihak J, Lösch U, Coltey M, Le Douarin NM (1989) Ontogeny of T cells and a third lymphocyte lineage in the chicken. In Progress in Immunology VII, Melchers F ed., Springer-Verlag, Heidelberg.
Guillemot F, Kaufmann JF, Skjoedt K, Auffray C (1989) The major histocompatibility complex in the chicken. Trends Genet 5:300-304
Matis LA, Cron R, Bluestone JA (1987) Major histocompatibility complex-linked specificity of $\gamma\delta$ receptor-bearing T lymphocytes. Nature 330:262-264
Rivas A, Koide J, Cleary M, Engleman EG (1989) Evidence for involvement of the $\gamma\delta$ T cell antigen receptor in cytotoxicity mediated by human alloantigen specific T cell clones. J Immunol 142:1840-1846
Sowder JT, Chen CH, Ager LL, Chan MM, Cooper MD (1988) A large subpopulation of avian T cells express a homologue of mammalian T$\gamma\delta$ receptor. J Exp Med 167:315-322
Spits H, Paliard, Engelhard VH, de Vries JE (1990) Cytotoxic activity and lymphokine production of T cell receptor (TCR)$\alpha\beta^+$ and TCR-$\gamma\delta^+$ cytotoxic T lymphocyte (CTL) clones recognizing HLA-A2 and HLA-A2 mutants. J Immunol 144:4156-4162
Wilhelm M, Pechumer H, Rank G, Kopp E, Riethmüller G, Rieber EP (1986) Direct monoclonal antibody rosetting. J Immunol Methods 90:89-96

Thymic Nurse Cells: a Site for Positive Selection and Differentiation of T Cells

G. WICK, THERESA RIEKER, and J. PENNINGER

Institute for General and Experimental Pathology, University of Innsbruck, Medical School, and Immunoendocrinology Research Unit of the Austrian Academy of Sciences, Fritz-Pregl-Straße 3, A-6020 Innsbruck, Austria

1. Introduction

The capacity to discriminate between "self" and "non-self" is acquired in the thymus where T cell progenitors develop to mature, antigen specific, self-tolerant T cells (Fink and Bevan 1978, Zinkernagel 1978). This process is based on the direct contact of developing T cells with stroma cells and extracellular matrix constituents and the action of humoral factors secreted by nonlymphoid and lymphoid thymic components (von Boehmer et al. 1989, Marrack and Kappler, 1988). The most widely - albeit not exclusively - accepted hypothesis postulates that thymocytes expressing T cell receptors (TCR) with affinity to MHC class I and II molecules displayed on cortical epithelial cells are positively selected (MHC restriction) (Benoist and Mathis 1989, Bill and Palmer 1989). In the medulla the interaction with (auto)antigen presenting bone marrow derived dendritic cells and/or macrophages entails clonal deactivation and/or deletion of T cells expressing potentially harmful TCR (negative selection) (Miller et al. 1989, von Boehmer et al. 1989). The exact sites of positive and negative selection are not yet known. However, thymic nurse cells (TNC) seem to provide an optimal microenvironment for the former (for review see Kyewski 1986). If they are also involved in negative selection is still an open question.

2. Thymic nurse cells

Thymic nurse cells are subcortically located multicellular complexes of single epithelial cells containing intact thymocytes enclosed withing vacuoles lined by the epithelial cell plasma membrane (Wekerle et al. 1980, Kyewski 1986). TNC have been described in mice and rats (Wekerle et al. 1980), humans (Ritter et al. 1981), chickens (Boyd et al. 1984) and frogs and tadpoles (Wick and Du Pasquier, unpublished). On average murine TNC contain 50 TNC-L, human TNC 20, chicken TNC 4 and frog/tadpole TNC 1-2 TNC-L. TNC are not phagocytic and it is not yet clear which thymocytes enter into these epithelial cells and how this ingression is brought about. The possible role of certain adhesion molecules during this process is currently under study in our laboratory.

Since thymic epithelial cells express MHC class I and class II antigens in high density it was suggested that TNC provide a specialized microenvironment where developing T cell come into close and prolonged contact with self MHC (Kyewski 1986). It has recently been shown by Lorenz and Allen (1989) and by Marrack et al. (1989) that thymic epithelial cells are able to present non-self and self antigens in the context of self

Current Topics in Microbiology and Immunology, Vol 173

MHC to T cell hybridomas and clones. We became interested in TNC through the observation that chickens of the Obese strain (OS) that develop a spontaneous autoimmune thyroiditis closely resembling human Hashimoto's disease show a severe deficiency of TNC (Boyd et al. 1984). This fact and the unique experimental possibilities in an avian system prompted our phenotypic and functional studies of chicken TNC that led to the concept of their role in self - non-self discrimination.

3. Phenotypic characterization of chicken TNC-L

The phenotypic characteristics of chicken TNC-L were determined by double staining immunofluorescence tests using mouse monoclonal antibodies against the chicken analogues of CD3, CD4, CD8, the T cell receptor 1 (TCR γ/δ), TCR 2 (TCR α/β) (kindly provided by Dr. M. D. Cooper) and the light chain of the interleukin-2 receptor (IL-2R) (Penninger et al. 1990a). TNC and extra-TNC thymocytes were prepared from the thymi of 4-12 week old white Leghorn chickens of the highly inbred congenic strains CB (MHC = B12B12) and CC (B4B4). Extra-TNC thymocytes were subjected to fluorescence activated cell sorter (FACS) analysis, TNC-L were assessed visually. Table 1 shows a simplified version of the results of 4 experiments a detailed account of which can be found elsewhere (Penninger et al. 1990a). From these and earlier data the following conclusions can be drawn: (i) $CD4^+ CD8^+$ (double positive) T cells are more numerous outside TNC, while $CD4^+ CD8^-$ and $CD4^- CD8^+$ (single positive) cells are encountered more frequently inside TNC. Thus, thymocytes seem to differentiate within TNC from double positive to single positive T cells. (ii) TNC-L consist of a larger proportion of TCR α/β^+ and a smaller proportion of TCR γ/δ^+ cells thus excluding a clonal origin of these cells. All TCR^+ cells coexpress CD3. (iii) Similar to the situation with the accessory molecules CD4 and CD8, $TCR^+ CD3^+$ cells are enriched inside TNC supporting the role of this microenvironment for T cell maturation and differentiation. (iv) As shown previously (Penninger er al. 1990a) but not mentioned in table 1, we constantly found a small but consistent population of $CD4^+$ or $CD8^+$ thymocytes expressing the TCR γ/δ. This is in contrast to the data of Sowder et al (1988) and Chan et al. (1988) but in agreement with recent similar findings in mice (Itohara et al. 1989).

Table 1. TNC-L and extra-TNCthymocyte subpopulations

Cell phenotype	Percent positive cells ±SEM Thymocytes	TNC-L
$CD4^+CD8^+$	**60.3 ± 4.4**	**46.4 ± 4.5**
$CD4^+CD8^-$	**5.8 ± 1.6**	**24.3 ± 4.1**
$CD4^-CD8^+$	**10.4 ± 2.0**	**27.3 ± 5.5**
$CD3^+$	**46.6 ± 4.0**	**70.0 ± 6.5**
$TCR\alpha\beta^+$	**36.4 ± 1.9**	**60.5 ± 2.6**
$TCR\gamma\delta^+$	**8.2 ± 1.9**	**20.3 ± 1.6**
$IL\text{-}2R^+$	**44.2 ± 1.3**	**23.2 ± 3.3**

4. Functional analysis of TNC-L

4.1. Alloreactivity of TNC-L

For functional analysis of the possible alloreactivity of TNC-L the chorionallantoic membrane (CAM) assay in avian systems provides an optimal and unique experimental tool (Burnet and Burnet 1960). In this assay T cells from immunocompetent mature donors are deposited onto the CAM of 10 day (ED10) old histoincompatible recipient embryos, which are evaluated macroscopically, (immuno)histologically and functionally on ED16. Since MHC antigens of the recipient are expressed on the CAM specific donor T cells proliferate in situ leading to the appearance of macroscopically visible foci, so called "pocks". Each pock is derived from a single T cell with graft-versus-host reaction (GvHR) potential and finally consits of both donor and recruited recipient T cells (Simonsen 1967). The number of pocks is proportional to the MHC disparity between donor and recipient.
The congenic donor-recipient combination in our experiments differed only at the MHC (B locus in chickens). Suspensions of peripheral blood lymphocytes (PBL), thymocytes and single, micromanipulated donor TNC (corresponding to approximatly 4 lymphocytes/TNC) from the same CC (B4B4) donor were transferred onto the CAM of CB (B12B12) recipient embryos (Wick and Oberhuber 1986, Penninger et al. 1989). Surprisingly, TNC-L displayed a considerably higher GvHR efficiency (1/18) than PBL ($1/10^4$) or thymocytes ($1/10^5$). Sex and syngeneic homing were excluded as possible factors contributing to this reactivity (Penninger 1990a). The cellular composition of such primary (1°) TNC-L induced as compared to PBL-induced pocks is shown in Table 2. The most salient points are: (i) the contribution of both donor and recruited recipient T cells to pock formation with the expected preponderance of recipient cells; (ii) the participation of both TCR α/β^+ and TCR γ/δ^+ cells in the reaction; (iii) the participation of both $CD4^+$ and $CD8^+$ donor and recipient cells; (iv) the expression of IL-2R not only by mature donor but also by embryonic T cells.

In order to prove the immunological specificity and thus true GvHR nature of this phenomenon serial tranfer experiments were performed (Penninger et al 1990a). For this purpose 1° pocks were produced by transfer of TNC-L, thymocytes or PBL from CC (B4B4) donors onto the CAM of CB (B12B12) recipients. Single 1° pocks were then excised on ED16 and transplanted in toto onto the CAM of ED10 secondary (2°) recipients comprising the MHC haplotypes B4 (like original donor), B12 (like 1° recipient) or B15 (unrelated third party). On ED14 these 2° pocks were evaluated. Further proliferation (increase in pock size) only occurred on B12 recipients thus suggesting immunological specificity of the reaction. Definitive proof for this suggestion came from the significant increase of the spleen weights of the secondary recipients which - in addition to the local proliferation within the 2° pocks - reflects a generalized GvHR (Simonsen, 1967). Data of one such experiment are given in Table 3.

Table 2. FACS analysis of 1° TNC-L or PBL induced pocks

Cell phenotype	Percent positive cells [a] PBL pocks CB-host	PBL pocks CC-donor	TNC-L pocks CB-host	TNC-L pocks CC-donor
TCR α/β	**10.7**	**11.5**	**6.8**	**3.3**
TCR γ/δ	**6.4**	**7.8**	**3.9**	**3.0**
CD4	**6.6**	**16.4**	**7.2**	**5.9**
CD8	**7.9**	**15.6**	**8.2**	**4.6**
IL-2R	**54.0**	**9.0**	**46.5**	**5.5**
surface Ig	**27**	**n. d.**	**21**	**n. d.**
MHC class I	**61.0**	**22.5**	**59.8**	**11.6**
MHC class II	**26.8**	**23.7**	**20.2**	**18.4**

a Percentages of positive cells were evaluated in immunofluorescence double staining by "gated analysis" using a FACS III. The mean value of positive cells from 5 experiments is shown. Standard deviations are omitted for clarity sake. TCR α/ß, TCR γ/δ, CD4, CD8 and IL-2 receptor light chain positive cells were visualized by means of monoclonal antibodies, surface Ig positive cells by a goat anti-chicken Ig affinity chromatography purified antibody, MHC class I and class II positive cells by alloantibodies. Double staining for donor and recipient type MHC class I carrying cells and the respective other markers were done by combining the appropriate directly FITC-labelled anti-MHC class I alloantibody with indirect immunofluorescence using a given monoclonal antibody visualized by a goat anti-mouse Ig PE conjugate. n.d. = not done.

Table 3. Serial transfer of 1° allogeneic pocks

Donor cells [a]	1° Host	2° Host	n	SI (X±SEM) [b]	RSI (X±SEM) [b]
CC-B4 TNC	CB-B12	CB-B12	5	**98.0 ± 5.8**	**1.30 ± 0.08** [c]
		CS-B4	5	**46.6 ± 1.3**	**0.99 ± 0.01**
		OS-B15	3	**43.7 ± 4.3**	**0.90 ± 0.04**

a Seven single living TNC from a CC-B4 donor were transplanted per CAM of 10 day old (ED10) congeneic CB-B12 embryos (n = number of recipients). On ED16 1° pocks were isolated and transferred in toto onto the CAM of ED10 secondary hosts. The 2° pocks were evaluated on ED14.

b SI = spleen index = [(mg spleen weight)/(g body weight)]x10^5; RSI = relative spleen index = mean SI transplanted animals/mean SI sham manipulated animals. RSI ≥1.3 indicates a specific GvHR.

c Significantly different from sham manipulated control (theoretical RSI = 1.0) and both B4 and B15 recipients ($p < 0.05$, paired student's t-test).

4.2. Syngeneic reactivity of TNC-L

Transfer of PBL or thymocytes from immunocompetent donors onto the CAM of syngeneic recipients did result in pock formation in a frequency of only 1/2 x 10^5 or

1/4 x 10^5 T cells, respectively. However, TNC-L were capable to induce pocks on syngeneic hosts with the astonishingly high efficiency of 1/35 - 1/50, i.e. only slightly lower than in allogeneic combinations (Wick and Oberhuber 1986, Penninger and Wick 1990b). Serial transfer experiments were then again performed to establish the specificity of this phenomenon. CB (B12B12) PBL, thymocytes or single TNC from the same donor were applied onto the CAM of CB hosts, the 1° pocks were excised and transplanted onto the CAM of 2° CB hosts or recipient embryos carrying other unrelated MHC haplotypes. Splenomegaly of 2° hosts was again taken as an indicator for a generalized GvHR. Table 4 presents the data of such an experiment showing that significant splenomegaly as determined by the relative spleen index (RSI) occurred only when the 2° recipient carried the same MHC haplotype as the 1° recipient. Thus, TNC-L seem to be in a stage of development where they are already positively selected for self-MHC restriction but not yet negatively selected for self tolerance.

Table 4. Serial transfer of 1° syngeneic pocks

Donor cells [a]	1° Host	2° Host	n	SI (X±SEM) [b]	RSI (X±SEM) [b]
a) CB-B12 TNC	CB-B12	CB-B12	16	97.3 ± 5.3	1.29 ± 0.07 [c]
		B7-B7	8	71.5 ± 3.6	0.95 ± 0.05
		CS-B4	6	59.2 ± 2.7	0.99 ± 0.04
b) CB-B12 PBL	CB-B12	CB-B12	4	96.6 ± 6.9	1.28 ± 0.09 [c]
		B7-B7	5	92.7 ± 1.3	1.24 ± 0.02 [c]
		CC-B4	5	83.3 ± 1.9	0.99 ± 0.02
		CS-B4	5	48.2 ± 5.7	0.85 ± 0.06

a Five single living TNC (group a) or 1x10^6 PBL from the same CB-B12 donor were transplanted and serially transferred as desribed in Table 3.

b SI and RSI as indicated in Table 3.

c Significantly different from sham operated controls (theoretical RSI = 1.0) and from recipients with different MHC ($p < 0.05$; paired student's t-test).

5. Conclusions

(a) TNC seem to provide an optimal microenvironment for T cell differentiation and positive selection
(b) TNC contain TCR α/β^+ and TCR γ/δ^+ cells
(c) TNC harbour T cells already displaying a mature phenotype and therefore seem to be a site for the differentiation of $CD4^+$ $CD8^+$ (double positive) to $CD4^+$ $CD8^-$ or $CD4^-CD8^+$ (single positive) T cells
(d) TNC-L are highly efficient in allogeneic GvHR
(e) TNC-L comprise a subpopulation of thymocytes which have already undergone positive selection

(f) TNC-L are at a pretolerant stage of maturation and have not yet undergone negative selection

6. Acknowledgements

This work was supported by the Austrian Research Council (project Nr. 7391). Dr. M. D. Cooper and Dr. J. Cihak generously supplied monoclonal antibodies to chicken T cell markers, Dr. K. Hála alloantibodies to chicken MHC antigens.

7. References

Benoist C, Mathis D (1989) Positive selection of the T cell repertoire: where and when does it occur? Cell 58: 1027-1033

Bill J, Palmer E (1989) Positive selection of CD4+ T cells mediated by MHC class II-bearing stromal cells in the thymic cortex. Nature 341: 649-651

Boyd RL, Oberhuber G, Hála K, Wick G (1984) Obese strain (OS) chickens with spontaneous autoimmune thyroiditis have a deficiency in thymic nurse cells. J. Immunol. 132: 718-724

Burnet FM, Burnet D (1960) Graft versus host reactions on the chorionallantoic membrane of the chicken embryo. Nature 188: 376-377

Chan MM, Chen CH, Ager LL, Cooper MD (1988) Identification of the avian homologues of mammalian CD4 and CD8 antigens. J. Immunol. 140: 2133-2138

Fink P, Bevan M (1978) H-2 antigens of the thymus determine lymphocyte specificity. J. Exp. Med. 148: 766-775

Itohara S, Nakanishi N, Kanegawa O, Kubo JR, Tonegawa S (1989) Monoclonal antibodies specific to native murine T cell receptor γ/δ analysis of γ/δ T cell during thymic ontogeny and in peripheral lymphoid organs. Proc. Natl. Acad. Sci. USA 86: 5094-5098

Kyewski BA (1986) Thymic nurse cells: possible sites of T cell selection. Immunol. Today 7: 374-379

Lorenz RG, Allen PM (1989) Thymic cortical epithelium lack full capacity for antigen presentation. Nature 340: 557-559

Marrack P, Lo D, Brinster R. Palmiter R, Burkly L, Flavell RH, Kappler JW (1988) The effect of thymus environment on T cell development and tolerance. Cell 13: 627-634

Marrack P, McCormack J, Kappler J (1989) Presentation of antigen, foreign major histocompatibility complex proteins and self by thymus cortical epithelium. Nature 340: 557-559

Miller JFAP, Morahan G, Allison J (1989) Immunological tolerance: new approaches using transgenic mice. Immunol. Today 10: 53-57

Penninger J, Klima J, Kroemer G, Dietrich H, Hála K, Wick G (1989) Intrathymic nurse cell lymphocytes can induce a graft-versus-host reaction with high efficiency. Dev. Comp. Immunol. 13: 313-327

Penninger J, Hála K, Wick G (1990a) Intrathymic nurse cell lymphocytes can induce a specific graft-versus-host reaction. J. Exp. Med. 172: 521-529

Penninger J, Wick G (1990b) Thymic nurse cell lymphocytes react against self-MHC. submitted for publication

Ritter MA, Sauvage CA, Cotmore SF (1981) The human thymus microenvironment: in vivo identification of thymic nurse cells and other antigenically distinct subpopulations of epithelial cells. Immunol. 44: 439-446

Simonsen M (1967) The clonal selection hypothesis evaluated by grafted cells reacting against their hosts. Cold Spring Harbor Symp. Quant. Biol. 32: 517-523

Sowder JT, Chen CH, Ager LL, Chan MM, Cooper MD (1988) A large subpopulation of avian T cells express a homologue of the mammalian T γ/δ receptor. J. Exp. Med. 167: 315-322

Von Boehmer H, Teh HS, Kisielow P (1989) The thymus selects the useful, neglects the useless and destroys the harmful. Immunol. Today 10: 57-61

Wekerle H, Ketelsen UP, Ernst M (1980) Thymic nurse cells. Lymphoepithelial cell complexes in murine thymuses: morphological and serological characterization. J. Exp. Med. 151: 925-944

Wick G, Oberhuber G (1986) Thymic nurse cells: a school for alloreactive and autoreactive cortical thymocytes? Eur. J. Immunol. 16: 855-858

Zinkernagel RM (1978) Thymus and lymphohematopoietic cells: their role in T cell maturation in selection of T cells, H-2-restriction specificity and H-2 linked IR gene control. Immunol. Rev. 42: 224-270

Marked Variations in γδ T Cell Numbers and Distribution Throughout the Life of Sheep

C. R. Mackay and W. R. Hein

Basel Institute for Immunology, Grenzacherstraße 487, Ch-4005 Basel

INTRODUCTION

A large proportion of T cells in sheep and cattle are γδ TCR+ (Mackay et al., 1989; Mackay and Hein, 1989; Hein et al., 1990a) and their function, as in other species, appears to be concerned mostly with the protection of epithelial surfaces (reviewed by Hein and Mackay, 1991). The large numbers of γδ T cells in sheep and cattle favor these species for the study of γδ T cells, particularly from the standpoint that these cells probably play a major role in cell-mediated immunity in these "γδ -high" species. In addition, the study of γδ T cells within various niches of vertebrate evolution might reveal how the γδ T cell system evolved, and its relationship to the αβ T cell system.

Here we wish to document one particularly pertinent feature of γδ T cells which may relate to their functional significance: that is, the striking prominence of γδ T cells in the circulation during late fetal and early post natal life. We will present only a brief outline of other aspects of the γδ T cell system in ruminants, since this has been the subject of recent reviews (Hein and Mackay, 1991; Mackay and Hein, 1991).

Distinctive features of γδ T cells in sheep and cattle

The features and function of γδ T cells may show heterogeneity amongst the vertebrate species; this is exemplified by interspecies variability with respect to numbers, concentrations at different epithelial sites, and the expression of cell surface molecules. Hence, γδ T cells may play different roles, or may have greater importance in some species compared with others. In sheep and cattle, γδ T cells comprise up to 60% of peripheral blood lymphocytes (PBL), depending on the age of the animal (see below), but the high level in the blood can be misleading since the blood contains only 2% of all lymphoid cells (Trepel, 1974) and is not necessarily representative of the total peripheral lymphocyte pool (Mackay et al., 1988). In fact, the most dramatic example of this is seen in sheep and cattle, since there is only ~1-5% γδ T cells in the mesenteric lymph node (LN) and about 2-10% in other LNs, yet there are about 60% γδ T cell among lymphocytes in the dermis and epidermis, and blood (see below). These differences suggest that γδ T cells specifically home and localize within certain tissues such as skin, and function there rather than in LNs (Mackay et al., 1988, 1989).

γδ T cells in the circulation

In our studies in sheep, we have tried to determine the differences between αβ and γδ T cells, so as to understand the roles of each type of T cell in immune responses. One striking difference between these subsets is the level and/or expansion of each cell

Current Topics in Microbiology and Immunology, Vol 173

type in the periphery over the lifetime of sheep (Figure 1). The proportions of γδ, CD4 and CD8 T cells in the circulation varies greatly during fetal and post-natal life (Figure 1). The first T cells to appear in the blood are not γδ T cells, but CD4⁺ and CD8⁺ T cells, at around 50 days of gestation. γδ T cells first appear around 69 days (Figure 1 and Maddox et al., 1987), although at this stage CD4⁺ and CD8⁺ cells comprise the majority of PBL. Between 100 and 140 days, this trend is dramatically reversed, since γδ T cells increase to around 60% of PBL by 140 days, and this high representation of γδ T cells in the blood continues for at least the first three months after birth (Figure 1; see also McClure et al., 1989 and Hein et al., 1990b). From 4 months to 2 years of age, the proportion of γδ T cells in the blood declines steadily, and in old age (5-8 years) γδ T cells stabilize at around 5-10%.

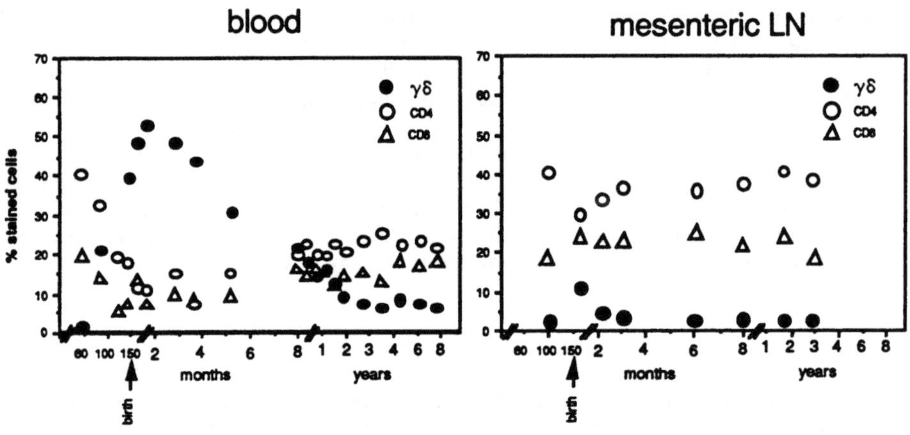

Fig. 1. γδ, CD4 and CD8 T cell percentages in the blood (left) and mesenteric lymph node (right) of sheep of various gestational and post-natal ages. γδ percentages were determined using a γδ-specific mAb 86D (Mackay et al., 1989).

Recirculation pathways for αβ and γδ T cells

The circulation of lymphocytes in blood and lymph serves to disseminate the functional subsets of the immune system to various tissues in the body. For αβ T cells, continual recirculation is necessary since the proportion of these cells which are reactive to any given antigen is very low. Hence, at some stage in evolution, strategically placed organized lymphoid tissue developed to allow for the collection and concentration of antigen, and for the percolation of large numbers of lymphocytes from the blood through such tissue. The lymphatic system also allows for the dissemination of antigenically experienced cells, generated at one site, to various other sites in the body.

The differential migration pathways for T cell subsets to lymphoid tissue is clearly obvious if one compares the levels of γδ T cells in the blood with those in LNs (Figure 1). γδ T cells usually comprise less than 2% of mesenteric LN lymphocytes. An exception is during the perinatal period when γδ T cells levels in the blood are at their highest, when the level of γδ T cells in LNs is around 10-15% (Figure 1). The general low levels of γδ T cells in LNs can be explained by the fact that γδ T cells do not localize or recirculate through LNs in the same way as CD4⁺ or CD8⁺ T cells, perhaps because γδ

T cells do not bind to LN high endothelial venules (HEV) as avidly as other types of T cells. On this point we have two pieces of conflicting data. The MEL-14/LAM-1 molecule, which is thought to be responsible for lymphocyte binding to HEV, is expressed at levels 5-8 fold higher on γδ T cells compared with $CD4^+$ or $CD8^+$ T cells. However, if a lymphocyte sample is labeled with FITC and injected back into the animal, then the capacity of γδ T cell migration to LNs from blood, over a 30 minute time period, is far less than that of $CD4^+$ T cells (Mackay unpublished). The vast majority of γδ T cells in sheep blood express a memory T cell phenotype i.e. $CD45RO^+$, which for αβ T cells signifies that they recirculate through peripheral tissues such as skin and gut rather than LNs (Mackay, 1991). Thus while it appears that the main recirculation pathway for most αβ T cells is by entering LNs through HEV, γδ T cells may be more prone to home to other tissues such as skin, although some γδ T cell localization in LNs does occur.

Table 1. Distribution of γδ T cells at epithelial sites during the life of sheep

Epithelial site	140 days of gestation	1 month (post-natal)	1 year
Intestine	+	+++	+++
Skin	-	+	+ → +++
Lung	+	+	+
Esophagus	-	++	+++
Tongue	-	++	+++
Trachea	-	++	+ → +++

γδ T cells in the sheep thymus

The numbers, distribution and phenotype of γδ T cells in the sheep thymus also varies according to fetal and post-natal age. $\gamma\delta^+$ thymocytes are apparent from a very early stage of fetal thymic development, ~40 days (gestation = 150 days, thymic lymphopoiesis begins at ~35 days). These "early" γδ thymocytes localize to the outer cortex and cortex. By day 60, a thymic medulla is clearly evident and γδ T cells begin to localize there in addition to the cortex. An interesting aspect of sheep γδ T cell development is the phenotypic change by γδ thymocytes after transition from the cortical-type "immature" γδ thymocytes to the medullary-type "mature-like" γδ thymocytes. Cortical γδ thymocytes are mostly MHC I^- and $T19^-$, whereas medullary γδ thymocytes are usually $T19^+$ and MHC I^+. Many of the medullary γδ thymocytes are closely associated with Hassall's corpuscles (Mackay et al., 1989; Hein and Mackay, 1991), and we have speculated that this association may play a part in their maturation in the thymus.

The large number of γδ T cells in the blood of late-term fetal and post-natal sheep implies a large rate of γδ T cell production, either within the thymus or in the periphery. γδ T cell numbers within the thymus do vary, but not by much, i.e. in the range of 0.5% to 4%. However, the levels of γδ T cells in the blood correlate with increased levels in the thymus, particularly within the medulla. γδ T cell numbers in the periphery will depend on their production in the thymus, their expansion in the periphery, and their sequestration or death within particular tissues. However, γδ T cells appear to have a limited capacity for expansion in the periphery since γδ T cells in thymectomized lambs are unable to reconstitute the peripheral T cell pool, whereas $CD4^+$ and $CD8^+$ T cells can do so (Hein et al., 1990b).

γδ T cells at epithelial surfaces

The presence of γδ T cells at different epithelial sites varies considerably for different species. In the sheep, γδ T cells are prevalent at most epithelial surfaces, although this prevalence depends very much on age (summarized in Table 1). In the intestine, γδ T cells are prevalent before and after birth; they localize to the lamina propria, as well as within the epithelium. However, birth leads to an increase in γδ T cells in the gut, presumably because of the effect of antigen. The situation in the gut differs from the other epithelial surfaces we have examined, since γδ T cells are sparse or absent from tongue, esophagus, skin, trachea and lung before birth, although after birth, and particularly by 1 year of age, they can be numerous at these sites (Table I and Hein and Mackay, 1991).

The distribution of γδ T cells at epithelial sites before and after birth would indicate that antigen has a pronounced effect on their localization to these sites. Perhaps an exception is intestinal epithelium which contains γδ T cells even before birth. However, this might reflect the presence of "antigen" in the form of cellular debris, mucin etc. in the gut lumen, and these constituents may not be regarded as self. The absence of γδ T cells from most epithelial surfaces before birth implies that such cells migrate to these sites after birth, although in mice,"families" of γδ T cells are thought to populate different epithelial sites at different stages of ontogeny, and each family consists of γδ T cells expressing particular Vγ or Vδ components of the γδ TCR (Carding et al., 1990; Itohara et al., 1990)

DISCUSSION

The γδ T cell system of the sheep and other ruminants has some unique features. First, the number of γδ T cells change markedly over the life-span of sheep, and this might be accounted for by several factors.

(a) Different generation kinetics for αβ and γδ T cells in the thymus during fetal life, and their rate of emigration to the periphery.
(b) The effect of antigen on the localization and expansion of γδ T cells at epithelial surfaces after birth.
(c) The long term maintenance of T cells, either by their intrinsic longevity or by constant cell division, probably differs for αβ and γδ T cells.

Secondly, the rapid localization and/or expansion of γδ T cells within epithelia after birth might be important as a first line of immune defence, since αβ T cells might have to undergo more rigorous selection and activation procedures, and hence may not be sufficiently "mature" in a short time period to cope with the massive antigenic flux immediately after birth.

The variation in certain features of the γδ immune system between species might relate to physiological differences between animals in different evolutionary pathways, or might be somewhat meaningless. Interestingly, those species which are precocious in their behavior at birth, such as chickens, sheep and cattle, have high γδ T cell levels, possibly because they have greater exposure to antigen at epithelial surfaces. Thus, γδ T cells may provide protection of epithelial surfaces particularly during early life, and this role might then be assumed by αβ T cells when this T cell system is more mature and established.

ACKNOWLEDGEMENTS

We thank W. Marston and L. Dudler for technical assistance. The Basel Institute for Immunology was founded and is supported by Hoffmann-La Roche, Ltd., Basel Switzerland.

REFERENCES

Carding, S.R., Kyes, S., Jenkinson, E.J., Kingston, R., Bottomly, K., Owen, J.J.T. and Hayday, A.C. (1990). Developmentally regulated fetal thymic and extrathymic T cell receptor γδ gene expression. *Genes and Development* **4**: 1304-1315.

Hein, W.R. and Mackay, C.R. (1990). Prominence of γδ T cells in the ruminant immune system. *Immunol. Today* **12**: 30-34.

Hein, W.R., Dudler, L., and Morris, B. (1990b). Differential peripheral expansion and in vivo antigen reactivity of αβ and γδ T cells emigrating from the early fetal lamb thymus. *Eur. J. Immunol.* 20: 1805-1813.

Hein, W.R., Dudler, L., Marcuz, A. and Grossberger, D. (1990a). Molecular cloning of sheep T cell receptor γ and δ chain constant regions: unusual primary structure of γ chain hinge segments. *Eur.J. Immunol.* **20**: 1795-1804.

Itohara, S., Farr, A.G., Lafaille, J.J., Bonneville, M., Takagaki, Y., Haas, W. and Tonegawa, S. (1990). Homing of a γδ thymocyte subset with homogenous T cell receptors to mucosal epithelia. Nature **343**: 754-756.

Mackay, C.R. and Hein, W.R. (1989). γδ T cells comprise a large proportion of bovine lymphocytes and show a distinct tissue distribution and surface phenotype. *Int. Immunol.* **1**: 540-545.

Mackay, C.R. and Hein, W.R. (1990). Analysis of the γδ T cell system in ruminants reveals further heterogeneity in γδ T cell features and function among species. *Research in Immunol.* **141**: 611-619.

Mackay, C.R., Beya, M.F. and Matzinger, P. (1989). γδ T cells express a unique surface molecule appearing late during thymic development. *Eur. J. Immunol.* **19**: 1477-1483.

Mackay, C.R., Kimpton, W.G., Brandon, M.R. and Cahill, R.N.P. (1988). Lymphocyte subsets show marked differences in their distribution between blood, and afferent and efferent lymph draining peripheral lymph nodes. *J. Exp. Med.* **167**: 1755-1765.

Mackay, C.R., Maddox, J.F. and Brandon, M.R. (1986). Three distinct subpopulations of sheep T lymphocytes. *Eur. J. Immunol.* **16**: 19-25.

Mackay, C.R. (1991) The nature and behaviour of memory T cells. *Immunol. Today (in press)*

Maddox, J.F. Mackay, C.R. and Brandon, M.R. (1987). Ontogeny of ovine lymphocytes. III. An immunohistological study on the development of T lymphocytes in sheep fetal lymph nodes. *Immunology* **62**: 113-118.

McClure, S.J., Hein, W.R., Yamaguchi, K., Dudler, L., Beya, M.-F. and Miyasaka, M. (1989). Ontogeny, morphology and tissue distribution of a unique subset of $CD4^-$ $CD8^-$ sheep T lymphocytes. *Immunol. Cell Biol.* **67**: 215-221.

Trepel, F. (1974). Number and distribution of lymphocytes in man. A critical analysis. *Klin. Wochenschr.* **52**: 511-516.

γ/δ T-Lymphocyte Subsets in Swine

M. J. Reddehase[1], A. Saalmüller[2], and W. Hirt[2]

[1] Department of Virology, Institute for Microbiology, University of Ulm, 7900 Ulm, FRG
[2] Federal Research Centre for Virus Diseases of Animals, POB 1149, 7400 Tübingen, FRG

1 Retrospect on Swine Immunology: The Null Lymphocyte

Evidence for fundamental differences in the overall organization of mammalian lymphoreticular systems dates back to the work of CHIEVITZ (1881) demonstrating microanatomical inversion of swine lymph node tissues in that B- and T-lymphocyte lodging areas are interchanged. Associated with this is an altered route of lymphocyte recirculation: entry and exit both take place via venules, causing efferent lymph to be almost acellular. The specialty of the porcine immune system extends also to the composition of the lymphocyte pool. In addition to B and classical T lymphocytes, a third major population of so-called 'null lymphocytes' was distinguished by the absence of surface Ig and by its inability to form xenoerythrocyte rosettes (BINNS 1982). Recent work of our group has classified these formerly enigmatic cells as a $CD2^-CD4^-CD8^-$ subpopulation of T lymphocytes, encompassing two subsets of γ/δ T lymphocytes (SAALMÜLLER et al. 1989; HIRT et al. 1990).

2 Swine Break with a Dogma: Coexpression of CD4 and CD8 Defines a Fourth Subpopulation of T Lymphocytes

Extrathymic circulating as well as resident T lymphocyte pools are composed of four subpopulations: $CD2^+CD4^-CD8^+$, $CD2^+CD4^+CD8^-$, $CD2^+CD4^+CD8^+$, and $CD4^-CD8^-$, the latter of which is subdivided into $CD2^+$ and $CD2^-$ subsets (SAALMÜLLER et al. 1989). A compilation of data from individual swine revealed high variance and an inverse relation in the numbers of $CD4^-CD8^-$ and $CD4^+CD8^+$ T lymphocytes, either of which can range from few percent up to more than half of the total T-lymphocyte pool. Whereas $CD4^-CD8^-$ T lymphocytes constitute a major subpopulation also in peripheral blood of other ungulate species (MACKAY et al. 1986), porcine $CD4^+CD8^+$ T lymphocytes are without precedent. There is preliminary evidence that these cells express the porcine homolog of T cell receptor (TcR) α/ß (HIRT, unpublished).

This work was supported by the Deutsche Forschungsgemeinschaft grant Re 712/1-2

Current Topics in Microbiology and Immunology, Vol. 173

3 Subsets of Circulating γ/δ T Lymphocytes Defined by Expression of a Common-Ungulate and a Private TcR γ/δ

Monoclonal antibody 86D that defines a phylogenetically conserved external epitope present on the TcR γ/δ of sheep and cattle (MACKAY et al. 1989) subdivides porcine $CD2^-CD4^-CD8^-$ peripheral blood T lymphocytes into $86D^+$ and $86D^-$ subsets, which make up ca. one third and two thirds of this subpopulation, respectively. A rabbit antiserum raised against synthetic peptide W6 representing a highly conserved region in human Cδ (BONYHADI et al. 1987) precipitated CD3-associated, disulfide-bonded heterodimers from the surface of sorted $86D^+$ and $86D^-$ cells: in the 86D-type TcR γ/δ of swine a 40 kDa^R δ chain is linked to a 38 kDa^R second chain, tentatively designated γ. In the 86D epitope-negative TcR the γ chain has a slightly lower apparent molecular mass of 37 kDa^R. All chains are N-glycosylated and migrate at ca. 35 kDa^R after deglycosylation (HIRT et al. 1990).

4 Lymphoid Homing of γ/δ T Lymphocytes: A Third Type of Porcine TcR γ/δ Expressed by the $CD2^+CD4^-CD8^-$ Subset

Homing to lymphoid tissues is usually not a prominent behaviour of γ/δ T lymphocytes. Specifically, the $86D^+$ γ/δ T lymphocytes of ruminant ungulates are sparse-to-absent in lymphoid tissue, even though they are frequent in peripheral blood (MACKAY et al. 1989). Also with respect to lymphoid homing of γ/δ T lymphocytes swine provide an exception: Interestingly, the 86D ungulate-type γ/δ T lymphocytes of swine share the ungulate-typic low-to-negative homing propensity, whereas the $86D^-$ γ/δ subset, which has no counterpart in other ungulates, is evenly distributed between circulating and resident pools and accounts for most of the $CD2^-CD4^-CD8^-$ T lymphocytes that lodge in lymphoid tissues of swine.

As mentioned above, the porcine $CD4^-CD8^-$ subpopulation splits into $CD2^-$ and $CD2^+$ subsets, the latter of which is minute in peripheral blood, but is highly enriched in lymphoid tissues (SAALMÜLLER et al. 1989). Expression of CD2 and the 86D epitope specifies four subsets of $CD4^-CD8^-$ T lymphocytes residing in the spleen: $CD2^-86D^+$ and $CD2^+86D^+$, which are both rare, as well as the major subsets $CD2^-86D^-$ and $CD2^+86D^-$. Precipitation with the anti δ chain reagent identified for the sorted $CD2^+86D^-$ subset a novel TcR γ/δ of 46/40 kDa^R (SAALMÜLLER et al. 1990). The 46 kDa^R γ chain differs from the 37 and 38 kDa^R γ chains in its protein core size, as deglycosylation reduced its molecular mass to only 43 kDa^R (HIRT, manuscript in preparation). Whether the $CD2^+86D^+$ γ/δ T lymphocytes express the 38/40 kDa^R receptor or carry the 86D epitope on a fourth type of TcR γ/δ remains open to question, because the low frequency of these cells has so far precluded an analysis. In conclusion, in porcine TcR γ/δ a 40 kDa^R δ chain is linked to different γ chains: $\gamma1$, $\gamma2$, and $\gamma3$ with molecular masses of 37, 38, and 46 kDa^R, respectively (Table).

T-lymphocyte subpopulations in swine

Subpopulation	Subset	TcR	Lymphoid homing
$CD4^-CD8^+$		α / β	++
$CD4^+CD8^-$		α / β	++
$CD4^+CD8^+$		α / β	++
$CD4^-CD8^-$	$CD2^-86D^+$	38/40kDa^R γ / δ	-
	$CD2^-86D^-$	37/40kDa^R γ / δ	+
	$CD2^+86D^-$	46/40kDa^R γ / δ	++

5 Prospect on γ/δ T Lymphocyte Function: Hog Cholera, a Viral Disease Associated with Loss of γ/δ T Lymphocytes

Considering the high frequency and the highly developed subset organization and homing pattern of the porcine γ/δ T lymphocytes, it is reasonable to expect a function in immune surveillance. We have preliminary evidence that sorted $CD2^-$ γ/δ T lymphocytes do not respond to alloantigens, but can be stimulated by ConA and acquire CD8 upon activation (SAALMÜLLER, manuscript in preparation). A clue to function may come from Hog Cholera, a natural disease of swine caused by a pestivirus. We have found that at a terminal stage of lethal disease γ/δ T lymphocytes are selectively depleted (SUSA, manuscript in preparation). Future research will have to show whether this is just an epiphenomenon of infection or whether γ/δ immunodeficiency contributes to the pathogenesis, possibly by a breakdown of immune control at epithelial frontiers.

The help of Mrs. Ingrid Bennett with preparation of the manuscript was greatly appreciated.

REFERENCES

Binns RM (1982) Organisation of the lymphoreticular system and lymphocyte markers in the pig. Vet Immunol Immunopathol 3:95-146

Bonyhadi M, Weiss A, Tucker PW, Tigelaar RE, Allison JP (1987) Delta is the C_x gene product in the γ/δ antigen receptor of dendritic epidermal cells. Nature 330:574-576

Chievitz JH (1881) Zur Anatomie einiger Lymphdrüsen im erwachsenen und fötalen Zustande. Arch Anat Physiol:347-370

Hirt W, Saalmüller A, Reddehase MJ (1990) Distinct γ/δ T cell receptors define two subsets of circulating porcine $CD2^-CD4^-CD8^-$ T lymphocytes. Eur J Immunol 20:265-269

Mackay CR, Maddox JF, Brandon MR (1986) Three distinct subpopulations of sheep T lymphocytes. Eur J Immunol 16:19-25

Mackay CR, Beya M-F, Matzinger P (1989) γ/δ T cells express a unique surface molecule appearing late during thymic development. Eur J Immunol 19:1477-1483

Saalmüller A, Hirt W, Reddehase MJ (1989) Phenotypic discrimination between thymic and extrathymic $CD4^-CD8^-$ and $CD4^+CD8^+$ porcine T lymphocytes. Eur J Immunol 19:2011-2016

Saalmüller A, Hirt W, Reddehase MJ (1990) Porcine γ/δ T lymphocyte subsets differing in their propensity to home to lymphoid tissue. Eur J Immunol, in press

Transgenic Mice
Manipulation *In Vitro*

Mechanisms of Development of αβ T Cell Antigen Receptor-Bearing Cells in γδ T Cell Antigen Receptor Transgenic Mice

A.L. Dent and S.M. Hedrick

Department of Biology and Cancer Center, University of California, San Diego, La Jolla, CA 92093, USA

Our lab has generated transgenic mice that carry the genes encoding a γδ T cell antigen receptor (γδ TCR) from a γδ T cell clone with known specificity . This γδ T cell clone, G8, was derived from a BALB/c nu/nu mouse by alloantigen stimulation (Matis et al. (1987)). G8 is specific for an MHC class-I-like gene that maps to the TL region of the H-2 locus (Bluestone et al. (1988)). G8 is typical of γδ T cells that derive from the adult thymus or spleen in that it is negative for the CD4 and CD8 T cell markers, and uses a Vγ2-Jγ1-Cγ1- encoded γ chain for its TCR. Recently (Dent et al. (1990)), we have used these γδ TCR transgenic mice to study whether self-reactive γδ T cells are tolerized in a similar way as T cells that bear the αβ T cell antigen receptor (αβ T cells). In transgenic mice that did not bear the TL-encoded ligand ($H\text{-}2^{d/d}$), CD4/CD8 negative γδ T cells bearing the G8 specificity were found in the lymph nodes and spleen. In transgenic mice that expressed the TL-encoded ligand ($H\text{-}2^{b/d}$), transgenic γδ T cells were eliminated from the peripheral lymphoid organs, although Vγ2-positive cells persisted in the thymus. This phenotype would imply that a deletion process is occuring similar to that shown for self-reactive αβ T cells.

An interesting issue that arose in this work was whether αβ T cells would develop normally in these transgenic mice, or whether some form of allelic exclusion would prevent the development of αβ T cells. As we reported previously (Dent et al.

Current Topics in Microbiology and Immunology, Vol. 173

(1990)), and as we have found in subsequent work, the number of αβ T cells that are present in a given transgenic mouse is somewhat variable. Thus, some transgenic mice have normal numbers of αβ T cells in the thymus, whereas other transgenic mice are almost completely depleted of αβ T cells in the thymus. The number of αβ T cells in the thymus always correlates with the number of αβ T cells that are found in the periphery. Thus, a mouse with a small thymus has a relatively small number of αβ T cells in the spleen and lymph nodes, whereas a mouse with a normal sized thymus has normal number of αβ T cells in the spleen and lymph nodes. Apparently the γδ transgenes could, in some instances, inhibit αβ T cell development. Interestingly, despite the variability in the development of αβ T cells, the absolute number of γδ T cells was relatively constant from one transgenic to another. Transgenic γδ T cells therefore developed independently of the αβ T cells, and were a consistently limited number of cells both in the thymus and periphery in all types of transgenic mice. This result is illustrated in Figure 1, which shows four thymuses taken from six week-old transgenic littermates. Two of the thymuses were much smaller than normal (type 1), and two of the thymuses were normal sized (type 2). Note that the thymus size does not depend on presence or absence of the ligand for the γδ TCR. Importantly, the number of γδ T cells in these thymuses is fairly constant: 5.8 million cells ± 2.7 million cells. The actual variation in the number of γδ T cells is even less if the tolerance process is taken into account. H-$2^{b/d}$ mice express the TL-encoded ligand, and therefore have fewer γδ T cells in the thymus, presumably due to the clonal deletion process. A consistent finding is that the levels of Vγ2 expressed by H-$2^{b/d}$ γδ T cells are significantly lower than the levels of Vγ2 expressed by H-$2^{d/d}$ γδ T cells, which is a hallmark of γδ T cell tolerance in these mice (Dent <u>et al.</u> (1990)). In contrast to the relatively constant number of γδ T cells, the number of αβ T cells varies dramatically in the different thymuses. Thus, as shown in Figure 1, so-called type 1 thymuses have

10- to 50-fold fewer mature αβ T cells than type 2 thymuses. An important point is that in type 1 transgenic mice, the number of γδ T cells did not increase in order to "fill up" the deficient αβ T cell compartments. This result implies a separate developmental lineage for γδ T cells. Also important is that fact that type 2 $H\text{-}2^{d/d}$ mice show a relatively normal development of αβ and γδ T cells in terms of proportion of cells and the expression of cell surface phenotype (CD4 and CD8 expression).

The variability in αβ T cell development in the γδ TCR transgenic mice was not dependent on transgene copy number, age, sex, or as shown, the presence of the TL-encoded ligand. On the other hand, the presence of αβ T cells was dependent on the founder line of the transgenic mice, and in some cases by the strain of mouse to which the transgenes were bred. Six original transgenic founder lines were bred for analysis of the progeny. The progeny of all six founder lines have the ability to make αβ T cells to some extent, although only three lines have the ability to make normal numbers of αβ T cells. The results from the different founder lines are summarized in Table 1. Both transgenes apparently contain their endogenous enhancers (Redondo et al. (1990); Bories et al. (1990); David Raulet, personal communication). Since all six transgenic founder mice were made with the same gene constructs, αβ T cell development is clearly affected by where the transgenes integrate. The γ and δ transgenes have apparently cointegrated in all six transgenic founder lines, since the two genes are always transmitted together. This integration effect may involve either where the transgenes integrate into the genome, or how the transgenes integrate with respect to one another. How the transgene integration could affect αβ T cell development is not known, but it is likely that normal regulation of the γ and δ genes is somehow affected. One reasonable hypothesis is that the ability to turn off expression of the γδ transgenes determines whether αβ T cells can develop or not.

In normal $\alpha\beta$ T cell development, the TCRδ gene is deleted before or during TCRα rearrangement (Chien et al. (1987); deVillartay et al. (1988); Hockett et al. (1988)), and transcription of Cγ1-containing γ genes is turned off (Garman et al. (1986)). It is not known if these events are causative or subsequent to the decision to become an $\alpha\beta$ T cell. Nonetheless, it may be that $\alpha\beta$ T cells can only develop normally if there is a specific shut off of γ and δ gene expression. If there were no regulation of $\alpha\beta$ and $\gamma\delta$ gene expression then the presence of rearranged γ- and δ-chain genes in transgenic mice could result in the complete absence of $\alpha\beta$ expression if "allelic exclusion" acted on $\alpha\beta$ gene rearrangements, or the co-expression of both $\alpha\beta$ and $\gamma\delta$ T cell receptors. Therefore, we tested expression of the transgenes in $\alpha\beta$ T cells. Initially we stained the thymuses of either type 1 or type 2 transgenic animals for co-expression of TCR$\alpha\beta$ and Vγ2 (data not shown). In type 2 thymuses, where normal numbers of $\alpha\beta$ T cells developed, the $\alpha\beta$ TCR-staining cell population was completely separate from the Vγ2-staining cell population. Thus, in type 2 thymuses, there appeared to be a $\alpha\beta$ T cell lineage-specific shut off of transgene expression. In type 1 transgenic thymuses, 30-50% of the cells staining with the TCR$\alpha\beta$ antibody also stained with the Vγ2 antibody. This lack of transgene down-regulation is a tempting explanation for why $\alpha\beta$ T cellls fail to develop normally in type 1 thymuses.

An important question is whether this transgene down-regulation occurs at the mRNA level. To address this issue, we grew out $\alpha\beta$ T cells from both type 1 and type 2 non-tolerizing transgenic spleens by solid-phase TCR$\alpha\beta$ antibody and IL-2. Total cellular RNA was prepared from these cells and assayed by Northern blot for transgene expression. Transgenic $\gamma\delta$ T cells grown out with solid-phase Vγ2 and IL-2 produce high levels of both Vγ2 and Cδ containing message. By contrast, the transgenic $\alpha\beta$ T cells express several-fold lower levels of transgene message. Interestingly, $\alpha\beta$ T cells grown from a type 2 spleen express negligible amounts of transgene message, whereas $\alpha\beta$ T cells grown from type 1 spleens express higher levels of transgene message. Since one explanation for higher levels of transgene expression in the type 1 αβ T cell lines is a greater chance of contaminating γδ T cells, we are currently confirming these results with hybridoma clones. Preliminary evidence supports the contention that type 1 αβ T cells have a defect in transgene down-regulation. Therefore, the difference between type 1 and type 2 transgenic animals may be due to a difference in transgene mRNA expression. The inability to down-regulate transgene message at some stage during thymus development could thus cause an inhibition in αβ T cell maturation.

The previous results involved transgene down-regulation at the mRNA level, since transgene DNA was still present in the αβ T cells (data not shown). A separate mechanism confirmed the importance of transgene down-regulation for the development of αβ T cells. About four out of every five transgenic mice derived from founder #73 have a severe defect in αβ T cell development, but the remaining one in five transgenic mice have more normal numbers of αβ T cells. Surprisingly, these latter mice have undergone a deletion of the Jδ1 portion of the δ *transgene* in the thymus (data not shown). This deletion is specific to αβ T cells, since γδ T cells grown from the same animals retain the complete δ transgene. Interestingly, the deletion may involve deletion of the δ enhancer, which has been mapped to the Jδ1-Cδ intron (Redondo et al. (1990); Bories et al. (1990)). Therefore, this result appears to be another mechanism used by developing αβ T cells to down-regulate transgene expression, and implies that shutting off δ transgene expression promotes αβ T cell maturation. This mechanism of transgene deletion appears to be unique to transgenic founder line #73.

We have also produced γ chain-only transgenic mice, using the same G8 γ transgene construct. These mice have normal numbers of αβ T cells, and Vγ2 expression is undetectable by flow cytometry. Interestingly, the αβ T cells from these mice express the γ transgene message at high levels (data not shown). These results have several implications. First, expression of the γ transgene by itself must not be detrimental to $\alpha\beta$ T cell development-- therefore the inhibition must also require the δ chain, or more likely the expression of a $\gamma\delta$ heterodimer as a cell-surface receptor. Also, the δ transgene would also appear to be necessary for the development of large numbers of $\gamma\delta$ T cells. Finally, the γ transgene may be dependent on the δ transgene for down-regulation of its expression in $\alpha\beta$ T cells.

Tonegawa's group has made $\gamma\delta$ TCR transgenic mice, and have argued for a γ silencer element that would control the development of $\alpha\beta$ T cells in these mice (Bonneville et al. (1989); Ishida et al. (1990)). With a short genomic γ construct, missing the putative silencer element, there is an inhibition of $\alpha\beta$ T cell development. With a long γ genomic construct, containing the γ silencer element, normal numbers of $\alpha\beta$ T cells develop. Therefore, Tonegawa's transgenic mice exhibit both a type 1 and and a type 2 $\alpha\beta$ T cell phenotype, although these phenotypes are dependent on the gene constructs. Interestingly, using our one genomic γ construct, which is a length in between Tonegawa's long and short γ constructs, we find two different $\alpha\beta$ T cell phenotypes in the transgenic mice. It would seem that other factors besides the length of the γ transgene construct can control $\alpha\beta$ T cell development.

What are the other factors that appear to play a role in $\alpha\beta$ T cell development in the transgenic mice? The offspring of founder #75 can be either type 1 or type 2, as shown in Figure 1. Interestingly, a mouse strain background gene appears to affect the development of $\alpha\beta$ T cells. Specifically, if the #75 transgenes are bred successively to BALB/c mice, the offspring all have the type 1 phenotype. If the #75 transgenes are bred to C57BL/10-derived strains, the offspring bear the type 2 phenotype. The nature of this polymorphism is not known, though neither the BALB/c effect nor the C57BL/10 effect is dominant. This lack of dominance implies a multi-gene effect. Since the background gene effect appears to be limited to founder line #75, the effect would seem to be dependent on the precise site of transgene integration, i.e., the level and temporal control of trangene expression.

One possibility is that development of the αβ lineage requires that the γδ transgenes are not expressed. The developing αβ T cell can down regulate the transcription of γδ genes, or actually delete the δ-chain genes. If the γδ transgenes incorporate in such a way that expression cannot be silenced and the transgenes cannot deleted, then αβ development would not progress. The development of αβ T cells may be drastically altered by just a slight alteration in transgene expression. Another possibility is that there is a positive feed-back mechanism whereby the first αβ T cells that arise in the thymus produce a factor that promotes further αβ T cell development. Thus, an inhibition of αβ T cell development, even partially, could drastically alter the outcome of the thymus populations. Such a mechanism may explain why normal T cell development in the thymus is skewed towards the αβ T cell pathway. If the outcome of αβ and γδ gene rearrangements randomly determined the T cell phenotype, the prediction would be that the percentage of γδ T cells would be greater than that which is actually observed.

In conclusion, in our transgenic model, we find that the lineage relationship between αβ T cells and γδ T cells is a very subtle and complex one. There are several factors that determine the development of αβ T cells in the γδ TCR transgenic mice. The most important factor appears to be a down-regulation of transgene expression, although there may be several different mechanisms that cause this end result. Down-regulation of endogenous γ and δ gene expression probably play a role in normal αβ T cell development-- it remains to be seen how closely our transgenic model reflects normal development. The molecular biology of γ and δ TCR gene regulation has not been well characterized as yet. The precise factors that control αβ T cell development thus await further research on the factors that control the expression of the γ and δ TCR genes.

REFERENCES:

Bluestone JA, Cron RQ, Cotterman M, Houlden BA, Matis LA (1988) Structure and specificity of T cell receptor gamma/delta on major histocompatibility complex antigen-specific CD3+, CD4-, CD8- T lymphocytes. J Exp Med 168(5):1899-1916

Bonneville M, Ishida I, Mombaerts P, Katsuki M, Verbeek S, Berns A, Tonegawa S (1989) Blockage of αβ T-cell development by TCR γδ transgenes. Nature 342:931-934

Bories JC, Loiseau P, dAuriol L, Gontier C, Bensussan A, Degos L, Sigaux F (1990) Regulation of transcription of the human T cell antigen receptor δ chain gene. A T lineage-specific enhancer element is located in the Jδ3-Cδ intron. J Exp Med 171(1):75-83

Chien YH, Iwashima M, Kaplan KB, Elliott JF, Davis MM (1987) A new T-cell receptor gene located within the alpha locus and expressed early in T-cell differentiation. Nature 327:677-682

de Villartay JP, Hockett RD, Coran D, Korsmeyer SJ, Cohen DI (1988) Deletion of the human T-cell receptor δ-gene by a site-specific recombination. Nature 335(6186):170-174

Dent AL, Matis LA, Hooshmand F, Widacki S, Bluestone JA, Hedrick SM (1990) Self-reactive γδ T cells are eliminated in the thymus. Nature 343:714-719

Garman RD, Doherty PJ, Raulet DH (1986) Diversity, rearrangement, and expression of murine T cell gamma genes. Cell 45(5):733-742

Hockett RD, de Villartay JP, Pollock K, Poplack DG, Cohen DI, Korsmeyer SJ (1988) Human T-cell antigen receptor (TCR) delta-chain locus and elements responsible for its deletion are within the TCR alpha-chain locus. Proc Natl Acad Sci U S A 85(24):9694-9698

Ishida I, Verbeek S, Bonneville M, Itohara S, Berns A, Tonegawa S (1990) T-cell receptor γδ and γ transgenic mice suggest a role of a γ gene silencer in the generation of αβ T cells. Proc Natl Acad Sci U S A 87(8):3067-3071

Matis LA, Cron R, Bluestone JA (1987 Nov 19-25) Major histocompatibility complex-linked specificity of γδ receptor-bearing T lymphocytes. Nature 330(6145):262-264

Redondo JM, Hata S, Brocklehurst C, Krangel MS (1990) A T cell-specific transcriptional enhancer within the human T cell receptor δ locus. Science 247(4947):1225-1229

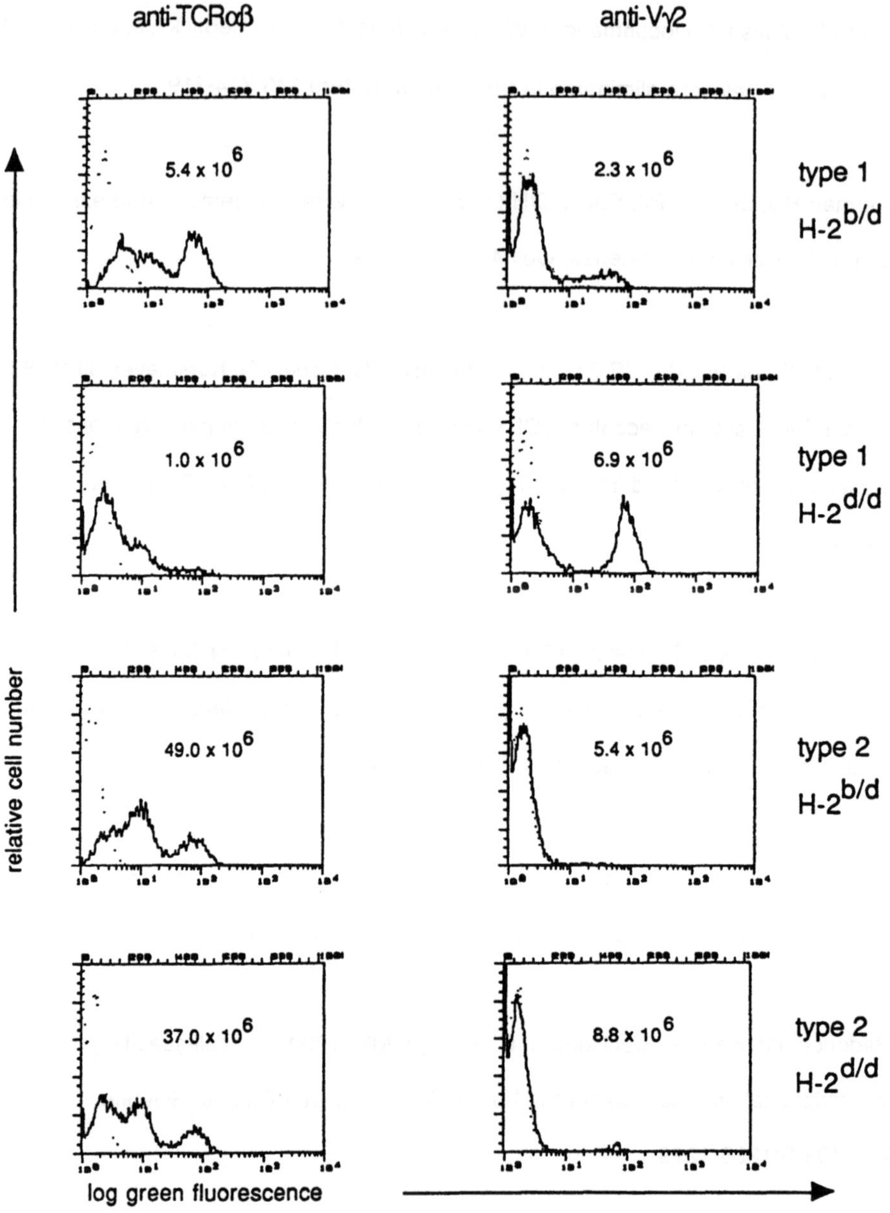

Figure 1: Comparison of type 1 and type 2 transgenic thymuses

Histograms represent 5000 thymocytes stained with either the anti-TCRαβ monoclonal H57 or the anti-Vγ2 monoclonal 10A6. Numbers in the boxes refer to total TCRαβ-bright or Vγ2-positive cells in each thymus. Type 1 thymuses had about 15 million cells each, type 2 thymuses had about 200 million cells each.

TABLE 1: Phenotypes of different $\gamma\delta$ TCR transgenic founder lines

Founder line	$\alpha\beta$ T cell development
73	INHIBITED (unless δ transgene deleted)
74	INHIBITED
75	BOTH NORMAL AND INHIBITED (mouse strain background gene effect)
86	INHIBITED
93	NORMAL
98	INHIBITED

Ligand Specificity and Repertoire Development of Murine TCRγδ Cells

J. A. Bluestone[1], R. Q. Cron[1], B. Rellahan[2], and L. A. Matis[2]

[1] Committee on Immunology, University of Chicago, Chicago, IL, USA
[2] Division of Cytokines, Food and Drug Administration, Bethesda, MD, USA

INTRODUCTION

The physiological role for, and the ligand(s) of, TCRγδ cells remain intense fields of investigation. The selective tissue localization of some subsets of TCRγδ cells (Allison *et al.* 1988, Tonegawa *et al.* 1989, Goodman and Lefrancois 1988, Bucy *et al.* 1988) as well as the relatively restricted T cell receptor usage in those tissues has led to the suggestion that this specialized T cell subset recognizes non-polymorphic MHC or conserved antigens (Asarnow *et al.* 1988). In fact, the identification of bacterial antigens (Janis *et al.* 1989, Kabelitz *et al.* 1990), more specifically heat shock proteins (Born *et al.* 1990, Rajasekar et al. 1990, Holoshitz *et al.* 1989), as potential ligands for TCRγδ cells has solidified the notion that these cells may indeed play a unique role in the immune system. However, many of these studies have been limited to an analysis of selected TCRγδ populations often localized in epithelial tissue or newborn thymus. In this brief review we would like to summarize our data with regards to the potential diversity of splenic and lymph node-derived TCRγδ cells, compare the general observations of ligand specificity for these cells and those of TCRαβ cells, and finally, to present one approach towards beginning to determine the effect of MHC and non-MHC genes on TCRγδ repertoire development.

GENERATION AND CHARACTERIZATION OF MHC-SPECIFIC TCRγδ CELLS

Early studies suggested that TCRγδ cells were tumor-specific CTL or NK cells which played a role in the immunosurveillance of cancer (Brenner et al. 1987, Borst *et al.* 1987, Moingeon *et al.* 1987, Ang *et al.* 1987). However, the ligand(s) of the TCRγδ heterodimer was still unknown because of the lack of any demonstration of a specific target antigen for these TCRγδ cells. In fact, as it turns out, one of the original human TCRγδ cell turned out to be specific for the human CD1c protein (Porcelli et al. 1989) Since a high percentage of TCRαβ cells are capable of recognizing allogenic MHC proteins, and because TCRγδ cells appeared phenotypically and functionally similar to TCRαβ cells, efforts were made to generate allo-MHC-reactive TCRγδ cells. BALB/c *nu/nu* mice ($H\text{-}2^d$) were primed with allogeneic antigen presenting cells (APC) (B10.BR - $H\text{-}2^k$). Athymic mice were used because TCRγδ cells represented a large percentage of the T cells in these animals. A T cell clone, G8, was derived from primed mice. This cell line proliferated to and lysed $H\text{-}2^k$-expressing, but not $H\text{-}2^d$-bearing, cells (Matis *et al.* 1987). Supernatants from G8 cells, stimulated with allogeneic (B10.BR or BALB.K) accessory cells, but not MHC-syngeneic (BALB/c or B10.D2) accessory cells, supported the growth of the lymphokine dependent cell line, HT-2, in a dose dependent manner. Further studies defined this lymphokine as GM-CSF. In addition, γ-IFN was released by G8 under the same conditions. Thus, when G8 was stimulated with anti-CD3-ε mAb or B10.BR accessory cells, the supernatants were found to contain IL-3, γ-IFN, GM-CSF and TNF, but no detectable IL-2 or IL-4, demonstrating that a TCRγδ-expressing T cell clone derived from lymph nodes was capable of producing some of the known T cell-derived lymphokines upon triggering through its antigen receptor. Thus, the target antigen recognized by this cell line was the first demonstration of any specific ligand for the TCRγδ (Matis *et al.* 1987).

Current Topics in Microbiology and Immunology, Vol. 173

Detailed mapping has localized the ligand, a non-classical Class I alloantigen, for these cells to a portion of the TL region linked to the T11-T15 genes (Houlden *et al.* 1989). It was determined that the G8 TCRγδ expressed a Vγ2-Cγ1 chain in association with a δ chain composed of a Vα11 variable gene element. Subsequently, other laboratories identified T cells specific for other non-classical MHC or MHC-like gene products, including, Qa-1 (Vidovic *et al.* 1989) and human CD1c (Porcelli *et al.* 1989). In fact, a thymus-derived autoreactive TCRγδ hybridoma sharing a similar, or possibly identical, specificity as the G8 clone was derived by Tonegawa and his colleagues (Ito *et al.* 1990). This led us and others (Janeway *et al.* 1988) to hypothesized that TCRγδ cells recognize conserved MHC class I proteins. This hypothesis fit nicely with the notion that the TCRγδ population had a more restricted and limited repertoire than TCRαβ counterparts especially in light of the limited number of Vγ and Vδ gene elements identified.

However, over the past two years, we have begun to question whether TCRγδ cells have a specialized repertoire. Of the many alloreactive TCRγδ cells we have generated, all have been shown to recognize classical class I and class II MHC molecules (Bluestone *et al.* 1988, Matis *et al.* 1989), rather than the non-polymorphic class I-like MHC molecules. In addition to the MHC-alloreactive TCRγδ cell G8, 3 other allo-MHC reactive TCRγδ cells have been well characterized. One cell line, LBK, which was generated from B10 *nu/nu* lymph nodes cells ($H\text{-}2^b$) that had been primed with B10.BR APC ($H\text{-}2^k$), reacts specifically with a classical MHC class I antigen, $H\text{-}2D^k$. This cell line was demonstrated by immunoprecipitation analysis to express a Cγ4-encoded TCRγ chain. However, not all alloMHC reactive TCRγδ cells recognize MHC class I molecules. One cell line, LBK5 and its clone G11, which were also derived from a B10 *nu/nu* mouse primed with B10.BR APC, recognizes a classical MHC class II Ag, $I\text{-}E^k$ (Matis *et al.* 1989). Another CTL clone, LKD1, recognizes the class II ligand, $I\text{-}A^d$ (Rellahan *et al.*, submitted). Interestingly, the γ chain used in these clones was encoded by a Vγ1.1-Jγ2 rearrangement, a gene combination that is expressed in small amounts throughout ontogeny and in the adult thymus and is maintained as a minor subset within the circulating δ pool in the lymphoid system. In addition, these clones share a common Vδ usage, Vδ5. However, nucleotide sequence analysis of the V(D)J junction shown an extensive amino acid diversity. This potential for the extensive diversity, especially at the junction, may have several implications for the MHC-specific repertoire of these cells. First, is the expression of the Vγ1.2 and Vδ5 genes linked in some way to the MHC class II specificity? It is worth noting that unlike many of the other TCRγ and δ chains, these chains do not appear to have a tissue-specific or developmentally regulated appearance. The Vγ1.2-Cγ2 gene product is expressed on a small subset of TCRγδ cells throughout fetal thymic ontogeny and remains as a small population throughout adult life in both the spleen and lymph node of normal and athymic mice (Houlden *et al.* 1989). Similarly, the Vδ5-Cδ1 chain is expressed on the majority of both thymic and peripheral TCRγδ cells. Second, recent studies suggest that, at least some, alloreactive TCRαβ cells recognize self-peptides in the context of allogeneic MHC class I or class II molecules. The recognition of peptides in the groove of the MHC molecule suggests similarities to the recognition of foreign peptidic antigens in the context of self MHC molecules by antigen-specific TCRαβ cells. Bjorkman and Davis have suggested that peptide recognition depends on the diversity of the amino acid sequence encoded by the CDR3 or V(D)J junctions. Thus, the dramatic effect of the junctional amino acid sequences of the two alloreactive class II TCRγδ cells suggests that these cells recognize a peptide in the context of the class II molecule.

The relative ease in generating "classical" alloreactive TCRγδ cells, has been confirmed and extended in other species including man. There have been several reports of both class I-reactive and class II-reactive human TCRγδ cells (Ciccone *et al.* 1988, Paliard *et al.* 1989, Rivas *et al.* 1989). In fact, in at least one instance, a nominal antigen, tetanus toxoid, has been shown to be recognized by a human TCRγδ T cell clone in an MHC class II-restricted manner (Kozbor *et al.* 1989). Therefore, it is now clear that MHC-reactive TCRγδ cells exist. Furthermore, these cells use a variety of TCRγ and δ proteins and seem to be capable of recognizing a broad range of MHC-encoded antigens. However, one criticism of the possibility that γδ cells have an MHC-specific repertoire has been the difficulty in routinely generating alloreactive TCRγδ cells. In many laboratories, unlike TCRαβ cells, randomly generated TCRγδ cells have not been found to react with allogeneic MHC molecules. Furthermore, the ability to generate antigen-specific, MHC-restricted TCRγδ cells has been extremely difficult, with only a few rare examples. In fact, the most common antigen-specific TCRγδ cells, those which recognize bacterial products such as heat shock proteins and toxins, frequently have been reported as non-MHC restricted in their specificity. Yet, there are still several unresolved issues. For instance, there is still a minimal understanding of the growth conditions for this cell population. The possibility of distinct lymphokines and other secondary signalling structures on accessory cells that may be necessary for triggering these cells. For instance, IL-7 has been shown to be potent lymphokine for growing TCRγδ thymocytes (Watson *et al.* 1989). In addition, the ability of human TCRγδ cells to recognize Staphylococcal enterotoxin can be shown by activation of lytic activity not proliferation (Rust *et al.* 1990). Finally, the diversity among the Vγ and Vδ chains used by TCRγδ cells may reflect different repertoires. For instance, the human TCRγδ cells generated against alloreactive cells often use the Vδ1 gene element (Rivas *et al.* 1989), whereas, the bacterial antigen-specific human TCRγδ cells use Vδ2 (Fisch *et al.* personal communication). These different TCRγδ cells may be developmentally regulated, positively selected in the periphery or localized to different anatomical sites. Thus, the antigen-specific repertoire of TCRγδ cells may be far more diverse than previously appreciated. Therefore, it is critical to determine the diversity of the TCRγδ gene expression.

T CELL RECEPTOR DIVERSITY OF SPLENIC TCRγδ CELLS

One approach towards examining the potential diversity of the TCRγδ repertoire has been an examination of the diversity of the δ and γ chains expressed on lymph node and splenic TCRγδ cells. Although TCRγδ cells selectively localized in epithelial tissues have restricted γ and δ gene segment usage (Asarnow *et al.* 1988, Tonegawa *et al.* 1989) and, in some instances, express a highly conserved TCRγ and δ VDJ junctional sequence, the circulating TCRγδ cells express a quite diverse potential repertoire (Cron *et al.* 1989a, Cron *et al.* 1989b, Cron *et al.* (1990), Ezquerra *et al.* 1990).

A total of 15 TCRγδ-expressing T cell hybridomas were generated from activated $CD3^+$,$CD4^-$,$CD8^-$ B6 and B10 splenocytes to gain a better understanding of the diversity, preponderance, and pairing of the different TCRγ and δ proteins in the murine peripheral lymphoid organs. The TCR on these hybridomas were analyzed for the type of γ and δ chains expressed using biochemical, molecular and phenotypic analysis with several recently derived hamster anti-TCRγδ mAb. One of these, UC7-13D5, reacts with all murine TCRγδ heterodimers, while the other mAbs, UC3-10A6 and GL2, specifically detect Vγ2-expressing and Vδ4-expressing TCRγδ cells, respectively. Of the 15 hybridomas, 7 expressed Vγ1-Cγ4 proteins, 5 expressed Vγ2-Cγ1, and 3 expressed Vγ1-Cγ2 proteins. This ratio of the 3 peripheral TCRγ chains roughly corresponds to the relative intensities of these proteins as observed on 2-D SDS-PAGE gels of anti-CD3-ε mAb-immunoprecipitations of bulk $CD3^+$,$CD4^-$,$CD8^-$ spleen cell lysates.

The TCRδ genes expressed by these hybridomas were determined in collaboration with Drs. Angel Ezquerra and John Coligan by northern blot and PCR analyses. Eight of the 15 splenic TCRγδ-expressing hybridomas were found to express Vδ5, by far the most commonly used Vδ gene segment. The three Vγ1-Cγ2-expressing hybridomas expressed only Vδ5 proteins, whereas, of the eight Vγ1-Cγ4-expressing hybridomas, 2 expressed Vδ2, 2 used Vδ5, 2 used Vδ6 proteins, 1 expresses Vδ4 and 1 expressed a Vδ7 (Vα10) encoded TCRδ chain (Ezquerra *et al.* 1990). To analyze the junctional diversity of these δ proteins, the rearranged TCR Vδ genes were sequenced. The amino acid sequence results confirmed the TCRδ designation of the hybrids. However, each junctional region sequence was unique due to differential joins, Dδ usage, and nucleotide additions between the various gene segments. In some instances, both D regions as many as three N regions have been identified to create an extremely diverse junctional sequence. Taken as a whole these results would suggest a considerable potential for diversity within the TCRγδ repertoire. In fact, the limited diversity seen in peripheral epithelial tissues may reflect a positive selection of a diverse population of TCRγδ cells due to antigenic exposure in those tissues (Lafaille *et al.* 1990).

ANALYSIS OF THE TCRγδ REPERTOIRE USING ANTI-TCR MAB: EVIDENCE FOR MHC AND NON-MHC ASSOCIATED DIFFERENCES IN TCRγδ EXPRESSION

The development of a panel of mAb which detect different subsets of murine TCRγδ heterodimers has allowed for more detailed analyses of the TCRγδ repertoire. However, the percentage of TCRγδ cells in spleen and lymph node is quite small. Therefore, in order to analyze the repertoire of TCRγδ cells using these reagents, splenic TCRγδ cells were expanded by plating unseparated splenocytes into tissue culture wells coated with the anti-pan TCRγδ mAb, UC7-13D5, in the presence of IL-2. After 7 days in culture, greater than 90% of the cells were TCRγδ$^+$. This method offered several advantages over the previous technique of examining CD4$^-$CD8$^-$CD3$^+$ cell populations. First, a potential subset of naive CD8$^+$ or CD4$^+$ TCRγδ-expressing T cells were not eliminated from the subsequent analyses, since it was no longer necessary to remove the CD4$^+$ and CD8$^+$ cells prior to culturing. Second, any deleterious or selective effects of C' treatment were eliminated. Third, contamination of TCRαβ cells was minimal after 1 wk of activation. Finally, treatment with anti-TCRγδ mAb plus IL-2 for 1 wk yielded about a 100 fold expansion of splenic TCRγδ cells with a diversity of TCRγδ chain usage similar to the naive population. By 2-D SDS-PAGE analysis, there appears to be no appreciable differences in these populations from the CD3$^+$,CD4$^-$,CD8$^-$ splenocytes. This suggested that the activating pan-reactive TCRγδ mAb did not preferentially activate subsets of TCRγδ-expressing splenocytes.

Bulk populations of splenic TCRγδ cells from a variety of inbred and MHC-congenic strains of mice were obtained and analyzed. The percentages of the different Vγ and Vδ subsets differed among the strains. Taken as a percentage of the TCRγδ$^+$ cells, 29.0%, 21.4%, and 14.1% of the splenocytes were Vγ2$^+$ (UC3-10A6 mAb) in the B10, B10.BR, and CBA/J strains, respectively. The percentages of TCRγδ$^+$ splenocytes which expressed Vδ4 (GL2 mAb) were 40.2%, 46.6%, and 30.2% in the B10, B10.BR, and CBA/J strains, respectively. FCF analyses were conducted on over 10 separate occasions on a variety of different inbred and congenic strains of mice.

From these analyses, one prominent observation was that in all 8 B6 or B6 MHC-congenic mice examined including B6.H-2^k, greater than 50% of the TCRγδ-expressing splenocytes were Vγ2$^+$. No other strain of mouse examined averaged over 32% Vγ2$^+$ splenocytes. The strain which expressed the second highest percentage of Vγ2$^+$ cells was B10. These 2 strains share the same MHC and Mls loci, thus, some other gene locus(i) must be accounting for the difference in Vγ2 expression between these strains. One known genetic difference between these strains which is related to the immune system is the TCRγ locus polymorphism. Therefore, mice which possessed presumably similar TCRγ loci were examined to address the question of whether or not some regulatory gene present within the TCRγ locus was influencing the high percentage of Vγ2$^+$ cells in B6 mice. Based on the TCRγ polymorphisms described earlier and confirmed by RFLP analysis , BALB/c should be more similar to B6 mice than B10 mice at their TCRγ loci. Yet, less than 20% of the TCRγδ-expressing splenocytes in BALB/c strains were Vγ2$^+$. Thus, a polymorphism at the TCRγ locus which is unique to B6 mice or a combination of factors may be responsible for the observed relatively high percentage of Vγ2$^+$ B6 splenocytes.

At the other end of the spectrum from B6, less than 15% percent of the splenic TCRγδ cells were Vγ2$^+$ in certain strains of mice (AKR/J and CBA/J). Although minor lymphocyte stimulating antigen (Mls)-encoded gene products have been shown to be involved in thymic selection of TCRαβ cells (137-139), the differences in Mls haplotype could not account for the difference between strains which expressed relative high and low levels of Vγ2$^+$ cells, since less than 10% of the TCRγδ-expressing splenocytes were Vγ2$^+$ in both CBA/J mice (Mls^d = Mls^a + Mls^c) and CBA/CaJ (Mls^b = Mls$^-$) mice (data not shown).

Although no detectable difference of the splenic TCRγδ repertoire could be observed between B6 (H-2^b) and B6.H-2^k mice, Lefrancois et al. have shown that the Vδ4$^+$ IELs is influenced by I-E^k (Lefrancois *et al.* 1990). Thus, it would appear that both MHC and non-MHC genes influence the TCRγδ repertoire.

There is no doubt that the TCRγδ cells are likely to be critical and important in the immune response. However, the distinct populations of TCRγδ cells may have different repertoires. Such variables as tissue localization, V gene usage and antigenic exposure undoubtably will play a fundamental role in determining the physiologic effects of TCRγδ cell activation. Yet, our studies and others reinforce the potential of this population to distinguish self from non-self.

REFERENCES

Allison JP, Havran WL, Asarnow D, Tigelaar RE, Tucker PW, Bonyhadi M (1988). Gamma delta antigen receptors of Thy-1+ dendritic epidermal cells: implications for thymic differentiation. Immunol Res 7: 292-302

Ang SL, Seidman JG, Peterman GM, Duby AD, Benjamin D, Lee SL, and Hafler DA (1987) Functional gamma chain-associated T cell receptors on cerebrospinal fluid-derived natural killer-like T cell clones. J.Exp.Med.165:1453.

Asarnow DM, Kuziel WA, Bonyhadi M, Tigelaar RE, Tucker PW, Allison JP (1988). Limited diversity of gamma delta antigen receptor genes of Thy-1+ dendritic epidermal cells. Cell 55: 837-847.

Bluestone JA, Cron RQ, Cotterman M, Houlden BA, Matis LA (1988). Structure and specificity of T cell receptor gamma/delta on major histocompatibility complex antigen-specific CD3+, CD4-, CD8- T lymphocytes. J Exp Med 168:1899-1916.

Borst J, van de Griend RJ, van Oostveen W, Ang SL, Melief CJ, Seidman JG, and Bolhuis RL (1987) A T-cell receptor gamma/CD3 complex found on cloned functional lymphocytes. Nature 325:683.

Brenner MB, McLean J, Scheft H, Riberdy J, Ang SL, Seidman JG, Devlin P, and

Krangel MS (1987). Two forms of the T-cell receptor gamma protein found on peripheral blood cytotoxic T lymphocytes. Nature 325:689.

Born W, Hall L, Dallas A, Boymel J, Shinnick T, Young D, Brennan P, O'Brien R (1990). Recognition of a peptide antigen by heat shock-reactive gamma delta T lymphocytes. Science 249:67-69.

Bucy RP, Chen CL, Cihak J, Losch U, Cooper MD (1988). Avian T cells expressing gamma delta receptors localize in the splenic sinusoids and the intestinal epithelium. J Immunol 141:2200-2205.

Ciccone E, Viale O, Pende D, Malnati M, Battista Ferrara G, Barocci S, Moretta A, Moretta L (1988). Antigen recognition by human T cell receptor gamma-positive lymphocytes. Specific lysis of allogeneic cells after activation in mixed lymphocyte culture. J Exp Med 167:1517-1522.

Cron RQ, Gajewski TF, Sharrow SO, Fitch FW, Matis LA, Bluestone JA (1989a) Phenotypic and functional analysis of murine CD3+, CD4-, CD8- TCR-gamma delta-expressing peripheral T cells. J Immunol 142:3754-3762.

Cron RQ, Ezquerra A., Coligan JE, Houlden BA, Bluestone JA, Maloy WL (1989b). Identification of distinct T cell receptor (TCR)-gamma delta heterodimers using an anti-TCR-gamma variable region serum. J Immunol 143:3769-3775.

Cron RQ, Coligan JE, Bluestone JA (1990). Polymorphisms and diversity of T-cell receptor-gamma proteins expressed in mouse spleen. Immunogenetics 31:220-228.

Esquerra A, Cron RQ, McConnell TJ, Valas RB, Bluestone JA, Coligan JE (1990) T cell receptor delta gene expression and diversity in the mouse spleen. J Immunol 145:1311-1317.

Goodman T, Lefrancois L (1988). Expression of the gamma-delta T-cell receptor on intestinal CD8+ intraepithelial lymphocytes. Nature 333:855-858.

Holoshitz J, Koning F, Coligan JE, DeBruyn J, Strober S (1989) Isolation of CD4- CD8- mycobacteria-reactive T lymphocyte clones from rheumatoid arthritis synovial fluid. Nature 339:226-229.

Houlden BA, Matis LA, Cron RQ, Widacki SM, Brown GD, Pampeno C, Meruelo D, and Bluestone JA (1989). A TCR gamma/delta cell recognizing a novel TL-encoded gene product. Cold Spring Harb Symp Quant Biol 54:45.

Houlden BA, Cron RQ, Coligan JE, and Bluestone JA (1988). Systematic development of distinct T cell receptor-gamma delta T cell subsets during fetal ontogeny. J.Immunol. 141:3753.

Ito K, Van Kaer L, Bonneville M, Hsu S, Murphy DB, Tonegawa S (1990). Recognition of the product of a novel MHC TL region gene (27b) by a mouse δ T cell receptor. Cell 62:549-561.
Janeway CA, Jones B, Hayday A (1988). Specificity and function of T cells bearing δ receptors. Immunology Today 9:73-77.

Janis EM, Kaufmann SH, Schwartz RH, Pardoll DM (1989) Activation of gamma delta T cells in the primary immune response to Mycobacterium tuberculosis. Science 244:713-716.

Kabelitz D, Bender A, Schondelmaier C, Schoel B, Kaufamnn SHE (1990). A large fraction of human peripheral blood δ T cells is activated by Mycobacterium tuberculosis but not by its 65-kD heat shock protein. J Exp Med 171:667-679.

Kozbor D, Trinchieri G, Monos DS, Isobe M, Russo G, Haney JA, Zmijewski C and Croce CM (1989) Human TCRγδ+,CD8+ T lymphocytes recognize tetnus toxoid in an MHC-restricted fashion. J Exp Med 169:1847-1851.

Lafaille JJ, Haas W, Coutinho A, Tonegawa S (1990). Positive selection of δ T cells. Immunology Today 11:75-78.

Lefrancois L, LeCorre R, Mayo J, Bluestone JA and Goodman T (1990) Extrathymic positive selection of TCR gamma-delta$^+$ T cells by class II histocompatibility molecules. Cell, in press.

Matis LA, Cron R, Bluestone JA (1987). Major histocompatibility complex-linked specificity of δ receptor-bearing T lymphocytes. Nature 330:262-264.

Matis LA, Fry AM, Cron RQ, Cotterman MM, Dick RF, Bluestone JA (1989). Structure and specificity of a class II MHC alloreactive gamma delta T cell receptor heterodimer. Science 245:746-749.

Moingeon P, Jitsukawa S, Faure F, Troalen F, Triebel F, Graziani M, Forestier F, Bellet D, Bohuon C, and Hercend T (1987). A gamma-chain complex forms a functional receptor on cloned human lymphocytes with natural killer-like activity. Nature 325:723.

Paliard X, Yssel H, Blanchard D, Waitz JA, DeVries JE, Spits H (1989). Antigen specific and MHC nonrestricted cytotoxicity of T cell receptor alpha beta+ and gamma delta+ human T cell clones isolated in IL-4. J Immunol 143:452-457.

Porcelli S, Brenner MB, Greenstein JL, Balk SP, Terhorst C., and Bleicher P (1989) Recognition of cluster of differentiation 1 antigens by human CD4-CD8- cytolytic T lymphocytes. Nature 341:447-450.

Rajasekar R, Sim G-K, Augustin A (1990). Self heat shock and gamma T-cell reactivity. Proc Natl Acad Sci USA 87:1767-1771.

Rivas A, Koide J, Cleary ML, Engleman EG (1989). Evidence for involvement of the gamma, delta T cell antigen receptor in cytotoxicity mediated by human alloantigen-specific T cell clones. J Immunol 142:1840-1846.

Rust CJ, Verreck F, Vietor H, Koning F (1990). Specific recognition of staphylococcal enterotoxin A by human T cells bearing receptors with the V9 region. Nature 346:572-574.

Tonegawa S, Berns A, Bonneville M, Farr A, Ishida I, Ito K, Itohara S, Janeway CA Jr, Kamagawa O, Katsuki M, Kubo R, LaFaille J, Mombaerts P, Murphy D, Nakanishi N, Takagaki Y, Van Kaer L, Verbeek S (1989). Diversity, development, ligands, and probable functions of δ T cells. Cold Spring Harb Symp Quant Biol 54:31-44.

Watson JD, Morrissey PJ, Namen AE, Conlon PJ, and Widmer MB (1989). Effect of IL-7 on the growth of fetal thymocytes in culture. J Immunol 143:1215-1222.

Vidovic D, Rogli CM, McKune K, Guerder S, Mackay C, Dembic Z (1989). Qa-1 restricted recognition of foreign antigen by a γδ T-cell hybridoma. Nature 34):646-650.

Antigen/Heat Shock Proteins

Specificity of Mycobacteria/Self-Reactive γδ Cells

Rebecca L. O'Brien[1] and W. Born[1,2]

[1] National Jewish Center for Immunology and Respiratory Medicine, Division of Infectious Diseases, Department of Medicine, 1400 Jackson Street, Denver, Colorado 80206, USA
[2] University of Colorado Health Sciences Center, Department of Microbiology, Denver, Colorado, USA

INTRODUCTION

The γδ T cell receptor (TCR), although structurally quite similar to the αβ TCR, may differ in terms of the ligand recognized, MHC restriction, and the immunological role of the cells that express it. αβ TCR-expressing cells are biased to recognize allogeneic MHC molecules, presumably because the allogeneic molecules three-dimensionally resemble self MHC molecules containing a bound foreign peptide. This implies that if γδ TCRs recognize the same kind of ligand, they should produce a similar frequency of alloreactivity. Although when cultured under conditions that select for cells that respond to allogeneic MHC molecules, γδ cells can be isolated with alloantigenic specificities (Bluestone et al., 1988; Matis et al., 1989; Rivas et al., 1989; Spits et al., 1990), γδ cells isolated under conditions not involving such selection usually show no alloreactivity (Haregewoin et al., 1989; Holoshitz et al., 1989; Modlin et al., 1989; O'Brien et al., 1989). Moreover, alloreactivity by γδ cells often shows a broad cross-reactivity uncommon among αβ T cells (Bluestone et al., 1988; Matis et al., 1989). Also, in many cases, non-classical MHC or MHC-like molecules such as mouse TL (Bonneville et al., 1989; Dent et al., 1990), mouse Qa-1 (Vidovic et al., 1989), and human CD1 (Porcelli et al., 1989), have been implicated as γδ TCR ligands, although classical MHC class I (Bluestone et al., 1988; Rivas et al., 1989; Spits et al., 1990) and II (Bluestone et al., 1988; Kozbor et al., 1989) molecules have been reported as well. Thus, while molecules that serve as αβ TCR ligands can be recognized by at least some γδ TCRs, the recognition usually seems to be less specific in terms of polymorphic determinants, and frequently involves molecules that are rarely found as αβ TCR ligands. These observations may indicate that the antigens recognized by γδ T cells also tend to be of a different sort than those recognized by αβ T cells.

Current Topics in Microbiology and Immunology, Vol. 173

Our studies on mycobacteria-reactive γδ T cells present in the newborn mouse thymus may hint at the nature of this difference. For reasons outlined below, we speculate that γδ TCRs may be specialized for recognition of a class of self proteins known as heat shock proteins, that are induced or over-expressed under conditions of cell stress. Perhaps γδ T can in this way circumvent infections by intracellular pathogens that "hide" within cells and produce little external foreign antigen of their own.

MYCOBACTERIA-REACTIVE γδ T CELLS

Newborn mouse thymus contains a fairly high percentage of γδ T cells. We fused C57BL/10 newborn thymocytes to a thymoma cell line lacking functional TCR genes of its own (BW/$\alpha^-\beta^-$, (White et al., 1989)), to create T cell hybridomas expressing the γδ TCR. We began screening these γδ hybridomas for alloreactivity on spleen cells from other mouse strains. Although we found no alloreactivity among any of the 51 γδ T cell hybridomas (even though 2 out of 16 αβ T cell hybridomas from the same fusions were alloreactive), about 1/3 of the γδ hybridomas "spontaneously" produced IL-2, when simply cultured overnight in tissue culture medium, without any other presenting cells (O'Brien et al., 1989). Because about 70% of these cells expressed Vγ6 mRNA, because anti-CD3 monoclonal antibody blocked the IL-2 production, and because γδ TCR-loss variants of these cells lost reactivity, we concluded that the IL-2 production was not "spontaneous" at all, but depended upon specific stimulation of a certain type of γδ TCR.

Upon screening of the same panel of newborn thymus γδ hybridomas with various bacterial antigens, we found that the same hybridomas produced even higher levels of IL-2 when cultured in the presence of purified protein derivative (PPD) of Mycobacterium tuberculosis (O'Brien et al., 1989). This elevated response was likewise receptor dependent, but was specific to this group of γδ cells, in that other hybridomas bearing different γδ TCRs, when stimulated submaximally with plate-bound anti-CD3 antibody, showed no augmentation of response if PPD was also added.

Because each and every cell responding "spontaneously" also responded to PPD, we speculated that both responses involved a cross-reactive antigen. The "spontaneous" reactivity was most likely caused by either an antigenic component in the culture medium (such as a component of fetal bovine serum) or an autoantigen produced by the hybridomas themselves. In either case, the antigen was likely to be derived from a mammalian source, and should also cross-react with an antigen from a bacterial source. Because heat shock proteins (HSPs) are known to be highly conserved in all life forms, and because HSP-65 is a known antigenic component of PPD (Young, 1990), we tested recombinant purified HSP-65 of mycobacteria for its ability to stimulate this group of hybridomas (O'Brien et al., 1989). Many of the PPD-reactive hybridomas also responded to purified HSP-65, although in general the responses were weaker than those to PPD. We are at present still uncertain as to why the purified HSP-65 is only weakly stimulatory, but in light of other evidence described in the next section, we conclude that HSP-65 indeed comprises an antigenic component in PPD.

PEPTIDE REACTIVITY OF γδ T CELLS

We tested synthetic peptides representing stretches of HSP-65 sequences that had previously been shown to stimulate mycobacteria reactive αβ T cells (kind gifts of Doug Young and Tom Shinnick). A peptide representing residues 180-196 strongly stimulated some of the PPD-reactive γδ hybridomas, although most responded only weakly to this peptide (Born et al., 1990). Additionally, a shorter peptide, representing residues 181-195, strongly stimulated one of the hybridomas. No other peptides tested stimulated any of the panel of γδ hybridomas.

The ability of a peptide to stimulate γδ T cells suggests two implications. First, γδ T cells, like αβ T cells, may generally recognize processed fragments (peptides) of naturally occurring proteins. Second, in analogy with αβ T cells, the ability of a peptide to act as antigen suggests that a "presenting" molecule binds peptide and may be co-recognized by the γδ TCR. We have at present no clues as to the possible identity of such a presenting molecule for HSP-65 reactive γδ T cells. Such experiments are difficult because these γδ hybridomas are apparently capable of presenting antigen to themselves, so that we cannot add or subtract the presenting molecule at will. The hybridomas do not express class II molecules, and none of the antibodies against known Class I or Class I-like antibodies so far tested has shown blocking ability in this system, however (O'Brien et al., In preparation; O'Brien et al., 1989).

"SPONTANEOUS" REACTIVITY ANTIGEN

The antigen responsible for "spontaneous" reactivity has not yet been determined. We have attempted to test fetal bovine serum (FBS) as a source of this antigen by adapting the cells to serum-free growth, using serum replacement media. When adapted from 10% FBS to growth in 1% FBS, the "spontaneously" reactive hybridomas show a greatly reduced "spontaneous" production of IL-2 (Born et al., 1990). When such cells are then restored to 10% FBS-containing medium, they show no increase in IL-2 production, although they still respond well to PPD. This could indicate that high FBS levels do not supply a stimulatory antigen, but instead induce expression of a stimulatory autoantigen.

We attempted to test this hypothesis by stimulating cells with a likely autoantigen, a synthetic peptide representing a region of the mouse homologue of HSP-65, that corresponds to the stimulatory region of mycobacterial HSP-65 (Born et al., 1990). This self peptide showed weak stimulation on two of the "spontaneously" reactive hybridomas. We are still uncertain whether the mouse version of HSP-65 is indeed an autoantigen for these cells, or whether instead, is only related to another, more stimulatory self-antigen.

γδ RECEPTORS OF PPD REACTIVE NEWBORN THYMUS HYBRIDOMAS

Protein gels of anti-CD3 immunoprecipitates of the PPD-reactive γδ hybridomas indicated that many bear similar γ and δ chains, and further, that all such cells expressed a certain γ chain with a comparatively high molecular weight (Happ et al., 1989). We deduced the amino acid sequences of the γ and δ chains expressed by most of these cells, and drew the following conclusions:

1. All 28 PPD-reactive γδ hybridomas in this collection expressed an in-frame Vγ1-Jγ4-Cγ4 chain, with limited junctional diversity. An occasional rare non-PPD reactive hybridoma also expressed this γ chain, indicating that it alone is not sufficient to confer PPD and autoreactivity.
2. About 70% of the PPD-reactive γδ hybridomas expressed an in-frame δ chain with a Vδ6 variable region. Some of these used a different Vδ6, about 65% identical to the more prevalent Vδ6 found in the cells in this collection. Again, a few non-PPD reactive hybridomas also expressed a Vδ6-containing δ chain, indicating that a Vδ6+ chain alone does not confer PPD reactivity. Delta chains with a different variable region, in conjunction with the Vγ1-Jγ4-Cγ4 chain, are expressed by a few of the PPD-reactive hybridomas.

PERIPHERAL AUTOREACTIVE $\gamma\delta$ CELLS

Because $\gamma\delta$ cells, like $\alpha\beta$ cells, for the most part develop within the thymus, we wondered whether this autoreactive subset of $\gamma\delta$ cells could be found in peripheral tissues of the mouse. It should be noted at this point that the autoreactivity displayed by the newborn thymus hybridomas may well be an in vitro artifact due to the culture conditions or to fusion with a tumor cell, and might not reflect a tendency of normal $\gamma\delta$ cells to stimulate themselves. Rather, the normal $\gamma\delta$ cells may only respond to infected or malignant cells expressing stress proteins or peptides on their surfaces. Even so, the "spontaneous" autoreactivity of $\gamma\delta$ cell hybridomas presents a functional assay for specificity.

We have begun our search for peripheral $\gamma\delta$ T cells of this type by isolating $\gamma\delta^+$ hybridomas from adult mouse spleen fusions. To our surprise, we found that about 1/2 of the $\gamma\delta^+$ spleen hybridomas showed some degree of "spontaneous" IL-2 production. We have begun receptor analysis on these cells, and so far, they all express the Vγ1-Jγ4-Cγ4 chain (O'Brien et al., In preparation), as do all of the newborn thymus autoreactive hybridomas. We have found so far two differences between the adult spleen derived hybridomas and the newborn thymus hybridomas. First, the adult cells, although their γ chains are encoded by the same genetic elements, are derived from many precursors, as is evident from junctional sequences. These γ chain juntions show a considerably greater degree of junctional variability than do those derived from newborn thymus. Second, the autoreactive $\gamma\delta$ cells from the adult mouse spleen frequently show no positive response to PPD, and sometimes even show a depression of "spontaneous" IL-2 production when cultured in the presence of PPD. Whether this difference in reactivity is related to differences in the receptors expressed has yet to be determined. The adult spleen cells, if originally derived from cells that develop in the newborn thymus, may show a selection for or expansion of cells with a certain reactivity.

CONCLUSIONS

We have shown that a subset of $\gamma\delta$ T cells responds to a mycobacterial heat shock protein, and our results implicate the mouse version of the homologous protein as an autoantigen for the same cells. Moreover, a synthetic peptide representing a segment of the mycobacterial heat shock protein is strongly stimulatory for some of these $\gamma\delta$ cells, perhaps indicating that TCR γδ recognizes peptide antigens as does TCR αβ, via specialized presenting molecules that are co-recognized by the TCR. This recognition of a heat shock protein, and in particular a self heat shock protein, may be a common feature in antigen recognition by γδ TCRs. Murine skin γδ cells provide additional evidence for this hypothesis, because these cells respond to heat-shocked self keratinocytes (Wendy Havran and Jim Allison, personal communication); additionally, some human γδ T cell clones seem to recognize the human HSP-65 homologue (Fisch et al., 1990; Haregewoin et al., 1990). Thus, recognition of this specific antigen type may be a key difference between γδ and αβ T cells, and may offer clues as to the functional role of γδ T cells.

We have wondered whether peptide recognition by γδ T cells as presented here might fail to be generalizable, if for instance we happened to have used a peptide that is capable of assuming a 3-dimensional structure that mimics the stimulatory portion of the intact protein. Recently, however, the keratinocyte produced antigen recognized by a subset of mouse γδ cells, and the mycobacterial antigen recognized by a subset of human γδ cells, have both been shown to be active when supplied in a degraded peptide form (Jim Allison, personal communication; and Robert Modlin, personal communication). We expect, then, that the ability to recognize peptide antigens will be a general characteristic of TCR γδ-expressing T lymphocytes.

REFERENCES

Bluestone JA, Cron RQ, Cotterman M, Houlden BA, Matis LA (1988) Structure and specificity of T cell receptor γ/δ on major histocompatibility complex antigen-specific CD3+, CD4-, CD8- T lymphocytes. J Exp Med 168: 1899-1916

Bonneville M, Ito K, Krecko EG, Itohara S, Kappes D, Ishida I, Kanagawa O, Janeway JCA, Murphy DB, Tonegawa S (1989) Recognition of a self major histocompatibility complex TL region product by γδ T cell receptors. Proc Natl Acad Sci USA 86: 5928-5932

Born W, Hall L, Dallas A, Boymel J, Shinnick T, Young D, Brennan P, O'Brien R (1990) Recognition of a peptide antigen by heat shock reactive γδ T lymphocytes. Science 249: 67-69

Dent AL, Matis LA, Hooshmand F, Widacki SM, Bluestone JA, Hedrick SM (1990) Self-reactive γδ T cells are eliminated in the thymus. Nature 343: 714-719

Fisch P, Malkovsky M, Klein BS, Morrissey LW, Carper SW, Welch WJ, Sondel PM (1990) Human Vγ9/Vδ2 T cells recognize a groEL homolog on Daudi Burkitt's lymphoma cells. Science submitted for publication:

Happ MP, Kubo RT, Palmer E, Born WK, O'Brien RL (1989) Limited receptor repertoire in a mycobacteria-reactive subset of γδ T lymphocytes. Nature 342: 696-698

Haregewoin A, Singh B, Gupta RS, Finberg RW (1990) A mycobacterial heat shock protein responsive γδ T cell clone responds to human homologous heat shock protein: a possible link between infection and autoimmunity. J Infect Dis in press:

Haregewoin A, Soman G, Hom RC, Finberg RW (1989) Human γδ T cells respond to mycobacterial heat-shock protein. Nature 340: 309-312

Holoshitz J, Koning F, Coligan JE, De Bruyn J, Strober S (1989) Isolation of CD4- CD8- mycobacteria-reactive T lymphocyte clones from rheumatoid arthritis synovial fluid. Nature 339: 226-229

Kozbor D, Trinchieri G, Monos DS, Isobe M, Russo G, Haney JA, Zmijewski C, Croce CM (1989) Human TCR-γ+/δ+, CD8+ T lymphocytes recognize tetanus toxoid in an MHC-restricted fashion. J Exp Med 169: 1847-1851

Matis LA, Fry AM, Cron RQ, Cotterman MM, Dick RF, Bluestone JA (1989) Structure and specificity of a class II alloreactive γδ T cell receptor heterodimer. Science 245: 746-749

Modlin RL, Pirmez C, Hofman FM, Torigian V, Uyemura K, Rea TH, Bloom BR, Brenner MB (1989) Antigen-specific T cell receptor γδ bearing lymphocytes accumulate in human infectious disease lesions. Nature 339: 544-548

O'Brien RL, Cranfill R, Lang J, Dallas A, Born W (In preparation) Structure and specificity of peripheral γδ TCRs that recognize a self heat shock protein.

O'Brien RL, Happ MP, Dallas A, Palmer E, Kubo R, Born WK (1989) Stimulation of a major subset of lymphocytes expressing T cell receptor γδ by an antigen derived from Mycobacterium tuberculosis. Cell 57: 667-674

Porcelli S, Brenner MS, Greenstein JK, Balk SP, Terhorst C, Bleicher P (1989) Recognition of cluster of differentiational antigens by human CD4-CD8- cytolytic T lymphocytes. Nature 341: 447-450

Rivas A, Koide J, Cleary ML, Engleman EG (1989) Evidence for involvement of the γ,δ T cell antigen receptor in cytotoxicity mediated by human alloantigen-specific T cell clones. J Immunol 142: 1840-1846

Spits H, Paliard X, Englehard VH, De Vries JE (1990) Cytotoxic activity and lymphokine production of T cell receptor αβ+ and TcR γδ+ cytotoxic T lymphocyte (CTL) clones recognizing HLA-A2 and HLA-A2 mutants. Recognition of TcR-γδ+ CTL clones is affected by mutations at positions 152 and 156. J Immunol 144: 4156-4162

Vidovic D, Roglic M, McKune K, Guerder S, MacKay C, Dembic Z (1989) Qa-1 restricted recognition of foreign antigen by a γδ T-cell hybridoma. Nature 340: 646-650

White J, Blackman M, Bill J, Kappler J, Marrack P, Gold D, Born W (1989) Two better cell lines for making hybridomas expressing specific T cell receptors. J Immunol 143: 1822-1825

Young RA (1990) Stress proteins and immunology. Ann Rev Immunol 8: 401-420

Modulation of Murine Self Antigens by Mycobacterial Components

R. MANN[1], E. DUDLEY[1], Y. SANO[3], REBEECA O'BRIEN[2], W. BORN[2], CH. JANEWAY, JR[3], and A. HAYDAY[1]

[1] Dept. of Biology, Yale University, New Haven, CT 06511, USA
[2] National Jewish Center, Denver, CO 80206, USA
[3] Dept. of Immunology, Yale University, CT 06510, USA

INTRODUCTION

Responses of γδ T cells to antigen are poorly understood. Murine double negative γ(+) cells generally increase in number late in mixed lymphocyte reactions (Jones *et al.*, 1988). In humans and mice, there are instances of allo-reactive γδ T cells directed at polymorphic products of the major histocompatibility complex (MHC) (Bluestone *et al.*, 1988; Ciccone, *et al.*, 1989). In addition, other γδ cells recognise, or are restricted by Class I-MHC-like molecules, TL, CD1, and Qa (Bonneville *et al.*, 1989, Porcelli *et al.*, 1989; Vidovic *et al.* 1989). These observations encourage the hypothesis that γδ cells recognise presenting molecules similar to, if not the same as, those recognised by αβ T cells. However, it is unclear whether *in vivo* , populations of γδ cells exploit high junctional diversity to recognise myriad peptides on such presenting molecules. Alternatively, populations of γδ cells with T cell receptors (TCRs) encoded by the same Vγ and Vδ genes may be selected primarily to recognise self (presenting) molecules. Such self molecules could be induced by stress, such as cell transformation or infection (Janeway *et al*, 1988).

A set of newborn, mouse, thymic γδ hybridomas that spontaneously secretes IL-2 has been described (O'Brien *et al.*, 1989). From experiments in which the growth media of the cells was manipulated, it appears that this secretion reflects a bona fide autoreactivity (O'Brien *et al.*, 1989; and unpublished). These γδ hybrids secrete IL-2 even more vigorously in response to mycobacterial extract, in the form of exogenously-supplied PPD (O'Brien *et al.*, 1989). A prominent component of the PPD is mycobacterial hsp65. At least some of these γδ hybrids respond to a peptide of this protein (Born *et al.*, 1990), presumably presented by a self molecule. These cells also show response to an equivalent peptide from the highly conserved mammalian counterpart, hsp63. This could serve as an explanation for the cells' autoreactivity.

There may, however, be an added dimension to the cells' recognition of PPD, and to their auto-reactivity. For instance, the panel of γδ

Current Topics in Microbiology and Immunology, Vol. 173

hybridomas that has these responses has an almost uniform γδ receptor, most often encoded by Vγ1-Vδ6, but with differences in the junctional sequences (Happ *et al.*, 1989; our unpublished data). Similarly, in humans, a large set of γδ T cells that bear a Vγ9-Vδ2 receptor all recognise the Burkitt lymphona cell line, Daudi (Sturm *et al.*, 1990). In this case, the antigen may again be hsp63 (Fisch *et al.* 1990). The same γδ cells also recognise antigen presenting cells pulsed with mycobacterial extract. However, in this case, recognition of hsp65 seems to be rare. Typically, recognition appears to be of low molecular weight compounds (Pfeffer *et al.*, 1990), presented by viable antigen presenting cells [APCs](Pfeffer-pers. comm.).

The recognition of mycobacterial components and of mammalian cells in the absence of mycobacteria by sets of γδ T cells with limited diversity might be reconciled if the mycobacterial components induce or stabilise on the surface of APCs self antigens that are recognised by γδ T cells. Such antigens would be expressed constitutively by T cell hybridomas and some other tumor cell lines. Because of the allo reactivities and restriction of γδ T cells discussed above, candidate self antigens would be components of the antigen presentation machinery. We therefore investigated the effect of mycobacterial extracts, in the form of PPD, on the cell surface expression of such components.

RESULTS and DISCUSSION

The murine γδ T cell hybridomas, BNT 19.8.2 and 13.7.12 were incubated overnight in the presence or in the absence of 100μg/ml exogenous PPD. The next day each culture was split into two. From one half, the cell supernatant was removed and assayed for IL-2 using an IL-2 dependent cell line, CTLL. Both 19.8.2 and 13.7.12 secrete IL-2 in response to PPD (Fig. 1), as previously described(O'Brien *et al.*, 1989). From the other half of each culture, cells were washed thoroughly, counted, and equal numbers of viable cells iodinated using ^{125}I and Iodogen (Pierce). The labeled material was immunoprecipitated using antibody Ly-m11(Tada *et al.* 1980) [a gift from Peter Robinson, CRC], that is directed against beta 2-microglobulin (b2-m) and that is efficacious at co-precipitating proteins to which b2-m is complexed. The products were resolved on 2D gels (Fig. 2). A large number of proteins of M.W. between approximately 40kD and 150kD are co-precipitated with b2-m. The identity of only some of the spots has been determined. One spot of around 43kD is 'contaminant' actin, but its intensity is relatively weak, implying that mostly cell surface rather than cytosolic proteins are detected. None of the proteins detected was a component of PPD. Most strikingly, the intensity of most of the proteins precipitated is, for both lines, greater after incubation with PPD. This experiment has been repeated four times for line 19.8.2 and twice for 13.7.12. Although the intensity of the increase varied, it was reproducibly obtained.

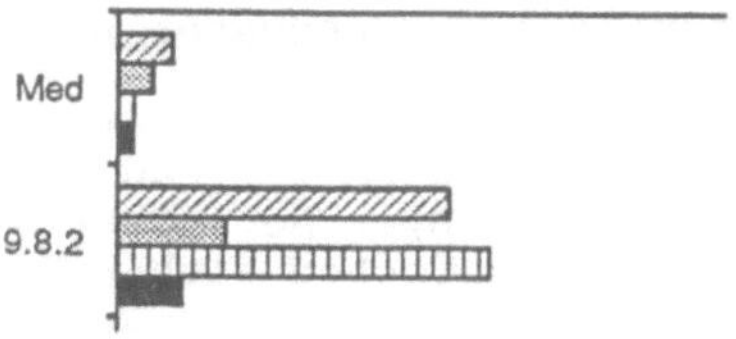

Fig.1 Response of BNT 19.8.2 cells to PPD x-axis: incorporation of tritiated thymidine into IL-2 dependent CTLL cells. CTLL cells were treated with supernatants from 19.8.2 cells incubated overnight with a) nothing (black); b) PPD, 100ug/ml (vertical hatching); c) syngeneic mitomycin C treated spleen cells (shaded); d) syngeneic spleen cells plus PPD (cross hatching). In the set of experiments labeled medium, CTLL cells were treated with the equivalent supernatants from incubations in the absence of 19.8.2 cells

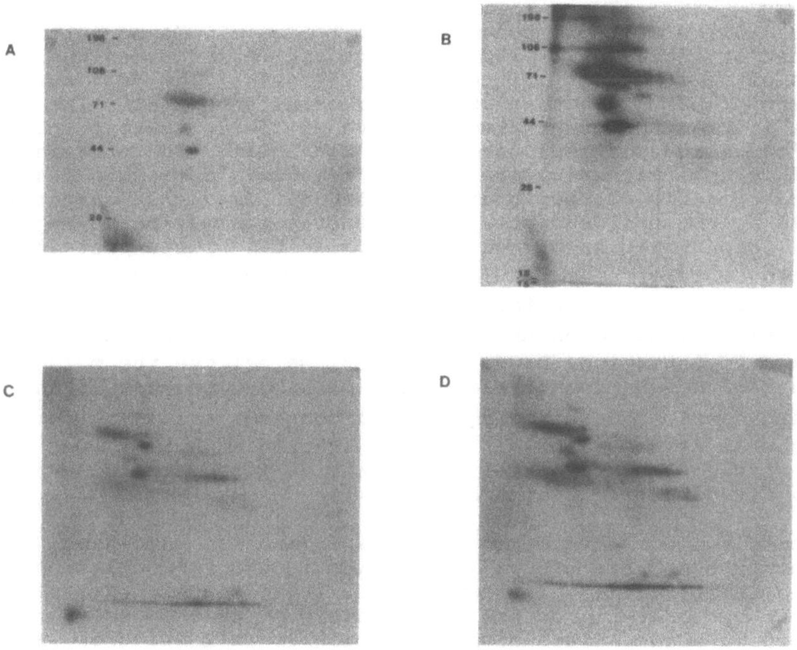

Fig.2 2-D gel analysis of ^{125}-I labeled cell surface proteins co-precipitated with beta 2 microglobulin from A. 19.8.2 cells; B. 19.8.2 cells incubated overnight with 100ug/ml. PPD; C. 13.7.12 cells; D. 13.7.12 cells incubated overnight with PPD. Separation in first dimension by iso-electric focusing (pH 3.5 -10); separation in second dimension by SDS poly acrylamide gel electrophoresis (size markers in kiloDaltons determined from co-electrophoresed markers). Equal numbers of viable cells were iodinated in each case. 50% of precipitated material was analysed. After this, the remaining 50% was analysed to confirm reproducibility (not shown).

A reproducible increase in b2-m and associated proteins was also detected by FACS analysis of viable cells (fig.3A), using the anti b2-m antibody followed by FITC rabbit anti-mouse IgG. The increases in protein expression were specific. No significant increases were detected in TCR. Use of an antibody,Y3, directed against polymorphic class-I antigens indicates that, as expected, at least some of the increased protein expression in response to PPD can be attributed to conventional class-I molecules(Fig. 3B). No increase in b-2m and associated proteins was reproducibly detected either by gels or by FACS when two other cell lines 33BTE.67.1, and RMA-S, were used. Neither of these cell lines (respectively, a murine γδ T cell hybrid, and a mutant murine B cell lymphoma that can present peptide antigens with which it is pulsed) is functionally responsive to PPD. B2-m and associated proteins could be up-regulated on all lines by alpha interferon. (Mann, Dudley,& Hayday -unpublished).

These data allow a simple conclusion to be drawn. Components of PPD can elicit an increase in b2-m and associated proteins on the cell surfaces of some but not all cell lines (Fig.4). The cell surface expression of numerous other proteins remains unchanged. The increase in b2-m and associated proteins may be a marker for an up-regulation of the antigen presentation machinery. The proteins involved in this machinery are likely numerous. Some such proteins, [X,Y, and Z in Fig.4], may ordinarily be only rarely or unstably expressed at the cell surface. Their expression may be up-regulated or stabilised by PPD or by cell transformation. These proteins would then be targets of γδ cells. X,Y, and Z may not be conventional class-I proteins, since b2-m deficient mice are not substantially depleted of γδ cells, and human γδ cells can respond to the b2-m deficient line, Daudi. However, we believe the molecules to be part of the antigen presentation machinery because of the modulation descibed here; because Daudi does not stimulate human γδ cells as well as does a Daudi line transfected with b2-m; and because viable APCs are better presenters of mycobacterial extracts than are fixed APCs (P.Fisch pers. comm.)

It is conceivable that mycobacterial hsp65 and its mammalian counterpart, hsp63 are centrally involved in the up-modulation described here (Fig. 4). As hsp65 is delivered in high amounts to cells (via mycobacteria or their extracts), or as endogenous hsp63 is induced by stress (including adaptation to cell culture), so the hsp may be re-compartmentalised from its conventional location within bacterial or mitochondrial matrices. It may gain entry to the antigen presentation machinery, which it may up-regulate through formation of molecular complexes based on its chaperonin properties (Ellis, 1987). Our sequence of murine hsp63 (Kyes *et al.*, unpublished) shows that unlike, for example, BiP it has a conserved N-glycosylation site. This is at least compatible with a capacity of hsp63 to appear at the cell surface with other proteins (e.g. X,Y, or Z).

Alternatively, or additionally the up-modulation or stabilisation of presenting molecules may be facilitated by the binding of abundant mycobacterial peptides, in a manner similar to that described by Townsend et al , 1989. (fig 4.). This would be consistent with the recognition of cell-bound hsp65 peptides by several γδ cell hybrids

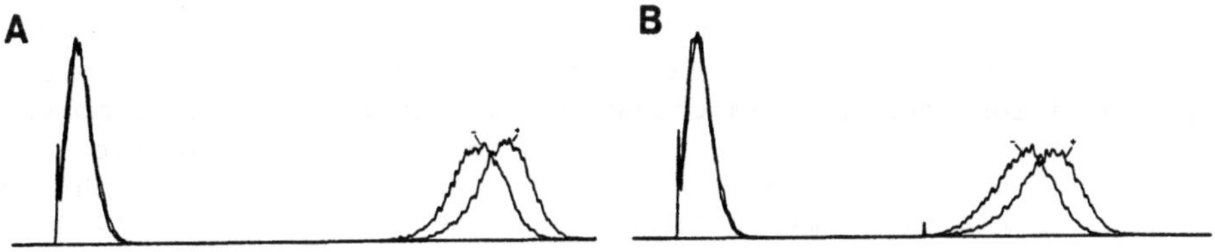

Fig.3A FACS analysis of beta 2-microglobulin expression on 19.8.2 cells without (-) , and with (+) prior overnight incubation with 100ug. /ml. PPD.

Fig.3B FACS analysis of polymorphic class I MHC antigens detected with antibody Y3 on 19.8.2 cells without (-) or with (+) PPD.

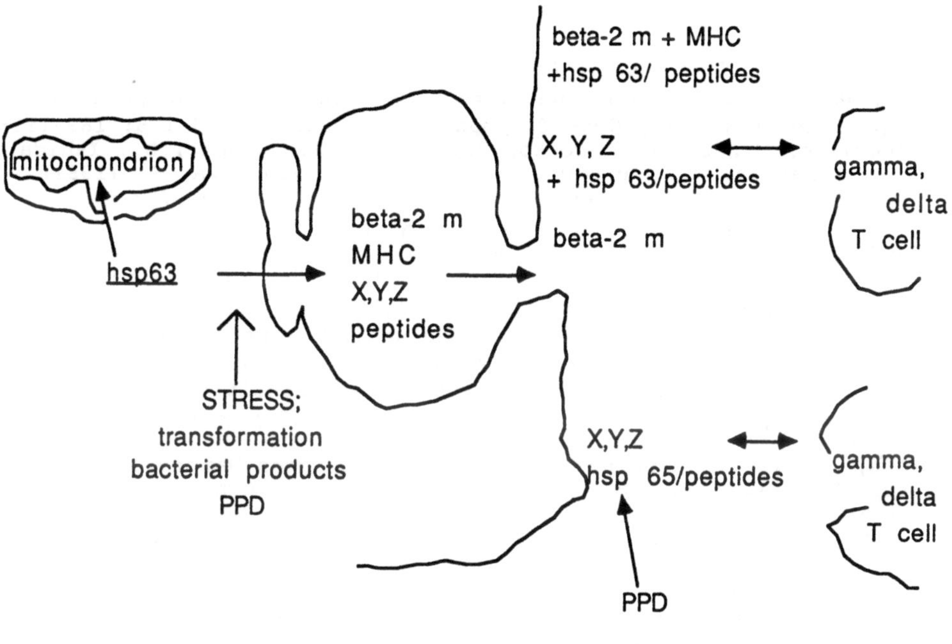

Fig.4 Surface modulation of Class I and other antigens by stress. A stressed state induced by cell transformation (e.g. Daudi cells), or by high concentrations of bacterial products induces re-compartmentalisation of hsp63 into the antigen presentation pathway. By direct protein associations, or as peptides, hsp63 is expressed at the cell surface with components of the antigen presentation machinery, thereby inducing or stabilising higher level expression of those components at the surface. Such components/complexes may also be stabilised to a lesser degree by direct binding of proteins or peptides from exogenously-added PPD. Either way, the resultant complexes are recognised by gamma delta T cells of particular subsets.

(Born et al., 1990). At minimum, these data go further in suggesting that γδ cells recognise products channeled through the antigen presentation machinery (although not necessarily the same molecules) as is recognised by αβ T cells. The modulation of specific self molecules within that machinery, either through stress or through action of bacterial products may be a biologically-significant stimulus for populations of γδ cells.

ACKNOWLEDGMENTS

We thank S.Kyes for advice, and Tom Taylor for work on the FACS. Work supported by NIH grants GM37759 (A.H.) and AI27785 (C.J./A.H.), and by an NSF training grant to E.D.

REFERENCES

Bluestone, J., Cron, R.Q., Cotterman, M., Houlden, B., Matis, L. (1988). Structure and specificity of T cell receptor γδ on MHC-specific CD3+,CD4-, CD8- T lymphocytes. J.Exp. Med 168: 189-1916

Bonneville, M., Ito,K., Krecko, E., Itohara,S., Kappes, D., Ishida, I., Kanegawa, O., Janeway, C., Murphy, D., Tonegawa, S. (1989) Recognition of a self MHC TL region product by γδ T cell receptors. Proc.Natl. Acad. Sci. USA 86: 5928-5932

Born,W., Hall, L., Dallas, A., Boymel, J., Shinnick, T., Young, D., Brennan, P., and O'Brien R. (1990) Recognition of a peptide antigen by heat shock reactive γδ lymphocytes Science 249: 67-69

Cicconi, E., Vialo, O., Pende, D., Malnati, M., Battista Ferrara, G.,Barocci, S., Moretta, A., Moretta, L.(1988) Antigen recognition by human TCR γ(+) lymphocytes. Specific lysis of allogeneic cells after activation in MLC. J.Exp.Med 167: 1517-1522

Ellis, J. (1987) Proteins as molecular chaperones Nature 328:378-379

Fisch, P., Malkovsky, M., Kovats, S., Sturm, E., Braakman, E., Klein, B., Voss, S., Morrisey, L., DeMars, R., Welch, W., Bolhuis, R., Sondel, P. (1990) Recognition by human Vγ9,Vδ2 T cells of a groEL homolog on Daudi cells Science -in the press

Happ, M.P., Kubo, R.T., Palmer, E., Born, W., O'Brien, R. (1989) Limited receptor repertoire in a mycobacteria-reactive subset of γδ T lymphocytes Nature 342: 696-699

Janeway, C., Jones, B., Hayday, A. (1988) Specificity and function of T cells bearing γδ receptors Immunol Today 9: 73-76

Jones, B., Carding, S., Kyes, S., Mjolsness, S., Janeway, C.,Hayday, A. Molecular analysis of T cell receptor γ gene expression in allo-activated splenic T cells of adult mice. Eur. J. Immunol. 18: 1907-1915

O'Brien, R., Happ, M.P., Dallas, A., Palmer, E., Kubo, R.,Born, W. (1989) Stimulation of a major subset of lymphocytes expressing T cell receptor γδ by an antigen derived from mycobacterium tuberculosis. Cell 57: 667-674

Pfeffer,K., Schoel, B., Gulle, H., Kaufman, S., Wagner, H. (1990) Primary responses of human T cells to mycobacteria: a frequent set of γδ T cells are stimulated by protease resistant ligands. Eur. J. Immunol. 20: 1175-1179

Porcelli, S., Brenner, M., Greenstein, J., Balk, S., Terhorst, C., Bleicher, P. (1989) Recognition of CD1 by human CD4-, CD8- cytolytic T cells. Nature 341: 447-449

Sturm, E., Braakman, E., Fisch, P., Vreugdenhil, R., Sondel, P., Bolhuis, R. (1990) Human Vγ9-Vδ2 lymphocytes show specificity to Daudi cells. J.Immuol. - in the press

Tada, N., Kimura, S., Hatzfeld, A., Hammerling, U.(1980) Ly-m11: The H3 region of mouse chromosome 2 controls a new surface alloantigen. Immunogenetics 11: 441-449

Townsend, A., Ohlen, C., Bastin, J., Ljunggren, H-G., Foster, L., Karre, K. (1989) Association of class I MHC heavy and light chains induced by viral peptides Nature 340: 443-448

Vidovic, D., Rogli, C., McKune, K., Guerder, S., MacKay, C., and Dembic, Z. (1989) Qa-1 restricted recognition of foreign antigen by a γδ T cell hybridoma Nature 340: 646-650

In Vitro Activation of Human γ/δ T Cells by Bacteria: Evidence for Specific Interleukin Secretion and Target Cell Lysis

M. E. Munk, A. J. Gatrill, and S. H. E. Kaufmann

Department of Medical Microbiology and Immunology, University of Ulm, Albert-Einstein-Allee 11, D-7900 Ulm, FRG

INTRODUCTION

Nontoxigenic bacteria can be roughly divided into two major groups: extracellular bacteria, which primarily replicate in the extracellular milieu, the host responding to them with a purulent reaction; and intracellular bacteria, which primarily multiply inside host cells, the host response being characterized by a granulomatous reaction (Hahn and Kaufmann 1981). Protective immunity against extracellular bacteria is mediated by protective antibodies which facilitate bacterial uptake and subsequent killing by professional phagocytes. Thus, the role of T cells in immunity to extracellular bacteria is thought to be restricted to their function as helpers in the generation of antibody-secreting B cells. This group of pathogens includes streptococci and staphylococci. In contrast, protective immunity against intracellular bacteria is mediated by T lymphocytes which alter macrophage functions. It is supposed that macrophage activation through interleukins produced by helper T cells, and lysis of infected macrophages by cytotoxic T cells, both participate in an effective immune response to intracellular bacteria. This group of pathogens includes *Listeria monocytogenes* and the pathogenic mycobacteria, *Mycobacterium tuberculosis* and *M. leprae*. Such immune responses (help or cytotoxicity) have been attributed to the major T cell population expressing a T cell receptor (TCR) composed of α- and β-chains.

Recent data have revealed that a less frequent set of T cells expressing a γ/δ TCR have a particular predilection for mycobacteria (Raulet 1989; Brenner et al. 1988; Modlin et al. 1989; O'Brien et al. 1989; Holoshitz et al. 1989; Janis et al. 1989; Haregewoin et al. 1989; Augustin et al. 1989). We have analyzed *in vitro* responses of γ/δ T cells of normal healthy individuals not only to mycobacteria, but also to gram-positive extracellular bacteria. We find that both types of bacteria stimulate γ/δ T cells in many individuals and that such activated T cells produce interleukin 2 (IL-2) and express cytolytic activity. Furthermore, evidence is presented for some kind of specificity of these γ/δ T cells.

RESULTS AND DISCUSSION

The *in vitro* effect of bacterial antigens on γ/δ T lymphocytes of normal individuals was assessed. Peripheral blood mononuclear cells (PBMC) from healthy adults with no history of mycobacterial disease were purified over Ficoll density gradients and 5 x 10^6 cells/ ml were cultured with different bacterial preparations. Samples were removed at day 0 and at day 7-10 and T cells were purified using nylon wool columns, then stained for analysis on a flow cytometer

Current Topics in Microbiology and Immunology, Vol. 173

cell sorter (FACS) for T cell subsets (Munk *et al.* 1990). The relative percentage of γ/δ T cells was between 1-9% in 12 donors examined at day 0 and was markedly increased (12-42%) after 7-10 days in culture with *M. tuberculosis* (strain H37Ra) in 8/12 donors, the other 4 donors showing little or no increase. The α/β T cell population showed no such increase after culture, remaining constant or decreasing slightly. *M. leprae* stimulated γ/δ T cells in an equivalent number of donors to *M. tuberculosis* (6/8 donors and 6/9 donors, respectively), though not always in the same individuals (Table 1). Group A streptococci stimulated an increase in γ/δ T cells in 7/9 donors, *Listeria monocytogenes* in 4/8 donors and *Staphylococcus aureus* in 2/6 donors. The different donors showed distinctly different patterns of response to this panel of antigens, arguing against a non-specific mitogenic stimulation.

Table 1: Increased γ/δ TCR expression after *in vitro* stimulation with different microorganisms[a]

Stimulus	% γ/δ TCR Expression								
Donor:	1	2	3	4	5	6	7	8	9
M. tuberculosis 50 μg/ml	**33**	**34**	**29**	4	**5**	**22**	**50**	1	**13**
M. leprae 50 μg/ml	ND[b]	**32**	**18**	3	**21**	**11**	3	**28**	**18**
L. monocytogenes 10^8/ml	ND	**14**	3	3	**6**	**30**	**36**	7	5
Group A streptococci 10^6/ml	8	**31**	5	**24**	**8**	**44**	**19**	8	**10**
S. aureus 2.5 x 10^7/ml	ND	4	**20**	3	**4**	ND	ND	7	2
Nil (d 0)	3	6	9	9	1	8	2	6	4
Nil (d 7)	2	7	2	4	1	4	1	4	3

[a] PBMC were incubated with the different microorganisms, T cells were enriched, labelled with TCRδ-1-FITC mAb and analyzed on d 7-10. Alternatively, T cells were analyzed on d 0.
[b] ND, not determined.
From Munk et al. 1990.

To verify the assumption that the increase in the γ/δ T-cell population is due to expansion of these cells, we negatively selected γ/δ T lymphocytes by FACS after the initial bulk culture period of 7-10 days with *M. tuberculosis*. Cells negative for the α/β TcR and for the natural killer (NK) cell marker CD56 were restimulated with the same antigen, or with other bacteria and irradiated autologous accessory cells (3000 R). Table 2 shows that the purified γ/δ T cells did proliferate upon restimulation with *M. tuberculosis*, as well as with other bacteria, indicating that expansion of this population does indeed account for the increases seen in bulk culture. However, it should be cautioned here that the broad reactivity pattern of *M. tuberculosis* activated γ/δ T cells towards different bacteria indicates some "antigen-independent" responsiveness of *M. tuberculosis* activated γ/δ T cells which currently we do not fully understand.

Table 2: Proliferation of *M.tuberculosis*-activated γ/δ T cells in response to different bacterial antigens[a]

Stimulus	Proliferative Response [cpm ^{3}H-TdR uptake]		
	Donor: 10	11	12
M.tuberculosis 25 μg/ml	1500	1810	1459
M.leprae 25 μg/ml	280	6606	100
L.monocytogenes 10^8/ml	633	266	2602
Group A streptococci 10^6/ml	608	1476	263
S.aureus 2.5 x 10^7/ml	817	1471	2022
r-IL-2 50 U/ml	6531	1799	37383
Nil	691	98	82

[a] PBMC were stimulated with *M.tuberculosis* for 7 to 10 d, afterwards γ/δ T cells were selected and restimulated with the microorganisms indicated.
[b] ND, not determined.
From Munk et al. 1990.

Expression of the IL-2 receptor (CD25) and the NK cell marker (CD56), was assessed in purified, *M.tuberculosis*-activated γ/δ T-cell populations from 3 donors. The majority (> 70%) of these γ/δ T cells expressed the IL-2 receptor, as assessed by staining with the anti-TAC monoclonal antibodies (mAb; kindly provided by O. Acuto; data not shown). CD56 was expressed to varying degrees (on 26-54% of the cells). IL-2 production by purified, *M.tuberculosis* activated γ/δ T cells was assessed after restimulation with irradiated autologous accessory cells and *M.tuberculosis* antigen in 4 donors. Supernatants were removed after 4 days restimulation and assayed for IL-2 activity on the IL-2 dependent cell line CTLL. Table 3 shows that the γ/δ T cells from all 3 donors produced IL-2, albeit to varying degrees. These data indicate that the *M.tuberculosis*-

Table 3: Specific IL-2 production by *M.tuberculosis*-activated γ/δ T cells[a]

Stimulus	IL-2 Activity [cpm ^{3}H-TdR Uptake by CTLL]		
	Donor: 13	14	15
γ/δ T cells + *M.tuberculosis* 25 μg/ml	95,430	29,404	107,637
γ/δ T cells + Medium	55	ND	46
IL-2 control 50 U/ml	89,786	80,828	93,520
Nil	23	5,721	318

[a] Selected γ/δ T cells were restimulated with *M.tuberculosis* and supernatants collected after 4 d. IL-2 activities were assessed on IL-2 dependent CTLL line.
[b] ND, not determined.
From Munk et al. 1990.

activated γ/δ T cells can both produce and respond to IL-2. Our findings would be consistent with the participation of an autocrine IL-2 pathway in the activation of γ/δ T cells by mycobacteria. Others have also shown secretion of IL-2 and other interleukins after various types of stimulation (Raulet 1989).

γ/δ T cells are known to express killer activity although the ligands involved in recognition of target cells are unknown (Raulet 1989; Phillips et al. 1987). We investigated the killer potential of *M.tuberculosis*-stimulated γ/δ T-cell populations in ^{51}Cr-release assays (detailed in Munk *et al* 1990) using a γ/δ mAb-facilitated kill to bypass the need for specific target cell recognition, as well as against antigen primed target cells. Targets were autologous cells, either isolated adherent cells from peripheral blood or EBV-transformed B cells. For the mAb-facilitated kills, whole, unsorted T-cell populations were used as effectors both before and after 7-10 days bulk culture stimulation with *M.tuberculosis* antigen. Prior to culture, T cells failed to lyse the targets in the presence of either an anti-α/β mAb (BMA031, kindly provided by Behringwerke, FRG) or an anti γ/δ mAb (TCRδ1). Following stimulation with *M.tuberculosis*, however, while the anti-α/β mAb still failed to bring about lysis, a high level of cytolytic activity in the presence of anti-γ/δ mAb was seen by the cells of all four donors examined (varying from 35-93% specific lysis at effector to target ratio of 15:1).

For studying the cytolytic potential of γ/δ T cells against antigen-bearing targets, effector cells were again the negatively-selected γ/δ T-cell population following initial stimulation by *M.tuberculosis* in bulk culture. These cells were then restimulated for a further 14 days and purity of γ/δ T cells was checked by FACS analysis (>98%) . Autologous target cells were pulsed overnight with the antigens indicated in Table 4, then labelled with $Na^{51}CrO_4$, washed, the γ/δ T cells added and ^{51}Cr release determined after 6 hours incubation at 37°C. As shown in Table 4, the γ/δ T cells from 7/8

Table 4: Lysis of antigen pulsed targets by *M.tuberculosis*-activated γ/δ T cells[a]

Target Cell Priming	% Specific Lysis (E/T 9:1)							
Donor:	9	10	16	17	18	19	20	21
M.tuberculosis 25 μg/ml	71	41	66	0	21	21	60	28
Group A Streptococci 10^6/ml	ND[b]	0	ND	ND	ND	ND	ND	8
S.aureus 2.5 x 10^7/ml	0	ND	ND	ND	ND	ND	0	ND
L.monocytogenes 10^8/ml	0	ND	20	0	0	13	7	ND
K562	0	52	ND	ND	ND	ND	44	35
Nil	5	9	8	0	10	0	5	7

[a] Cytolytic activities of selected γ/δ T cells, restimulated with *M.tuberculosis* for 14 d were assessed on autologous EBV-transformed B cells (donors 9, 20) or autologous adherent cells (donors 10, 16, 17, 18, 19, 21) which had not been pulsed (Nil) or pulsed with different bacteria, or on K562 cells.

[b] ND, not determined.

From Munk et al. 1990.

donors tested lysed *M.tuberculosis*-pulsed targets but had little or no effect on unpulsed targets. Three out of 4 donors also lysed the NK susceptible K562 cell line and it should be noted that the donor lacking this NK-like activity (#9) still efficiently lysed the *M.tuberculosis*-primed targets, indicating a distinction between the two killer activities. In 2 of the donors thus tested (#16, #9), *M.tuberculosis*-activated γ/δ T cells also lysed significantly targets pulsed with a different bacterial antigen, *L.monocytogenes*. We would assume from this, as from the proliferation assays, that while some *M.tuberculosis*-activated γ/δ T cells recognize specific *M.tuberculosis* antigens, others see antigens common to different microorganisms.

Finally, we investigated further the question of specificity of these γ/δ T cells. Four more donors were selected who showed an expansion of γ/δ T cells during *in vitro* stimulation to both group A streptococci as well as to *M.tuberculosis*. For each donor, the populations of streptococcus-activated and *M.tuberculosis*-activated γ/δ T cells were negatively selected by FACS and used as effectors in a ^{51}Cr-release assay against autologous targets pulsed with either streptococci or *M.tuberculosis*. As shown in Fig. 1, killing of *M.tuberculosis*-pulsed targets was only brought about by *M.tuberculosis*-activated γ/δ T cells, likewise the streptococcus-pulsed targets were only lysed by streptococcus-activated effectors.

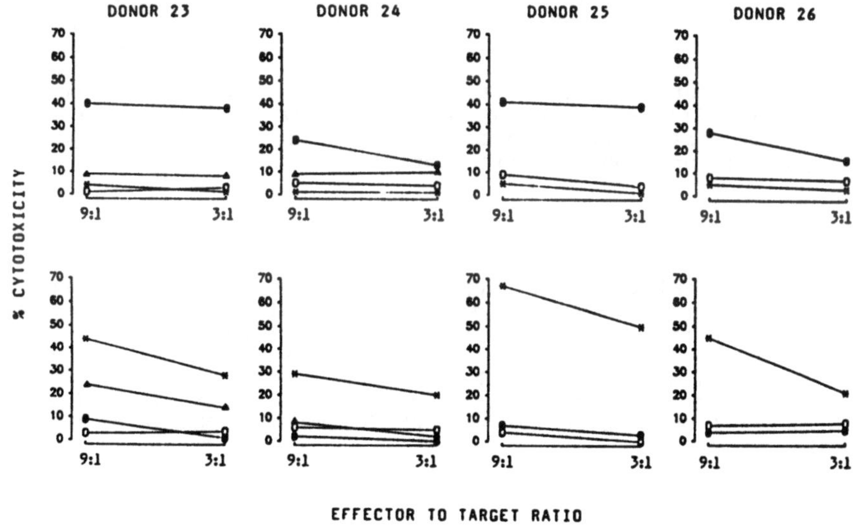

Fig. 1: Specific target cell lysis by *M.tuberculosis* and by streptococcus activated γ/δ T lymphocytes. PBMC were stimulated with either *M.tuberculosis* (upper row) or group A streptococci (bottom row) for 10 days and subsequently γ/δ T lymphocytes were selected. Their cytolytic activity was assessed on autologous adherent cells which had been pulsed with 25 µg/ml of killed *M.tuberculosis* (●), 2.5 x 10^7/ml group A streptococci (*), 1 x 10^8 *L.monocytogenes* (▲), or were left unpulsed (o). Mean values of triplicate wells; S.D. < 10%. From Munk et al. 1990.

For two of these donors, both types of effectors were also tested against *L.monocytogenes*-pulsed targets. In one case the streptococcus-activated γ/δ T cells also lysed these *L.monocytogenes*-pulsed targets, while the *M.tuberculosis*-activated γ/δ T cells from the same donor remained ineffective against this alternative antigen. Furthermore, anti-α/β mAb had no effect upon killing, whereas low concentrations of TCRδ1 mAb inhibited this cytolysis of antigen-pulsed targets (data not shown).

The data from these last experiments provide evidence for antigen-specific cytolytic activity of activated γ/δ T cells, although they do not rule out the possibility of cross-reactive antigens for γ/δ T cells shared between different microbial species (Kaufmann 1990). To our knowledge, this is the first demonstration of killer activity by γ/δ T cells against a nominal antigen, although others have already observed alloantigen-specific killer activity of γ/δ T cells (Raulet 1989; Brenner et al. 1988).

The present study has revealed that γ/δ T cells which recognize mycobacteria are cytolytic and produce IL-2. It remains to be established whether distinct subpopulations are responsible for these different functions, as is often the case with α/β T cells or whether a single population can express both activities. Our findings suggest a role for γ/δ T lymphocytes in immunity to mycobacterial pathogens. Since it has been suggested before that cytolytic and interleukin secreting α/β T cells contribute to immunity against mycobacteria (Kaufmann 1988), the functional role of γ/δ T cells seems similar to that of α/β T cells. Perhaps γ/δ T cells and α/β T cells differ more in their activation requirements, genetic restriction, tissue distribution and/or antigen recognition pattern than in their effector capacities. Because our data also show that, besides mycobacteria, different gram-positive bacteria can stimulate γ/δ T cells *in vitro*, it is tempting to speculate that γ/δ T cells perform a more general function in antibacterial resistance.

ACKNOWLEDGMENTS

This work received financial support to S.H.E. Kaufmann from: SFB 322, German Leprosy Relief Association, EC-India Science and Technology Cooperation Program, UNDP/World Bank/WHO Special Program for Research and Training in Tropical Diseases, and the A.Krupp award for young professors. M.E. Munk is supported by Conselho Nacional de Desenvolvimento Cientifico & Tecnologico (CNPq) Brazil, and A.J.Gatrill by a Wellcome research fellowship. Thanks to Mrs. R. Mahmoudi.

REFERENCES

Augustin A, Kubo RT, Sim GK (1989) Resident pulmonary lymphocytes expressing the γ/δ T-cell receptor. Nature 340:239-240

Brenner MB, Strominger JL, Krangel MS (1988) The γδ T cell receptor. Adv Immunol 34:133-192

Hahn H, and Kaufmann SHE (1981) Role of cell mediated immunity in bacterial infections. Rev Infect Dis 3:1221-1250

Haregewoin A, Soman G, Hom RC, Finberg RW (1989) Human $\gamma\delta^+$ T cells respond to mycobacterial heat shock protein. Nature 340:309-312 rheumatoid arthritis synovial fluid. Nature 339:226-229

Janis EM, Kaufmann SHE, Schwartz RH, Pardoll DM (1989) Activation of γ/δ+ T cells in the primary immune response to *Mycobacterium tuberculosis*. Science 244:713-716

Kaufmann SHE (1988) CD8+ T lymphocytes in intracellular microbial infections. Immunol Today 9:168-174

Kaufmann SHE (1990) Heat shock proteins and the immune response. Immunol. Today 11:129-135

Modlin RL, Pirmez C, Hofman FM, Torigian V, Uyemura K, Rea TH, Bloom BR, Brenner MB (1989) Lymphocytes bearing antigen-specific γδ T-cell receptors accumulate in human infectious disease lesions. Nature 339:544-548

Munk ME, Gatrill AJ, Kaufmann SHE (1990) Target cell lysis and Interleukin-2 secretion by γ/δ T lymphocytes after activation with bacteria. J Immunol 145:2434-2439

O'Brien RL, Happ MP, Dallas A, Palmer E, Kubo R, Born WK (1989) Stimulation of a major subset of lymphocytes expressing T cell receptor γδ by an antigen derived from *Mycobacterium tuberculosis*. Cell 57:667-674

Phillips JH, Weiss A, Gemlo BT, Rayner AA, Lanier LL (1987) Evidence that the T cell antigen receptor may not be involved in cytotoxicity mediated by γ/δ and α/β thymic cell lines. J Exp Med 166:1579-1584.

Raulet DH (1989) The structure, function, and molecular genetics of the γ/δ T cell receptor. Ann Rev Immunol 7:175-207

A Dichotomy Between the Cytolytic Activity and Antigen-Induced Proliferative Response of Human $\gamma\delta$ T Cells

J. HOLOSHITZ, N. K. BAYNE, D. R. MCKINLEY, and Y. JIA

Department of Internal Medicine, University of Michigan School of Medicine, Ann Arbor, Michigan 48109-0531, USA

INTRODUCTION

T cells bearing the $\gamma\delta$ T cell receptor have been commonly found to display non-MHC-restricted, NK-like cytolytic activity (Moingeon 1987, Bosnes 1989, Fisch 1990). Other subsets of $\gamma\delta$ T cells proliferate in response to mycobacterial antigens, in particular the 65 kd heat shock protein (HSP) of mycobacteria (Holoshitz 1989, Modlin 1989, Haregewoin 1989, O'Brien 1989, Janis 1989, Kabelitz 1990). We have previously reported four $CD4^{-}CD8^{-}$ $\gamma\delta$ T cell clones isolated from the synovial fluid of a patient with rheumatoid arthritis. These clones were selected by virtue of their proliferative response to mycobacteria and were found to recognize the mycobacterial HSP65. Their reactivity to mycobacteria was not restricted by MHC. In the present report we studied the cytolytic activity of these cells. We found that our $\gamma\delta$ T cells display "promiscuous" non-MHC-restricted cytolytic activity against neoplastic cells. The cytolytic activity was not directed against exogenously added HSP65 or endogenous HSPs induced by heat. These results suggest that the cytolytic activity and recognition of soluble antigens are two distinct functional properties.

MATERIALS AND METHODS

Antigens

Acetone precipitable fraction of *Mycobacterium tuberculosis* (AP-MT) was prepared from *M. tuberculosis* H37Ra (Difco Detroit, MI) as previously described (Holoshitz 1986). The 65 kd HSP of BCG (P64) and another mycobacterial protein (P32) were purified and kindly donated by Dr. J. DeBruyn, Brussels, Belgium (DeBruyn 1987, DeBruyn 1987a).

Current Topics in Microbiology and Immunology, Vol 173

Cells

Clone 1.4 bearing the phenotype CD4-CD8- γδ was isolated from the synovial fluid of a patient with early rheumatoid arthritis. The isolation and characterization of this and other clones have been described (Holoshitz 1989). Targets used for cytotoxic assays included JBEB, an autologous EBV-transformed B cell line, and DSEB, a heterologous EBV-transformed B cell line. Other target lines were obtained from ATCC or donated (U937) by Dr. B. Eisenstein, University of Michigan.

Cell Mediated Lysis (CML) Assays

Four-hour ^{51}Cr-release assays were carried out in 10% FCS/RPMI 1640 medium as described (Fisch 1990), using different ^{51}Cr-labelled targets and clone 1.4 effector cells at various effector:target ratios. In some experiments target cells were either pre-incubated overnight with 100 µg/ml of AP-MT or incubated for one hour in medium at 42°C before use. Target cells were subsequently collected, washed twice and used for CML assays. In other experiments, the assays were carried out in the presence or absence of AP-MT (10 µg/ml) throughout the length of the experiment.

Proliferation Assays

Proliferative responses were assayed as described (Holoshitz 1989) using $2x10^4$ T cells and 10^5 irradiated (3000R) peripheral blood mononuclear cells per well in the presence or absence of AP-MT (10 µg/ml), P64 (10 µg/ml) or P32 (10 µg/ml). Cultures were pulsed with [^{3}H] thymidine at 54 h and harvested at 72 h.

RESULTS

Promiscuous Cytolytic Activity of γδ T Cells

Cytolytic activity of clone 1.4 was tested with different long-term cell lines. As can be seen in Table 1, potent cytolytic activities were found with almost all malignant lines tested including Daudi (Burkitt's lymphoma), Molt 4, Jurkat, Peer, Molt 13 (T cell leukemias), U937 (monocytic line), and K562 (erythroleukemia). The cytolytic activity was particularly potent with Daudi and Jurkat cells (25% and 54% specific cytolysis, respectively, at E:T ratio of 0.1:1, not shown).

Interestingly, cytolytic activity against Raji cells (Burkitt's lymphoma) was significantly lower. This finding corroborates other studies which showed no cytolytic activity of γδ T cells against Raji (Fisch 1990). No cytolytic activity was found with autologous (JBEB) or heterologous (DSEB) EBV- transformed B cell lines, even with higher E:T ratios (not shown).

Table 1. Cytolytic activity of γδ clone 1.4 against various tumor cell targets

Cell Line	Percent specific lysis[a]
Daudi	77
Jurkat	73
U937	61
Molt 13	59
Molt 4	58
Peer	57
K562	50
Raji	19
JBEB	3
DSEB	1

[a] Effector:target ratio of 5:1

Antigenic Specificity of γδ T Cells

We have previously found that our γδ T cell clones proliferate in response to AP-MT, which contains the HSP65. Since the proliferative response to AP-MT and the cytolytic activity of the clones were both non-MHC-restricted, we wished to know whether the cytolytic activity of these cells is directed against HSPs. Figure 1 shows proliferative responses of clone 1.4. As can be seen, this clone responded to AP-MT and to P64, a biochemically purified mycobacterial HSP65, but not to another purified mycobacterial antigen P32. In contrast, neither JBEB, Molt 4, nor Daudi cells became more susceptible to lysis by clone 1.4 cells following preincubation with AP-MT. Experiments with continuous incubations of the antigen throughout the CML assay similarly did not show increased cytolysis (not shown). Endogenous HSPs, induced by preincubating targets at 42°C for 1 hour, also failed to trigger cytolysis of the EBV-transformed B cell line JBEB by clone 1.4 (Table 2).

Table 2. Cytolysis of clone 1.4 of AP-MT-pulsed targets and heat-treated targets

Cell line	Preincubation		Percent specific lysis at E:T ratio	
	Temp.	Antigen	12.5:1	6.25:1
JBEB	37°C	--	9	8
	37°C	AP-MT	10	8
	42°C	--	7	4
Molt 4	37°C	--	24	25
	37°C	AP-MT	23	21
Daudi	37°C	--	63	62
	37°C	AP-MT	68	67

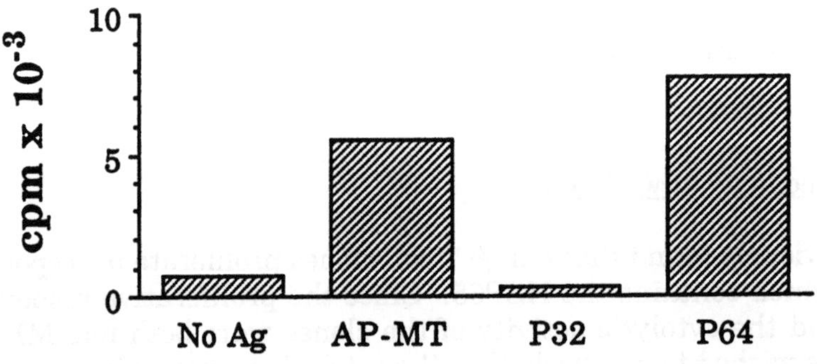

Fig. 1. Proliferative response of clone 1.4 in the presence of AP-MT, 10μg/ml, P64, 10μg/ml, and P32, 10μg/ml

DISCUSSION

In this study we have examined the cytolytic activity of HSP65-reactive γδ T cells. We found that these cells, which proliferate specifically in response to HSP65, displayed also promiscuous cytolytic activity against neoplastic cells. The two activities were not contributed by distinct subpopulation of cells, since TCR δ-chain Southern blot analysis, and TCR γ-chain cDNA sequencing data confirmed the clonality of clone 1.4 (unpublished data).

We studied whether killing by γδ T cells involves recognition of HSPs. To this end we found that externally added mycobacterial HSP65 did not produce any increase in the cytolytic activity of clone 1.4. Induction of endogenous HSP by heat also did not increase susceptibility of target cells to killing by γδ cells. Thus, recognition for killing involves different antigens than those recognized for specific proliferation. Our preliminary results suggest also that the two distinct recognition events are mediated by different cell surface molecules. Antigen-induced proliferation of the γδ cells could be effectively blocked by anti-TCR antibodies and by anti-LFA1-α antibodies. In contrast, cytolysis could not be blocked by these antibodies (in preparation).

The polyspecific killing of neoplastic cells bears some similarity to LAK activity (Grimm 1984). Our clones have been maintained in relatively low concentrations of IL-2 (10-20 u/ml of Cetus recombinant IL-2). However, some antigen-specific CD4+ T cell clones were found to display polyspecific cytolytic activity even with low doses of IL-2 (Gromkowski 1988). It is possible that γδ T cells have similar increased inherent tendency to get activated by lymphokines. Long-term culture of γδ T cells in IL-2 has been reported to broaden the specificity of their cytolytic activity (Bosnes 1989). Our recent finding (H. Koizumi et al., submitted) of nuclear inclusion bodies in 1.4 cells, similar to those found in LAK cells, gives further support to the possibility that γδ T cells acquire LAK activity *in vitro*.

ACKNOWLEDGEMENT

We thank Dr. J. DeBruyn for providing P64 and P32. Supported in part by NIH grants AR40544, AR07080, AR20557 and RR00042, and by the Michigan Chapter of the Arthritis Foundation. Dr. J. Holoshitz is also supported by an Arthritis Investigator Award and by the Searle Scholar Program/ The Chicago Community Trust.

REFERENCES

Bosnes V et al. (1989) Specificity of γδ receptor-bearing cytotoxic T lymphocytes isolated from human peripheral blood. Scand J Immunol 29: 723-731

De Bruyn J et al. (1987a) Purification, partial characterization, and identification of a skin-reactive protein antigen of *Mycobacterium bovis* BCG. Infect Immun 55: 245-252

De Bruyn J et al. (1987) Purification, characterization and identification of a 32 kDa protein antigen of *Mycobacterium bovis* BCG. Microb Path 2:351-366

Fisch P et al. (1990) γ/δ T cell clones and natural killer cell clones mediate distinct patterns of non-major histocompatibility complex-restricted cytolysis. J Exp Med 171: 1567-1579

Grimm EA, Rosenberg SA (1984) The lymphokine activated killer phenomenon. In: Pick, E (ed) The Lymphokine, Vol 9. Academic Press, New York, p 279

Gromkowski SH, Hepler KM, Janeway CA (1988) Low doses of interleukin 2 induce bystander cell lysis by antigen-specific $CD4^+$ inflammatory T cell clones in short-term assay. Eur J Immunol 18: 1385-1389

Haregewoin A et al. (1989) Human $\gamma\delta^+$ T cells respond to mycobacterial heat-shock protein. Nature 340: 309-312

Holoshitz J et al. (1986) T lymphocytes of rheumatoid arthritis patients show augmented reactivity to a fraction of mycobacteria cross reactive with cartilage. Lancet II 305-309

Holoshitz J et al. (1989) Isolation of $CD4^-$ $CD8^-$ mycobacteria-reactive T lymphocyte clones from rheumatoid arthritis synovial fluid. Nature 339: 226-229

Janis EM et al. (1989) Activation of γδ T cells in the primary immune response to mycobacterium tuberculosis. Science 244: 713-716

Kabelitz D et al. (1990) A large fraction of human peripheral blood $\gamma\delta^+$ T cells in activated by mycobacterium tuberculosis but not by its 65-kd heat shock protein. J Exp Med 171: 667-679

Modlin RL et al. (1989) Lymphocytes bearing antigen specific γδ T cell receptors accumulate in human infectious disease lesions. Nature 339: 544-548

Moingeon P et al. (1987) A γ-chain complex forms a functional receptor on cloned human lymphocytes with natural killer-like activity. Nature 325:723-726

O'Brien RL et al. (1989) Stimulation of a major subset of lymphocytes expressing T cell receptor γδ by an antigen derived from *Mycobacterium tuberculosis*. Cell 57: 667-674

Analysis of Primary T Cell Responses to Intact and Fractionated Microbial Pathogens

K. PFEFFER[1], B. SCHOEL[3], H. GULLE[3], HEIDRUN MOLL[2], SANDRA KROMER[1], S. H. E. KAUFMANN[1], and H. WAGNER[1]

[1] Institute of Medical Microbiology and Hygiene, Technical University of Munich, FRG
[2] Institute of Clinical Microbiology, University of Erlangen, FRG
[3] Institute of Medical Microbiology, University of Ulm, FRG

SUMMARY

Freshly isolated human T lymphocytes were tested for their response to mycobacteria, mycobacterial lysates, 2 dimensional (2D) PAGE separated mycobacterial lysates, leishmania and defined leishmanial antigen preparations. While $\gamma\delta$ T cells proliferated vigourously in the presence of mycobacteria and mycobacteria derived lysates, a significant stimulation from 2 D gel separated lysates was not detected. In addition $\gamma\delta$ T cells failed to respond towards leishmania or leishmanial components. In the $\alpha\beta$ T cell compartment some donors, presumably according to their state of immunity against mycobacteria, responded to mycobacteria, mycobacterial lysates and 2 D gel separated mycobacterial lysates. Neither freshly isolated $\gamma\delta$ T cells nor $\alpha\beta$ T cells from naive donors did mount a significant immune response against leishmania.

INTRODUCTION

Within the last few years studies revealed the existence of two distinct T lymphocyte subsets, characterized by expression of either an $\alpha\beta$ T cell receptor- (TCR) or a $\gamma\delta$ TCR heterodimer(Brenner et al. 1988; Davis and Bjorkman, 1988; Raulet, 1989; Strominger, 1989). In contrast to $\alpha\beta$ T cells, the biological role and the antigen specificity of $\gamma\delta$ T cells has not yet been elucidated(Brenner et al. 1988; Strominger, 1989; Bluestone and Matis, 1989). Recent observations, however, have suggested an involvement of $\gamma\delta$ T cells in mycobacterial infections (Janis et al. 1989; Modlin et al. 1989; Holoshitz et al. 1989; O'Brien et al. 1989; Augustin et al. 1989; Kabelitz et al. 1990). In particular, it has been shown that $\gamma\delta$ TCR expressing hybridomas derived from murine fetal thymuses recognize a 65 kDa heat shock protein (hsp65) (O'Brien et al. 1989). In addition human $\gamma\delta$ T cell lines or clones have been isolated responding to hsp65 (Holoshitz et al. 1989; Haregewoin et al. 1989).
This study was performed to characterize the responsiveness of freshly isolated human $\alpha\beta$ and $\gamma\delta$ T cells to molecular components derived from mycobacteria, mycobacterial lysates and 2-dimensional PAGE separated lysates. We found that $\gamma\delta$ T cells proliferate vigourously in response to mycobacterial lysates but not to the same lysates separated by 2 D gel electrophoresis. In contrast $\alpha\beta$ T cells respond both to unseparated and 2 D gel separated lysates.These data indicate that ligands recognized by $\gamma\delta$ and $\alpha\beta$ T cells are biochemically different. Further analysis revealed that the major proportion of $\alpha\beta$ T cells responded to components of 30-100 kDa MW, whereas the majority of $\gamma\delta$ T cells was stimulated by rather small molecular weight components of 1 - 3 kDa. We also analysed the responses of freshly isolated $\alpha\beta$ and $\gamma\delta$ T cells to another microbial pathogen, leishmania major. However, no stimulation of T cells, derived from healthy donors, was observed.

MATERIALS AND METHODS

$\alpha\beta$ and $\gamma\delta$ T cells were isolated from mononuclear cells (MNC) of healthy donors as described (Pfeffer et al. 1990). Briefly, adherent cells were removed from MNC by plastic adherence. E-rosetting cells were purified from nonadherent cells by E-rosette formation. For purification of $\alpha\beta$ T cells TCRδ1 (T Cell Sciences) positive and CD16 (Leu11, Becton Dickinson) positive cells were removed by fluorescence activated cell sorting (EPICS V, Coulter). For isolation of $\gamma\delta$ T cells, $\alpha\beta$ T cells were depleted on a MACS-system (Stefan Miltenyi, Biotechnische

Geräte,Bergisch Gladbach, FRG)) using BMA031 (kind gift of Dr. R. Kurrle, Behringwerke). Final purification of γδ T cells was achieved by removing CD16 positive cells by FACS sorting.

Lysis of mycobacteria (H37Rv), 2 dimensional PAGE separation, and electroblotting into 480 individual samples was done as described (Gulle et al. 1990), and gel filtration was performed accordingly to (Pfeffer et al. 1990). rhsp65 from E.coli clone M1103 (Thole et al. 1987) was purified as published (Kaufmann et al. 1987). Leishmania major and leishmanial antigens were prepared as published (Moll et al. 1989).

Replicate cultures of 5000 responder cells were incubated in the presence of 50,000 γ-irradiated autologous feeder cells in complete RPMI1640 (20mM glutamine, 10mM Hepes, antibiotics and 10% heat inactivated human serum) and antigenic preparations as indicated. rIL-2 (10U/ml, Eurocetus) was added unless other mentioned on day 5. Proliferation was measured by ^{3}H-thymidine uptake. Proliferative responses are shown as mean values in cpm of replicate cultures, standard deviations were less than 10% of mean values.

RESULTS

Proliferative responses of freshly isolated human αβ and γδ T cells to mycobacteria and mycobacterial lysates.

An efficient isolation procedure was established to purify peripheral γδ T cells from mononuclear cells by consecutive steps of plastic adherence, E-rosetting, magnet activated cell sorting and finally fluorescence activated cell sorting, yielding in 2 - 5 x 10^6 γδ T cells out of 500 ml peripheral blood.
The in vitro responsiveness of pan-T cells, αβ T cells and γδ T cells to graded amounts of mycobacterial lysates is detailed in Table 1.

Table 1 Responsiveness of T cell subsets to graded amounts of mycobacterial lysates.

Lysate		0	0.05	0.1	0.5	1.0	5.0 ug/ml
Exp.A	E-rosette$^+$ cells	800	48663	53668	60906	48221	66615
	αβ T cells	230	19138	36603	33675	19579	9053
	γδ T cells	55	9878	28516	35376	28954	25638
Exp.B	E-rosette$^+$ cells	240	10895	10221	20250	8158	867
	αβ T cells	126	711	3280	4861	708	696
	γδ T cells	100	28875	26564	49607	42508	8781

Mean values (cpm) of two representative experiments are shown.

Stimulations of αβ T cells varied from donor to donor depending on the immune status of the donor tested (H. Gulle, Klaus Pfeffer, unpublished results), whereas γδ T cells of almost all healthy donors responded vigourously to mycobacterial lysates (data not given). As indicated in Table 1 maximal proliferative responses were obtained in the range of 0.5 - 1 ug/ml of bacterial lysates.

Responses of T cell subsets stimulated with 2 D PAGE separated mycobacterial lysates.

To characterize the T cell immune response at a molecular level, mycobacterial lysates were separated by high resolution 2 dimensional PAGE electrophoresis using isoelectric focusing (IEF) in the first dimension and native PAGE electrophoresis in the second dimension. Consecutively 480 individual samples were obtained from 2D gels by electroelution, and afterwards analysed individually for their stimulating capacity towards both αβ and γδ T cells. Results are shown in Fig. 1. While αβ T cells vigourously responded to many fractions (A), no significant response of $\gamma\delta$ T cells could be obtained (B). Note that the very same $\gamma\delta$ T cells proliferated vigourously to unseparated lysates (Table 1, Exp.A).

Fig. 1 Analysis of 2 D gel separated mycobacterial lysates for stimulating capacity for either $\alpha\beta$ T cells (A) or $\gamma\delta$ T cells (B).

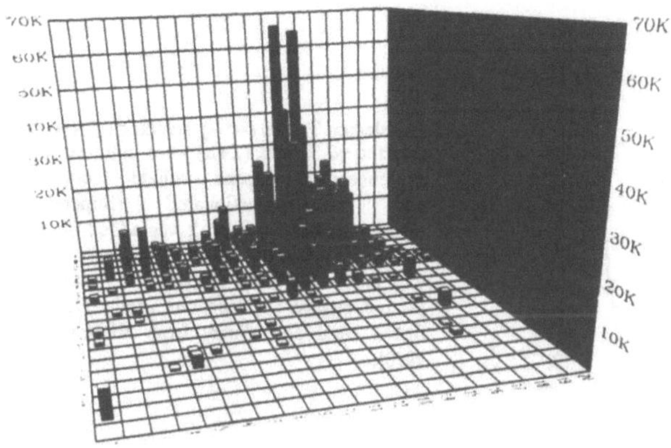

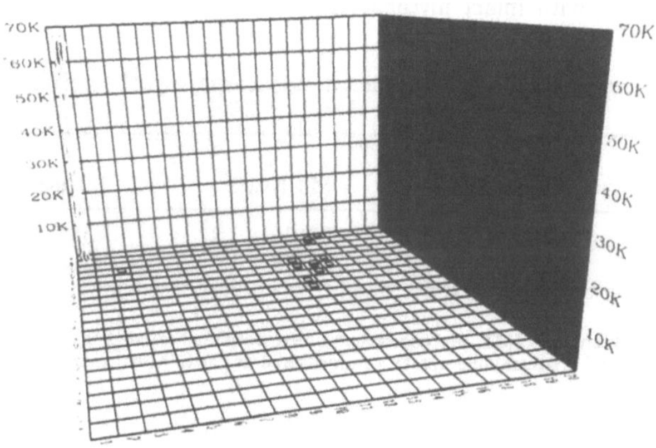

$\alpha\beta$ T cells (A) or $\gamma\delta$ T cells (B) (10^3 per culture), respectively, were incubated individually with 480 samples derived from electroeluted 2 D gels. ^{3}H-tyhimidine uptake was measured after 8 days. rIl-2 was added at day 5.

These results indicated that $\gamma\delta$ and $\alpha\beta$ T cells responded to distinct ligands differing at the level of their biochemical properties such as molecular weight and/or isoelectric point.

To clarify this question in detail mycobacterial lysates were separated according to size and distinct molecular weight fractions were analysed for their stimulating capacity to $\alpha\beta$ or $\gamma\delta$ T cells. As shown in Fig. 2 the major stimulation of $\alpha\beta$ T cells occured in the range of 30 - 100 kDa, while responses of $\gamma\delta$ T cells were mainly obtained in fractions containing mycobacterial components smaller than 10 kDa.

Fig. 2 Response of unselected and selected T cell populations to size-fractionated mycobacterial lysates.

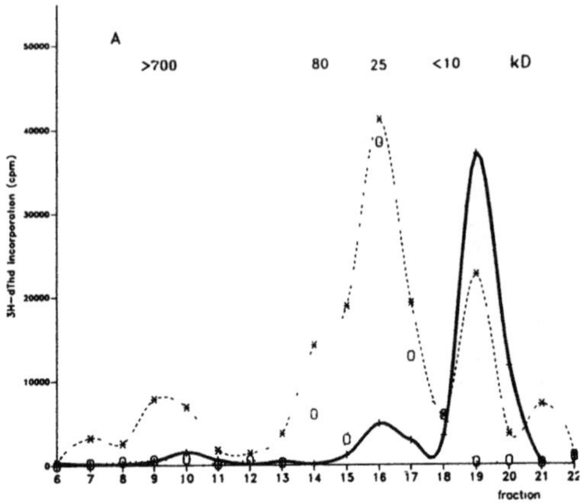

Unselected T cells (--*--), selected $\alpha\beta$ (··o··) or selected $\gamma\delta$ T cells (_+_) were cultered in the presence of autologous irradiated feeder cells and size- fractionated (Superose12) mycobacterial lysates. rIl-2 was added on day 5 and ^{3}H-Thymidine uptake was measured at day 8. Proliferative responses (y axis) are plotted against the fractions used. (Data and figure cited accordingly from Pfeffer et al. 1990).

To elucidate the role of hsp65 in primary immune responses to mycobacterial antigens freshly prepared $\gamma\delta$ T cells were cocultered with intact mycobacteria, mycobacterial lysates and rhsp65 (Table 2).

Table 2 Proliferative response of $\gamma\delta$ T cells against intact mycobacteria, mycobacterial lysates and rhsp65.

Antigen	control	intact MT	MT lysate	rhsp65
Exp.A	1431	3445	20430	1385
Exp.B	273	5341	21139	252

Freshly isolated $\gamma\delta$ T cells were cultered with antigens (MT: Mycobacterium tuberculosis) indicated (1ug/ml). rIl-2 was added from the beginning of the culture. ^{3}H-thymidine uptake was measured at day 8. Mean values of 12 cultures are shown.

Again $\gamma\delta$ T cells were stimulated by mycobacterial lysates and to a lesser degree by complete mycobacteria. But the response to rhsp65 was not significant, indicating that exogeneously added rhsp65 is poor as antigen in primary human $\gamma\delta$ T cell responses.

Finally we tested whether leishmania and leishmanial antigens have the capacity to stimulate $\gamma\delta$ T cells. These microbial pathogens are known to contain high levels of glycolipids and glycoproteins (Moll et al. 1989) and the possibility was considered that $\gamma\delta$ T cells may preferentially respond to such antigens (Pfeffer et al. 1990).

Table 3 Analysis of freshly isolated T cell populations in response to leishmania and different leishmanial derived preparations in comparison to mycobacterial lysates.

Antigen	control	leishmania	LPG	SLA	MT lysate
E-rosette$^+$ cells	398	445	146	574	5381
$\alpha\beta$ T cells	680	788	202	438	862
$\gamma\delta$ T cells	3343	3411	823	3669	14201

The T cell populations indicated were cultered in the presence of leishania (leishmania major, $4*10^4$ organisms/ml), leishmanial lipoglycan (LPG, 3 ug/ml), soluble leishmanial antigen (SLA, 1 ug/ml) and mycobacterial lysate (MT lysate, 1ug/ml).Mean values from quadruplicate cultures are shown. ^{3}H-thymidine uptake was determined at day 8. rIl-2 was supplemented at the beginning of the culture.

As depicted in Table 3 neither T cell population proliferated in the presence of leishmania or leishmanial antigen preparations such as lipoglycan (LPG) or disrupted leishmania (SLA). On the other hand $\gamma\delta$ T cells of this donor proliferated in the presence of mycobacterial lysate. These results provide evidence that leishmanial antigens are not able to evoke a significant immune reponse from unprimed human donors.

DISCUSSION

In vitro stimulation of freshly isolated human $\gamma\delta$ T lymphocytes with mycobacteria involves both $\gamma\delta$ and $\alpha\beta$ T cells. Lysis of mycobacterial organisms resulted in a significant enhancement of stimulation, presumably by overcoming limitations of antigen processing in APC's. 2D gel separation of mycobacterial lysates reveals that $\alpha\beta$ T cells respond to distinct and defineable mycobacterial components (Gulle et al. 1990), whereas proliferation of $\gamma\delta$ T cells is virtually absent. Subsequent studies showed that the major ligands for $\gamma\delta$ T cells derived from mycobacterial lysates reside in a molecular weight range between 1 to 3 kDa (Pfeffer et al. 1990), while ligands for $\alpha\beta$ T cells are contained in MW-fractions from 30 - 100 kDa. Electroelution of 2D gels is performed with an dialysis membrane with a molecular cutoff of 3.5 kDa (Gulle et al. 1990). Therefore ligands for $\gamma\delta$ T cells are not contained in 2D gel samples. We conclude from these data that ligands for $\alpha\beta$ and $\gamma\delta$ T cells derived from mycobacteria differ in size and biochemical properties.
Protease digestion of mycobacterial lysates results in complete abolishment of $\alpha\beta$ T cell responses, whereas $\gamma\delta$ T cells show virtually unaltered responses to protease-treated lysates (Pfeffer et al. 1990). These findings could be explained by either of two conclusions: the fraction stimulating $\gamma\delta$ T cells contains unusual peptides which are resistant to conventional protease digestion or, alternatively, $\gamma\delta$ T cells respond themselves to non-peptides.
Observation in mice indicate that within mycobacterial lysates hsp65 represents a major immunogen for $\gamma\delta$ T cells (O'Brien et al. 1989). The results presented here indicate that in primary responses human $\gamma\delta$ T cells do not respond to rhsp65. Possible explanations are that hsp65 has to be bound to cellular membranes, or that hsp65 has to be induced in the antigen presenting cells for $\gamma\delta$ T cell stimulation to occur, or that the frequency of freshly isolated $\gamma\delta$ T cells is rather low. Experiments evaluating these possibilities are currently performed.
Primary immune responses of $\gamma\delta$ T cells to defined glycolipids of leishmania (Moll et al. 1989), as well as lysed or intact leishmania were not observed. These results are in agreement to studies indicating that in leishmania infections $\gamma\delta$ T cells do not respond (Röllinghoff et al., see this issue of CTMI).

Acknowledgements
We thank Dr. R. Kurrle for mAb BMA031 and Dr. J.D.A. Van Embden for the recombinant E.coli clone M1103 expressing hsp65.This work was supported by the SFB 322, the BMFT, and the UNDP/World Bank/WHO Special Program for Research and Training in Tropical Diseases.

References

Augustin, A., Kubo, R.T. and Sim, G.K. (1989) Resident pulmonary lymphocytes expressing the gamma/delta T-cell receptor. Nature 340:239-241

Bluestone, J.A. and Matis, L.A. (1989) TCR gamma delta cells--minor redundant T cell subset or specialized immune system component?. J Immunol 142:1785-1788

Brenner, M.B., Strominger, J.L. and Krangel, M.S. (1988) The gamma delta T cell receptor. Adv Immunol 43:133-192

Davis, M.M. and Bjorkman, P.J. (1988) T-cell antigen receptor genes and T-cell recognition [published erratum appears in Nature 1988 Oct 20;335(6192):744]. Nature 334:395-402

Gulle, H., Schoel, B. and Kaufmann, S.H.E. (1990) Direct blotting with viable cells of protein mixtures separated by two-dimensional gel electrophoresis. J Immunol Methods 133:253-261

Haregewoin, A., Soman, G., Hom, R.C. and Finberg, R.W. (1989) Human gamma delta+ T cells respond to mycobacterial heat-shock protein. Nature 340:309-312

Holoshitz, J., Koning, F., Coligan, J.E., De Bruyn, J. and Strober, S. (1989) Isolation of CD4- CD8- mycobacteria-reactive T lymphocyte clones from rheumatoid arthritis synovial fluid. Nature 339:226-229

Janis, E.M., Kaufmann, S.H., Schwartz, R.H. and Pardoll, D.M. (1989) Activation of gamma delta T cells in the primary immune response to Mycobacterium tuberculosis. Science 244:713-716

Kabelitz, D., Bender, A., Schondelmaier, S., Schoel, B. and Kaufmann, S.H. (1990) A large fraction of human peripheral blood gamma/delta + T cells is activated by Mycobacterium tuberculosis but not by its 65-kD heat shock protein. J Exp Med 171:667-679

Kaufmann, S.H., Vath, U., Thole, J.E., van Embden, J.D. and Emmrich, F. (1987) Enumeration of T cells reactive with Mycobacterium tuberculosis organisms and specific for the recombinant mycobacterial 64-kDa protein. Eur J Immunol 17:351-357

Modlin, R.L., Pirmez, C., Hofman, F.M., Torigian, V., Uyemura, K., Rea, T.H., Bloom, B.R. and Brenner, M.B. (1989) Lymphocytes bearing antigen-specific gamma delta T-cell receptors accumulate in human infectious disease lesions. Nature 339:544-548

Moll, H., Mitchell, G.F., McConville, M.J. and Handman, E. (1989) Evidence of T-cell recognition in mice of a purified lipophosphoglycan from Leishmania major. Infect Immun 57:3349-3356

O'Brien, R.L., Happ, M.P., Dallas, A., Palmer, E., Kubo, R. and Born, W.K. (1989) Stimulation of a major subset of lymphocytes expressing T cell receptor gamma delta by an antigen derived from Mycobacterium tuberculosis. Cell 57:667-674

Pfeffer, K., Schoel, B., Gulle, H., Kaufmann, S.H. and Wagner, H. (1990) Primary responses of human T cells to mycobacteria: a frequent set of gamma/delta T cells are stimulated by protease-resistant ligands. Eur J Immunol 20:1175-1179

Raulet, D.H. (1989) The structure, function, and molecular genetics of the gamma/delta T cell receptor. Annu Rev Immunol 7:175-207

Strominger, J.L. (1989) The gamma delta T cell receptor and class Ib MHC-related proteins: enigmatic molecules of immune recognition. Cell 57:895-898

Thole, J.E., Keulen, W.J., De Bruyn, J., Kolk, A.H., Groothuis, D.G., Berwald, L.G., Tiesjema, R.H. and van Embden, J.D. (1987) Characterization, sequence determination, and immunogenicity of a 64-kilodalton protein of Mycobacterium bovis BCG expressed in escherichia coli K-12 [published erratum appears in Infect Immun 1987 Aug;55(8):1949]. Infect Immun 55:1466-1475

Function and Specificity of Human Vγ9/Vδ2 T Lymphocytes

P. Fisch[1], Susan Kovats[3], Natalia Fundim[1], E. Sturm[6], E. Braakman[6], R. DeMars[3], R. L. H. Bolhuis[6], P. M. Sondel[1,2,3], and M. Malkovsky[1,4,5]

[1] Department of Human Oncology [2] Department of Pediatrics [3] Department of Genetics, University of Wisconsin, Madison, WI 73792, USA
[4] Department of Medical Microbiology and Immunology [5] Wisconsin Regional Primate Research Center, University of Wisconsin, Madison, WI 53706, USA
[6] Department of Immunology, The Dr. Daniel Den Hoed Cancer Center, 3075 EA Rotterdam, The Netherlands

The majority of T cells express the αβ T-cell receptor (TCR) that recognizes antigen in the context of the major histocompatibility complex (MHC). The repertoire of αβ T cells reflects the diversity of potential environmental antigens and the polymorphism of MHC molecules. A minor subset of T cells expresses the γδ TCR. Although human peripheral blood γδ T cells display great junctional diversity of the γ and δ chains, only a limited combinatorial diversity of variable (V) region usage has been recorded (reviewed by Triebel and Hercend 1989). Moreover, recent reports indicate that several human γδ T-cell lines and a large proportion of murine γδ T cells recognize heat shock proteins (hsps) in mycobacterial extracts (O'Brien et al. 1989; Holoshitz et al. 1989; Haregewoin et al. 1989). Therefore, it was suggested that the antigens and the antigen presenting molecules for γδ T cells could be relatively non-polymorphic (reviewed by Born et al. 1990).

To study the mechanisms of MHC-unrestricted cytotoxicity, we derived γδ T-cell clones, αβ T-cell clones and natural killer (NK)-cell clones from various normal donors. Over the last two years, we tested more than 3000 clones from various normal donors. These clones were derived after sorting human peripheral blood lymphocytes (PBLs) by flow cytometry for cells expressing the γδ or αβ TCR, CD8 or CD4, or the surface antigens CD16 or CD56. The positively selected cells were cloned, using irradiated autologous or allogeneic PBLs and lymphoblastoid B cell lines (LCLs) as feeder cells. The clones were expanded as described (Fisch et al. 1990a), and characterized for function and phenotype.

We found that all human γδ clones coexpressing Vγ9 and Vδ2 gene products (reviewed by Triebel and Hercend 1989) mediated a pattern of MHC-unrestricted cytolysis distinct from NK-cell and αβ T-cell clones (Fisch et al. 1990a). All Vγ9/Vδ2 clones strongly lysed the Daudi cell line, but not the Raji cell line (both Burkitt's lymphoma cell lines). In contrast, NK clones lysed both Daudi and Raji targets effectively. However, virtually all $CD4^+$ or $CD8^+$ αβ T-cell clones killed neither Daudi nor Raji. Most of the γδ, αβ and NK clones displayed MHC-unrestricted cytotoxicity against K562, Molt4, U937, TK6 and other susceptible target cells. The NK-cell clones and the Vγ9/Vδ2 T-cell clones were generally more potent killers of these tumor targets than the αβ T-cell clones. In contrast to alloreactive αβ T-cell clones, the γδ T-cell clones did not lyse the LCL feeder cells. Moreover, γδ T-cell clones that were isolated after "priming" of PBLs in bulk cultures to allogeneic LCLs from various donors, did not mediate allospecific lysis. Thus, we conclude that most γδ T cells do not recognize foreign HLA antigens. We believe, that lysis of a few LCLs by γδ T cells (such as TK6 or HLA class I-deficient LCL 721.221), could be due to an increased sensitivity of these target cells to MHC-unrestricted lysis (e.g., associated with the absence of HLA class I on the mutant LCL 721.221), rather than to alloreactivity.

However, the strong cytotoxic activity of Vγ9/Vδ2 T cells against Daudi was not associated with the absence of HLA class I on the cell surface of Daudi, since an HLA class I-expressing Daudi variant [transfected with the murine β_2-microglobulin (β_2m) gene] was strongly lysed by the Vγ9/Vδ2 T-cell

Current Topics in Microbiology and Immunology, Vol. 173

clones. Also, lysis of Daudi cells was still present, when the cloned Vγ9/Vδ2 cells were cultured in medium in the absence of stimulatory feeder cells and interleukin-2 (IL-2), whereas the lysis of K562, Molt4 and the other susceptible target cells gradually disappeared. Using cold target inhibition assays, we have found that Daudi cell lysis by the Vγ9/Vδ2 T-cell clones is mediated via high affinity receptors, whereas lower-affinity receptors bind Molt4, K562 and other target cells. These data support the idea that the recognition of Daudi cells by Vγ9/Vδ2 T cells is mediated through the TCR, whereas the recognition of other targets involves determinants with lower affinity.

Our subsequent studies (Fisch et al. 1990b) showed that all human Vγ9/Vδ2 T-cell clones but not γδ T-cell clones expressing Vδ1, mediated specific proliferative responses to Daudi cells but to none of the other target cells, such as Raji, K562, Molt4 and LCLs. However, the same clones responded to LCLs in the presence of certain mycobacterial extracts. The presentation of mycobacterial antigens to the human Vγ9/Vδ2 cells was not restricted by HLA class I, nor HLA class II determinants, since mutant LCLs, deficient for HLA class I (LCL 721.221) or class II (LCL 721.180) could present the mycobacterial antigens. Also, the proliferative response of Vγ9/Vδ2 lymphocytes to the β_2m-deficient Daudi cells was enhanced in the presence of mycobacterial antigens. This is incompatible with the possibility that nonpolymorphic β_2m-associated HLA class Ib molecules (reviewed by Strominger 1989) are the antigen-presenting molecules for human Vγ9/Vδ2 T cells. Moreover, the presentation of mycobacterial antigens did not depend on intracellular antigen processing. Paraformaldehyde-fixed LCLs were able to present the mycobacterial antigens, although the antigen presentation was somewhat reduced as compared to unfixed LCLs (Fisch et al. 1990b). It is noteworthy that only very few preparations of mycobacterial antigens are highly stimulatory for human γδ T cells.

The stimulation with Daudi cells induced expansion of initially undetectable levels of Vγ9/Vδ2 T cells in umbilical cord blood lymphocytes, so that after repeated stimulation over a 4 week interval the cultures contained 50 to 80% Vγ9/Vδ2 T lymphocytes (Fisch et al. 1990b). Thus, the Daudi-induced expansion of Vγ9/Vδ2 T cells in the umbilical cord blood could serve as an *in vitro* model for the *in vivo* expansion of the Vγ9/Vδ2 subset in the peripheral blood (but not in the thymus) in the first 10 years after birth (Parker et al. 1990). Also, human peripheral blood Vγ9/Vδ2 T cells were expanded *in vitro* by the stimulatory mycobacterial extracts and by some extracts from *E. coli*. These data suggest that the antigens for the Vγ9/Vδ2 subset are shared by mammalian cells and diverse microorganisms.

The proliferative response of human peripheral blood γδ T lymphocytes to Daudi cells (Sturm et al. 1990; Fisch et al. 1990b) (measured as an increase in the percentage of γδ TCR-positive cells), and to the bacterial extracts from mycobacteria and *E. coli* could be substantially inhibited (approximately 90% inhibition in some experiments) by a polyclonal rabbit antiserum, against the mammalian groEL homolog (hsp58 or hsp60) but not by other immune and nonimmune rabbit sera (Fisch et al. 1990b) The hsp58 antiserum did not inhibit the proliferation of human Vγ9/Vδ2 T cells induced by IL-2. Using this antiserum, we were able to immunoprecipitate a Daudi cell surface molecule with a relative molecular weight of 58-kD corresponding to the mammalian hsp58. This molecule was coprecipitated with a 66-kD molecule which could be a member of the hsp70 family. Our data suggest that human Vγ9/Vδ2 T cells recognize a ligand on Daudi cells that is homologous to the human and bacterial groEL-related proteins. Recognition of groEL determinants by mycobacteria reactive human γδ T cells is compatible with results of Holoschitz et al. (1989) and Haregewoin et al. (1989). Modlin et al. (1989), Kabelitz et al. (1990) and we (Fisch et al. 1990b) were unable to demonstrate direct stimulation of human γδ T cells by the purified mycobacterial groEL homolog, hsp65. Nevertheless, our results indicate that all the mycobacteria-reactive human γδ T cells coexpress Vγ9 and Vδ2 genes and that all these cells recognize a determinant on Daudi cells while other human γδ T cells do not respond to Daudi or mycobacteria. Our experience is that the Daudi- or mycobacteria-induced specific proliferation of Vγ9/Vδ2 clones may be difficult to demonstrate experimentally after a prolonged (many weeks) *in vitro* culture period. Therefore, the absence of proliferative responses to purified hsp65, could be related to suboptimal antigen presentation of some groEL preparations or other experimental difficulties. These hsp preparations may be recognized in a different conformation by $\gamma\delta$ T cells than by $\alpha\beta$ T cells.

The specific antigenic stimulation of human Vγ9/Vδ2 T cells resembles superantigen responses. The mycobacteria- and Daudi-induced reactivities could be analogous to the stimulation of murine Vβ8 T cells by staphylococcal enterotoxins and Mls, since the response to these antigens appears to be determined by the expression of certain V region genes, but not by the junctional diversity of the TCR. Moreover, the Vγ9/Vδ2 T-cell response to the bacterial antigens may not require intracellular antigen processing. However, unlike the superantigenic responses of murine and human Vβ8 (Marrack and Kappler 1990) and human Vγ9 cells (Rust et al. 1990), coexppression of both Vγ9 and Vδ2 seems to be required for strong antigen responsiveness. We found that some (but not all) clones that express Vγ9 and Vδ1 recognized Daudi cells in cytotoxic and proliferative responses. Single-color flow cytometry analysis of the V region repertoire of the PBLs stimulated for one week with Daudi or mycobacteria revealed that more expanded $\gamma\delta$ T cells expressed the Vγ9 chain than the Vδ2 chain. Two-color analysis of the $\gamma\delta$ T cells in PBL preparations stimulated with Daudi showed that most Vγ9 expressed Vδ2, but very few cells expressed Vδ1. Therefore, it is possible that the Vγ9 protein is the chain determining the superantigen reactivity, and that the superantigen-binding sequence is always accessible on Vγ9 molecules associated with Vδ2, but only rarely with Vδ1.

Our data suggest that the human Vγ9/Vδ2 T cells recognize homologs of the groEL family of heat shock proteins. GroEL hsps are intracellular proteins that are highly conserved in the phylogeny (reviewed in Young et al. 1987, Lindquist and Craig 1988; Welch 1990; Kaufmann 1990), but they have not been detected on the cell surface. However, these proteins are frequent targets of cell-mediated and humoral immunity during bacterial and parasitic infections (Young et al. 1988). The induction of hsps can occur in cells stressed by heat, viral infections or transformation (reviewed by Born et al. 1990; and Kaufmann 1990). Previous work from Ottenhoff et al. (1988) and Koga et al. (1989) showed that endogenous groEL proteins presented by HLA antigens could be target antigens for cytotoxic $\alpha\beta$ T cells on macrophages. Our data suggest that groEL molecules expressed on Daudi cells are the target antigens for Vγ9/Vδ2 T cells. Exogenous ligands from some bacterial extracts may bind to yet unknown antigen presenting molecules in a complex that is immunologically crossreactive with the epitope recognized by the Vγ9/Vδ2 T cells on Daudi. Such antigen presenting molecules could also be involved in the transport of endogenous groEL antigens to the cell surface in Daudi cells, and, possibly, under conditions of stress in other cells. However, on the basis of our findings, we cannot exclude the possibility that the hsp homologs on Daudi are themselves involved in the antigen presentation of endogenous peptide antigens, that are crossreactive with peptides present in crude bacterial extracts (Born et al. 1990). However, human $\gamma\delta$ T cells did not respond to the groEL peptides that are the antigens of murine $\gamma\delta$ T cell hybridomas (Fisch et al. 1990b).

Interestingly, Holoshitz et al. (1989) isolated Vγ9/Vδ2 T cell clones from rheumatoid arthritis synovium and one clone, indeed, responded to groEL hsps. Haregewoin et al. (1989) found that human mycobacteria-reactive polyclonal $\gamma\delta$ T cells expressing the disulphide-linked form of the $\gamma\delta$ TCR (the TCR structure of Vγ9/Vδ2 T cells) mediated strong responses to the mycobacterial groEL homolog, hsp65. These and our data indicate that human Vγ9/Vδ2 T cells could participate in diseases of autoimmune, infectious and neoplastic origins.

REFERENCES

Born W, Happ MP, Dallas A, Readon C, Kubo R, Shinnick T, Brennan P, O'Brien R (1990) Recognition of heat shock proteins and $\gamma\delta$ cell function. Immunol Today 11:40-43

Born W, Hall L, Dallas A, Boymel J, Shinnick T, Young D, Brennan P, O'Brien R (1990) Recognition of a peptide antigen by heat shock reactive $\gamma\delta$ lymphocytes. Science 249:67-69

Fisch P, Malkovsky M, Braakman E, Sturm E, Bolhuis RLH, Prieve A, Sosman J, Lam VA, Sondel PM (1990a) γ/δ T cell clones and natural killer cell clones mediate distinct patterns of non-major histocompatibility complex restricted cytolysis. J Exp Med 171:1567-1579

Fisch P, Malkovsky M, Kovats S, Sturm E, Braakman E, Klein BS, Voss SD, Morrissey LW, DeMars R, Welch WJ, Bolhuis RLH, Sondel PM (1990b) Recognition by human Vγ9/Vδ2 T ce lls of a groEL homolog on Daudi Burkitt's lymphoma cells. Science, in press

Holoshitz J, Koning F, Coligan JE, De Bruyn J, Strober S (1989) Isolation of mycobacteria-reactive T lymphocyte clones from rheumatoid arthritis synovial fluid. Nature 339:226-229

Haregewoin A, Soman G, Hom RC, Finberg RW (1989) Human γδ T cells respond to mycobacterial heat-shock protein. Nature 340:309-312

Kabelitz D, Bender A, Schondelmaier S, Schoel B, Kaufmann SHE (1990) A large fraction of human peripheral blood γ/δ T cells is activated by *Mycobacterium tuberculosis* but not by its 65-kD heat shock protein. J Exp Med 171:667-679

Kaufmann SHE (1990) Heat shock proteins and the immune response. Immunol Today 11:129-136

Koga T, Wand-Würtenberger A, DeBruin J, Munk M E, Schoel B, Kaufmann SHE (1989) T cells against a bacterial heat shock protein recognize stressed macrophages. Science 245:1112-1115

Lindquist S, Craig EA (1988) The heat shock proteins. Annu Rev Genet 22:631-677

Marrack P, Kappler J (1990) The staphylococcal enterotoxins and their relatives. Science248:705-711

Modlin RL, Pirmez C, Hofman FM, Torigian V, Uyemura K, Rea TH, Bloom BR, Brenner MB (1989) Lymphocytes bearing antigen specific γδ T-cell receptors accumulate in human infectious disease lesions. Nature 339:544-548

O'Brien RL, Happ MP, Dallas A, Palmer E, Kubo R, Born WK (1989) Stimulation of a major subset of lymphocytes expressing T cell receptor γδ by an antigen derived from mycobacterium tuberculosis. Cell 57:667-674

Ottenhoff THM, Ab BK, VanEmbden JDA, Thole JER, Kiessling R (1988) The recombinant 65-kD heat shock protein of *Mycobacterium Bovis* Bacillus Calmette-Guerin/*M. tuberculosis* is a target molecule for $CD4^+$ cytotoxic T lymphocytes that lyse human monocytes. J Exp Med 168:1947-1952

Parker CM, Groh V, Band H, Porcelli SA, Morita C, Fabbi M, Glass D, Strominger J, Brenner MB 1990) Evidence for extrathymic changes in the T cell receptor γ/δ repertoire. J Exp Med 171:1597-1612

Rust CJJ, Verreck F, Vietor H, Koning F (1990) Specific recognition of staphylococcal enterotoxin A by human T cells bearing receptors with the Vγ9 region. Nature 346:572-575

Strominger JL (1989) The γδ Receptor and class Ib MHC-related proteins:Enigmatic molecules of immune recognition. Cell 57:895-898

Sturm E, Braakman E, Fisch P, Vreugdenhil RJ, Sondel P, Bolhuis RLH (1990) Human Vγ9-Vδ2 TCR γδ lymphocytes show specificity to Daudi Burkitt's lymphoma cells. J Immunol, in press

Triebel F, Hercend T (1989) Subpopulations of peripheral T gamma delta lymphocytes. Immunol Today 10:186-188

Welch WJ (1990) The mammalian stress response:cell physiology and biochemistry of stress proteins. In: Morimoto R, Georgopoulos C, Tissieres A (Eds) Stress Proteins in Biology and Medicine, Cold Spring Harbor Laboratory Press, New York, pp 223-277

Young DB, Ivanyi J, Cox JH, Lamb JR (1987) The 65kDa antigen of mycobacteria- a common bacterial protein? Immunol Today 8:215-219

Young D, Lathigra R, Hendrix R, Sweetser D, Young RA (1988) Stress proteins are immune targets in leprosy and tuberculosis. Proc Natl Acad Sci USA 85:4267-4270

ACKNOWLEDGEMENT

The authors thank R. O'Brien, W. Born, W. Welch and S. Pierce for helpful discussions. Supported by grants from the NIH, the American Cancer Society, the World Health Organization, the Dutch Cancer Society and a Leukemia Society of America Fellowship to P. Fisch. This is publication no. 30-007 of WRPRC.

Daudi Cell Specificity Correlates With the Use of a Vγ9-Vδ2 Encoded TCRγδ

Els Sturm[1], E. Braakman[1], P. Fisch[2], P. M. Sondel[2], and R. L. H. Bolhuis[1]

[1] Dept. of Immunology, The Daniel Den Hoed Cancer Centre, Rotterdam, The Netherlands
[2] Dept. of Human Oncology, University of Wisconsin, Madison, WI, USA

TCRγδ cells in peripheral blood with different functional Vγ and Vδ gene rearrangements represent two non-overlapping subsets. The majority expresses a TCR encoded by the Vγ9-Vδ2 gene segments whereas the minor subset uses Vδ1 in its functional receptor rearrangement [1,2]. Upon in vitro activation, these two subsets of TCRγδ lymphocytes display MHC unrestricted lytic activity against a wide variety of tumor cells of distinct histologic origin [3,4]. Recently antigen-specific and/or MHC restricted TCRγδ lymphocyte functions have been documented [5,6]. We aimed at studying whether functional differences exist between the two subsets of TCRγδ lymphocytes and in particular whether the TCRγδ is involved in their distinct activation. Stimulation of fresh unfractionated PBL with irradiated Daudi cells, a Burkitt lymphoma derived cell line, resulted in the selective expansion of TCRγδ lymphocytes within the $CD3^+$ population in a seven day period. This selective expansion appeared strictly confined to the Vγ9-Vδ2 T lymphocytes. Stimulation with other B-LCL lines e.g. APD, BSM (in vitro epstein bar virus transformed B-LCL) or Raji (a Burkitt lymphoma derived cell line) did not result in such selective expansion of Vγ9-Vδ2 lymphocytes. Our results therefore point at a TCR Vγ9-Vδ2 mediated antigen-specific Daudi cell interaction. Analysis of the lytic activities of Vγ9-Vδ2 and Vδ1 cloned lymphocytes against a panel of tumor cells of distinct histologic origin revealed a striking difference. Daudi target cells were exclusively lysed by all Vγ9-Vδ2 T lymphocyte clones. This specific lysis of Daudi cells by Vγ9-Vδ2 clones even was found when these effector

Current Topics in Microbiology and Immunology, Vol 173

lymphocytes were generated in an non-Daudi B-LCL stimulator cell comprising culture system, used to establish the Vγ9-Vδ2 clones. Whereas none of the Vδ1 lymphocyte clones could lyse Daudi cells. The Vγ9$^-$, Vδ2$^-$,Vδ1$^-$ clone PJ4 and the thymus derived Vγ9$^-$, Vδ1$^+$ clones Thy-CL5 and Thy-CL10 did not lyse Daudi cells either. The resistance to lysis of Daudi cells by Vδ1$^+$ clones was not due to a lower capacity of conjugate formation to the Daudi target cells by Vδ1 clones. Efficiently crosslinking of the TCR/CD3 complex via the relevant FcRγ expressed on Daudi cells by means of an anti-CD3 or TCRγδ mAb resulted in their lysis, conclusively demonstrating that Daudi cells are not intrinsically resistant to lysis by Vδ1 clones.

The critical question remains whether the specific lysis of Daudi cells involves the TCRγδ because the Daudi cell line is also susceptible to MHC unrestricted lysis by IL-2 activated CD3$^-$ and CD3/TCRαβ$^+$ lymphocytes reportedly not involving the TCR structure [7-9]. In the latter case this difference in specificity of Daudi cell lysis for the Vγ9-Vδ2 versus the Vδ1 clones might reflect differential expression and/or involvement of multiple receptors, e.g. accessory molecules in target cell recognition. Although both subsets of TCRγδ lymphocytes express equal levels af adhesion molecules CD2, CD11a/18 and LFA-1, they have been reported to differ in their cell matrix composition, which is reflected by differences in their respective adherence and motility capacities [10]. These differences in cell matrix composition may also result in different target cell repertoires, f.i. Daudi cells. However, as mentioned Vδ1 clones form conjugates with Daudi cells equally well as the Vγ9-Vδ2 clones. Moreover, freshly isolated Vγ9-Vδ2 PBL, that do not display MHC restricted cytolysis, specifically proliferate to Daudi cells but not to other B-LCL or K562 cells [11]. These data support the concept of the TCRγδ involvement in the recognition of Daudi cells.

What is the nature of the antigen expressed on Daudi cells that is specifically recognized by Vτ9-Vδ2 lymphocytes?

MHC class-I and CD1 molecules, reported to serve as specific TCRτδ ligands are not expressed on Daudi cells [6,12-13]. Recently evidence has accumulated to indicate that TCRτδ can recognize mycobacterial antigens. Interestingly only Vτ9-Vδ2 but not Vδ1 T lymphoytes specifically proliferate to myobacterial antigens. These mycobaterial antigens can be presented by MHC class I and II negative B-LCL, and by Daudi cells [14]. Daudi cells showed no requirement for MHC antigens in the antigen presentation. Mixtures of anti-MHC class II mAb did not inhibit the lytic activities of Vτ9-Vδ2 clones towards Daudi cells. The evolutionary conserved nature of heat shock proteins from bacteria to man and the reported expression of the latter on the membrane of certain B-LCL [15], makes HSP or HSP related molecules good candidates for the cell structure expressed by Daudi cells, recognized by Vτ9-Vδ2 lymphocytes. The specific proliferation to and cytolysis of Daudi cells by the entire subpopulation of Vτ9-Vδ2 lymphocytes, irrespective of the junctional diversity at the TCRδ gene is reminiscent of a superantigen reponse. In "classic" superantigen stimulation, the Vβ8 or Vτ9 encoded TCR chain in itself is sufficient to impose superantigen specificity [16,17]. In case of analogy, the Vτ9 chain is expected to dictate Daudi cell specificity. However two Vτ9-Vδ1 clones of thymic origin did not lyse Daudi cells, implying that either the Vτ9 perse is not sufficient to dictate Daudi cell specificity or represents a physiological constraint due to their thymic origin. Therefore either the expression of a Vδ2 encoded TCRδ chain alone might be sufficient or the combination of Vτ9-Vδ2 encoded TCR chains is required.

Specifity of the Vτ9-Vδ2 lymphocyte subset in the periphery for autologous HSP would explain the overrepresentation of these lymphocytes at the sites of autoimmune reactions where HSP are abundantly present [5].

This work was supported by the Dutch Cancer Society KWF-Grants RRTI 85-13 and DDHK 89-09.

References

1. Faure, F., S. Jitsukawa, F. Triebel, and T. Hercend (1988). Characterization of human peripheral lymphocytes expressing the CD3/τδ complex with anti-receptor monoclonal antibodies. J. Immunol. 141:3357.
2. Sturm, E., E. Braakman, R. E. Bontrop, P. Chuchana, R. J. Van de Griend, F. Koning, M-P. Lefranc, and R.L.H. Bolhuis (1989). Coordinated Vτ and Vδ gene segment rearrangements in human T cell receptor τ/δ$^+$ lymphocytes. Eur. J. Immunol. 19:1261.
3. Borst, J., R. J. Van de Griend, J. W. Van Oostveen, S.-L Ang, C. J. M. Melief, J. G. Seidman, and R. L. H. Bolhuis (1987). A T-cell receptor τ/CD3 complex found on cloned functional lymphocytes. Nature 325:683
4. Jitsukawa, S., F. Triebel, F. Faure, C. Miossec, and T. Hercend (1988). Cloned CD3$^+$ TCRαβ$^-$ TiτA$^-$ peripheral blood lymphocytes compared to the TiτA$^+$ counterparts: Structural differences of the τ/δ receptor and functional heterogeneity. Eur. J. Immunol. 18:1671.
5. Holoshitz, J., F. Koning, J. E. Coligan, J. De Bruyn, and S. Strober (1989). Isolation of CD4$^-$ CD8$^-$ mycobacteria reactive T lymphocyte clones from rheumatoid arthritis synovial fluid. Nature 339:226.
6. Bluestone, J. A., R. Q. Cron, M. Cotterman,, B. A. Houlden, and L. A. Matis (1988). Structure and specificity of T cell receptor τδ on major histocompatibility complex antigen specific CD3$^+$, CD4$^-$, CD8$^-$ T lymphocytes. J. Exp. Med. 168:1847.
7. Phillips, J. H., A. Weiss, T. B. Gemlo, A. A. Rayner, and L. L. Lanier (1987). Evidence that the T cell antigen receptor may not be involved in cytotoxicity mediated by τδ and αβ thymic cell lines. J. Exp. Med. 169:1579.

8. Van de Griend, R. J., W. J. M. Tax, B. A. Van Krimpen, R. J. Vreugdenhil C. P. M Ronteltap, and R. L. H. Bolhuis. (1987). Lysis of tumor cells by $CD3^{+}$,4^{-},8^{-},16^{+} T cell receptor $\alpha\beta^{-}$ clones regulated via CD3 and CD16 activation sites, recombinant interleukin-2 and interferon β. J. Immunol. 138:1627.

9. Fisch, P., M. Malkovsky, E. Braakman, E. Sturm, R. L. H. Bolhuis, A. Prieve, J. A. Sosman, V. A. Lam, and P. M. Sondel (1990). Gamma/Delta T cell clones and natural killer cell clones mediate distinct patterns of non-MHC restricted cytolysis. J. Exp. med 171:1567.

10. Grossi, C. E., E. Ciccone, N. Migone, C. Bottino, D. Zarcone, M. C. Mingari, S. Ferrini, G. Tambusi, O. Viale, G. Casorati, R. Millo, L. Moretta, and A. Moretta (1989). Human T cells expressing the τ/δ receptor (TCR-1): Cτ1- and Cτ2 encoded forms of the receptor correlate with distinctive morphology, cytoskeletal organization, and growth characteristics. Proc. Natl. Acad. Sci. USA 86:1619.

11. Sturm, E., E. Braakman, P. Fisch, R .J. Vreugdenhil, P. M. Sondel, and R. L. H. Bolhuis (1990). Human Vτ9-Vδ2 TCRτδ lymphocytes show specificity to Daudi Burkitt's lymphoma cells. J. Immunol. in press.

12. Porcelli, S., M. B. Brenner, J. L. Greenstein, S. P. Balk, C Terhorst, and P. A. Bleicher (1989). Recognition of differentiation antigens by human $CD4^{-}$, $CD8^{-}$ cytolytic T lymphocytes. Nature 341:447.

13. Calabi, F., and C. Milstein (1986). A novel family of human major histocompatibility complex-related genes not mapping chromosome 6. Nature 323:540

14. Fisch, P., M. Malkovsky, E. Sturm, E. Braakman, B. S. klein, S. D. Voss, L. W. Morissey, R. DeMars, W. J. Welch, R. L. H. Bolhuis, and P. M. Sondel (1990). Recognition by human Vτ9/Vδ2 T cells of a groEl homolog on Daudi Burkitt's lymphoma cells. Science in press.

15.Vanbuskirk, A., B. L. Crump, E. Margiolash, S. K. Pierce (1989). A peptide binding protein having a role in antigen presentation is a member of the HSP70 heat shock family. J. Exp. Med. 170:1799.

16.Marrack, P. and, J. Kappler (1988). The T cell repertoire for antigen and MHC. Immunol. Today 9:308.

17.Rust, C. J. J., F. Verreck, H. Vietor, and F. Koning (1990). Specific recognition of Staphylococal enterotoxin A by Vτ9 bearing human T cells. Nature 346:572.

TCT.1: A Target Structure for a Subpopulation of Human γ/δ T Lymphocytes

F. Mami-Choubaid and T. Hercend

"Laboratoire d'Immunologie Cellulaire", INSERM, U333, Institut Gustave-Roussy, Villejuif, France

INTRODUCTION

We have previously shown that the largest number of human γ/δ T cell clones express a Vγ9/Vδ2$^+$ T cell receptor (TCR) recognized by both anti-TiγA and anti-TiVδ2 monoclonal antibodies (mAb) (Jitsukawa et al 1987; Triebel et al 1988 and Miossec et al 1990). These clones generally display a non-MHC (major histocompatibility complex) requiring cytotoxicity detectable against a wide variety of tumor cells, which is likely TCR independent (Bank et al 1986; Brenner et al 1987; Borst et al 1987 and Moingeon et al 1987). A minority of the clones were found to express the δTCS1 epitope which involves Vδ1-Jδ1 gene segments (Mami-Chouaib et al 1989). Among these clones we have identified allogeneic γ/δ cytotoxic T lymphocytes (CTL) recognizing either class II MHC gene product (Jitsukawa et al 1988; Mami-Chouaib and Hercend unpublished data) or MHC "class I like" molecule, CD1c (Faure et al 1990a).

To further assess the specificity of human γ/δ T cells we have generated several alloreactive clones in mixed lymphocyte reactions (MLR) against an Epstein Barr virus (EBV) transformed B cell line, designated E418. Two of these clones (E102 and E117) were studied in details. They were found to recognize and kill the E418 immunizing cells. A 43 Kd surface molecule, termed TCT.1 (i.e., T cell target), broadly distributed in the hematopoietic system, was shown to be recognized on the target cells in these particular cytotoxic interactions.

RESULTS AND DISCUSSION

Peripheral blood mononuclear cells (PBMC) were extracted by Ficoll Hypaque centrifugation and the γ/δ T cells were purified by an immuno-rosetting technique using BMA031 (anti-TCRα/β), anti-CD4, anti-CD8, anti-CD20 (B cell specific) anti-CD14 (monocytes specific) and anti-CD56/NKH1 (anti-N901) mAbs. Non-rosetting lymphocytes were cultured at 2 10^4 cells/well in the presence of irradiated (10^4 cells/well) EBV transformed B cells, termed E418. rIL2 was added every 3 days, starting from day 12. The polyclonal cell line was cloned by limiting dilution at 0.5 cell/well on a feeder layer containing both allogeneic PBL and the sensitizing E418 B cell line. A series of clones with cytolytic activity against the E418 cells were generated. Two of them, termed E102 and E117, were studied in details. Both clones were found to be CD3$^+$, BMA031$^-$ (TCRα/β$^-$), TCRδ1$^+$, δTCS1$^+$, A13$^+$, TiVδ2$^-$, TiγA$^-$, CD4$^-$, CD8$^-$ and NKH1$^-$. Note that anti-TCRδ1 mAb is specific for a constant determinant

Current Topics in Microbiology and Immunology, Vol. 173

of the TCRδ chain (Band et al 1987), anti-δTCS1 mAb for an epitope encoded by Vδ1-Jδ1 rearranged gene segments (Mami-Chouaib et al 1989), anti-A13 and -TiVδ2 for peptides encoded by Vδ1 and Vδ2 gene segments respectively (Miossec et al 1990) and anti-TiγA for the Vγ9 gene product (Jitsukawa et al 1987; Triebel et al 1988). Therefore the two clones possess a V$\gamma 9^-$/V$\delta 1^+$ receptor found in a very small γ/δ peripheral cell fraction (10%) of the individual studied here.

<u>Characterization of the TCR expressed by E102 and E117 clones</u>

A series of Southern blots were performed to characterize the TCRγ and TCRδ gene rearrangements in the E102 and E117 clones. Southern blot analysis of both clone DNAs with a Vδ1 probe showed a 3.3 kb EcoRI band known to include (Hata et al 1987; Triebel et al 1988) the Vδ1-Jδ1 rearrangement (data not shown). Regarding the γ chain rearrangements, the pH60 probe (a Jγ1 fragment) detected two KpnI bands at 4.7 kb and 8.5 kb. It has been previously shown that such fragments correspond to rearrangements of a member of either the VγI or the VγIII gene subfamily to JP2 and JP1 respectively (Huck et al 1987). Hybridizations performed with a VγI and a VγIII probes indicated that these bands correspond to Vγ3-JP2 and to Vγ8-JP1 recombinations respectively (data not shown). Immunoprecipitations performed with the anti-TCRδ1 and the anti-δTCS1 mAbs show that E102 and E117 express a disulfide unlinked receptor (data not shown) indicating that both clones use an unfrequent γ/δ heterodimer encoded by Vδ1-Jδ1-Cδ/Vγ3-JP2-Cγ2 heterodimer.

<u>Functional activiy of E102 and E117</u>

A

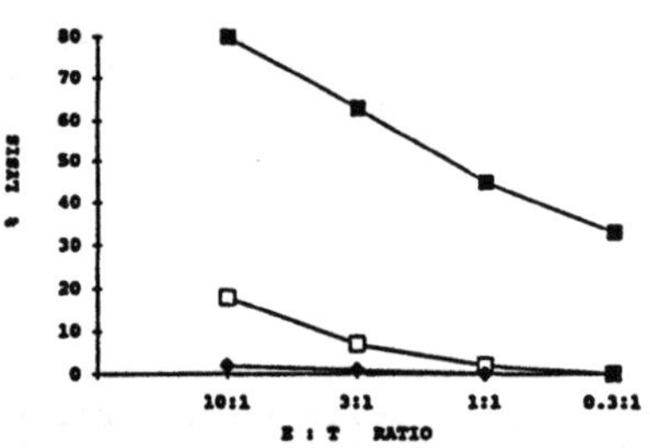

B

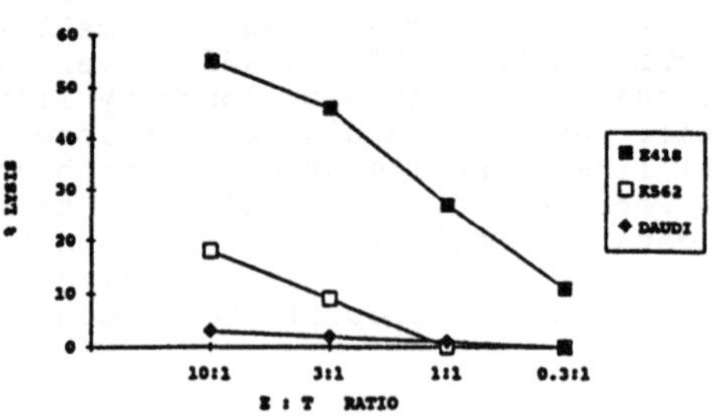

Fig. 1. Cytotoxic activity of E102 (A) and E117 (B) γ/δ T cell clones against E418 EBV transformed B cell line, DAUDI and K562 target cells. Effector/Target ratios were 10/1, 3/1, 1/1 and 0.3/1.

The lytic activity of the E102 and E117 cells was tested against a series of target cells including a panel of 8 EBV transformed B cell lines carrying various MHC class I and class II gene products as well as conventional "NK/LAK" target cells (K562, REX and DAUDI). There was little if any cytotoxicity against 6 of the B cell lines while two (F601 and KAS) were lysed more efficiently. The DAUDI "LAK" target cell was not killed and varying levels of lysis were found against K562 and REX (fig. 1).

The cytotoxicity of both clones against the E418 cell line was not altered by either anti-W6/32 (anti-class I) or 9-49 (anti-class II) mAbs. Therefore, E102 and E117 appeared to display a non-MHC Class I/II requiring cytotoxic activity (fig. 2).

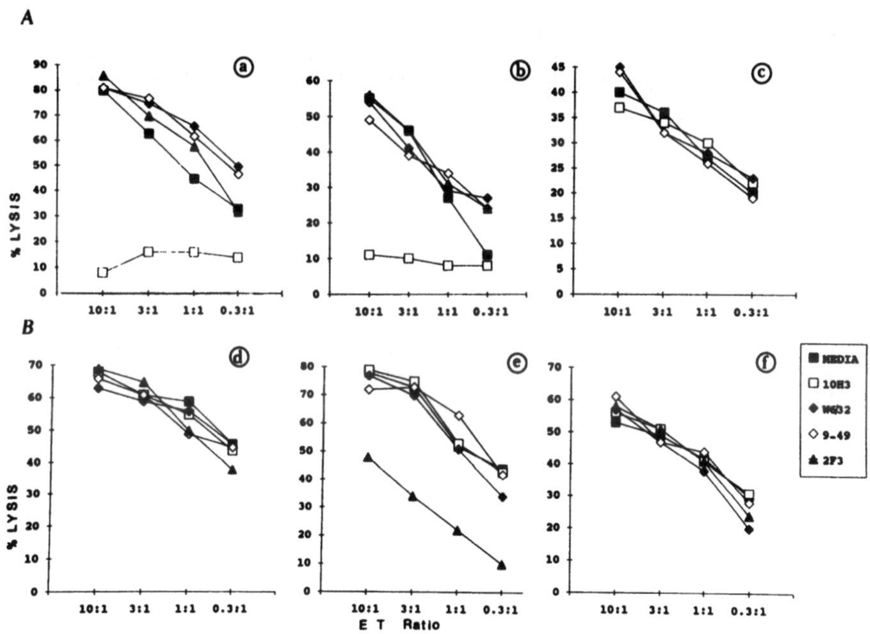

Fig. 2. Cytotoxic activity of E102 (a), E117 (b) and CD3$^-$.1 NK cell line (c) towards E418 target (A) and that of AB12 (d), JT9 (e) and the CD3$^-$.1 NK cells (f) towards DAUDI target (B). Cytolytic experiments were performed either in media or in the presence of mAbs. E418 and DAUDI target cells were preincubated for 2 hours with saturating concentrations of the 10H3, W6/32 (anti-class I), 9-49 (anti-class II) or 2F3 (anti TNKTar) mAbs and then effector cells were added.

To identify molecules potentially recognized by the clones, we have generated by immunizing Biozzi mice with E418 cells a mAb, termed anti-10H3, able to block the cytotoxic interaction of E102 and E117 clones with the E418 immunizing cells (Fig. 2Aa, b). The corresponding antigen, designated TCT.1 has been characterized as a 43 kd molecule. It was found to be broadly distributed in the hematopoietic system while cells from various other origins appeared to be negative (Table 1). Results obtained with the DAUDI cell line indicated that the TCT.1 protein does not require the ß2 microglobulin for its expression and is therefore distinct from the "class I-like" antigens.

TABLE 1 . reactivity of the anti-10H3 and the anti-W6/32 mAbs

A. Cell lines and clones

	Phenotype	W6/32 Reactivity	10H3 Reactivity
E418	$T3^-/9\text{-}49^+/B4^+$	90%$ (205)	90% (179)
DAUDI	$T3^-/9\text{-}49^+/B4^+$	0%	90% (105)
REX	$T3^+/9\text{-}49^-/B4^-$	95% (162)	95% (113)
AB12	$T3^+/9\text{-}49^+/B4^-$	94% (185)	94% (192)
E117	$T3^+/$ ND $/B4^-$	ND	97% (173)
K562	$T3^-/9\text{-}49^-/B4^-$	68% (88)	0%

B. Cell fraction :

	W6/32 Reactivity	10H3 Reactivity
PBL	97%$ (159)	93% (125)
Monocytes	98% (187)	89% (126)
Bone Marrow	91% (189)	82% (139)
Polymorphonuclear	95% (127)	46% (75)

C. Allogeneic tissues :

	W6/32 Reactivity	10H3 Reactivity
Liver	+	-
Kidney	+	-
Ovary	+	-
Placenta	+	-
Pancreas	+	-
Breast	±	-

Cells were analysed by indirect immunofluorescence (A and B) or by radioisotope assay on microfolds (C). $ = Percentage of positive cells. Number between parentheses correspond to fluorescence intensity mean. ND = not done.

Specificity of anti-10H3 mAb inhibitory function

Experiments performed with appropriate pre-incubation of either effector or target cells with anti-10H3 indicated that its inhibitory effect resulted from its binding to the membrane of target cells (data not shown).

In light of the broad distribution of the TCT.1 molecule, we tested the blocking activity of anti-10H3 in effector/target cell combinations distinct from E102 (or E117)/E418. The selected killer cells included AB12 (a γ/δ T cell clone with the predominant peripheral $Ti\gamma A^{+}/TiV\delta 2^{+}$ phenotype, Triebel et al 1988; Faure et al 1988 and Mami- Chouaib et al 1989), JT9 (an α/β T cell clone defined through the expression of the NKTa clonotypic determinant, Hercend et al 1983a; Hercend et al 1983b) and $CD3^{-}$.1, a polyclonal NK ($OKT3^{-}$, $NKH1^{+}$) cell line. The $10H3^{+}$ DAUDI cells were used as a target because of their susceptibility to the three types of effectors. As shown in fig.2, B, anti-10H3 had no effect at all in the cytotoxic reactions. Controls included W6/32 and 9-49 antibodies that were also inactive while, as described previously (Hercend et al 1984), anti-TNKtar blocked specifically the cytotoxicity mediated by JT9 cells (Fig.2, Be). Note, in addition, that anti-10H3 was unable to inhibit the cytotoxicity mediated by the $CD3^{-}$.1 NK cells against the E418 cell line (Fig.2, Ac).

In addition to E418, the functional activity of anti-10H3 was assessed against NK target cells (REX) and EBV transformed B cell lines (F601 and KAS), susceptible to the cytotoxic activity of E102 and E117. It was found to strongly inhibit the cytotoxicity against the F601 cells while being active, although less efficient, against the KAS cell line. In contrast, the antibody did not alter the interaction of E102 or E117 with the REX cell line (data not shown). These results strongly suggested that TCT.1 is not involved in a generally operating pathway of cell-cell interaction.

Concluding remarks

Together, the present results support the view that the E102 and E117 lymphocytes "see" the TCT.1 molecule on the surface of target cells probably through a TCR dependent pathway. Both clones also display a TCR independent NK-like function. The variability of the activity found against K562 and REX probably corresponds to the known dependence of NK/LAK lysis upon the effector cell status for IL2 induced activation (Faure et al 1990a; Hercend et al 1983a; Faure et al 1990b; Seeley et al 1989 and Rimm et al 1981). The absence of cytotoxicity against the DAUDI cell line which represents one of the conventional "LAK" target reflects the heterogeneity at the clonal level regarding the IL2-augmented NK function (Faure et al 1990a; Hercend et al 1983a; Faure et al 1990b and Fisch et al 1990).

The recognition of the TCT.1 molecule may allow to distinguish the TCR-dependent and the TCR-independent target cell recognition by the E102 and E117 lymphocytes. That TCT.1 is likely to be recognized via the γ/δ heterodimer is supported by several observations : (i) it has to be mentioned that the interaction between the clones and the E418 cells are inhibited by anti-TCR antibodies (data not shown); (ii) the blocking activity of the antibody is dependent upon the use of unique effector CTL, namely here E102 and E117; (iii) the blocking activity is dependent upon the use of individual target cells. If the TCT.1/ligand system would operate through a simple model of bimolecular interaction, one would expect that the specific antibody alters the cytotoxic interactions of E102 and E117 cells against all TCT.1$^+$ targets. The data obtained here particularly in the REX assays where anti-10H3 has no effect at all does not favor the latter possibility. A more likely hypothesis is that REX cells are recognized by a distinct mechanism than the EBV transformed B cell lines, namely in an "NK-like" TCR/TCT.1 independent fashion. In addition, the varying levels of functional inhibition obtained with the 3 B cell lines tested further suggest the existence of a complex pathway of TCT.1 recognition. Future studies will have to assess more directly the potential TCR/TCT.1 interaction and to establish whether TCT.1 may serve as a peptide presenting structure.

REFERENCES

Band, H., F.Hochstenbach, J.McLean, S.Hata, M.S.Krangel and M.B.Brenner. 1987. Immunochemical proof that a novel rearranging gene encodes the T cell receptor delta subunit. Science. 238:682.

Bank, I., R.A. Depinho, M.B.Brenner, J.Cassimeris, F.W.Alt and L.Chess. 1986. A functional T3 molecule associated with a novel heterodimer on the surface of immature human thymocytes. Nature. 322:179.

Borst, J., R.J.Van de Griend, J.W.Oostveen, S.L.Ang, C.J.Melief, J.G.Seidman and R.L.H.Bolhuis. 1987. A T-cell receptor gamma/CD3 complex found on cloned functional lymphocytes. Nature. 325:683.

Brenner, M.B., J.McLean, H.Scheft, J.Riberdy, S.L.Ang, J.G.Seidman, P.Devlin and M.S.Krangel. 1987. Two forms of the T-cell receptor gamma protein found on peripheral blood cytotoxic T lymphocytes. Nature. 325:689.

Faure, F., S. Jitsukawa, C. Miossec and T. Hercend. 1990a. CD1c, as a target structure for human T lymphocytes : analysis with peripheral blood γ/δ cells. Eur. J. Immunol. 20:703.

Faure. F., F. Triebel., Th. Hercend. 1990b. MHC-unrestricted cytotoxicity. Immunology Today. 11:108

Faure, F., S.Jitsukawa, F.Triebel and T.Hercend. 1988. Characterization of human peripheral lymphocytes expressing the CD3- gamma/delta complex with anti-receptor monoclonal antibodies. J. Immunol. 1 141:3357.

Fisch. P., M. Malkovsky., E. Braakman., E. Sturm., R. H. Bolhuis., A. Prieve., J. A. Sosman. V. A. Lam and P. M Sondel. 1990. T cell clones and Natural killer cell clones mediate distinct patterns of non-major histocompatibility restricted cytolysis. J. Exp. Med. in press.

Hata, S., M.B.Brenner and M.S.Krangel. 1987. Identification of putative human T cell receptor delta complementary DNA clones. Science. 238:678.

Hercend, T., E. L. Reinherz., S.C Meuer., S. F. Schlossman and J. Ritz. 1983a. Phenotypic and functional heterogeneity of human cloned natural killer cell lines. Nature. 301:158.

Hercend, T., S.Meuer, A.Brennan, M.A.Edson, O.Acuto, E.L.Reinherz, S.F.Schlossman and J.Ritz. 1983b. Identification of a clonally restricted 90 kD heterodimer on two human cloned natural killer cell lines : Its role in cytotoxic effector function. J. Exp. Med. 158:1547.

Hercend, T., R.Schmidt, A.Brennan, M.A.Edson, E.L.Reinherz, S.F.Schlossman and J.Ritz. 1984. Identification of a 140-kda activation antigen as a target structure for a series of human cloned natural killer cell lines. Eur. J. Immunol. 14:844.

Huck, S., and M.P.Lefranc. 1987. Rearrangements to the JP1, JP and JP2 segments in the human T-cell rearranging gamma gene (TRG gamma) locus. FEBS Letters. 224:291.

Jitsukawa, S., F.Faure, M.Lipinski, F.Triebel and T.Hercend. 1987. A novel subset of human lymphocytes with a T cell receptor-gamma complex. J. Exp. Med. 166:1192.

Jitsukawa, S., F.Triebel, F.Faure, C.Miossec and T.Hercend. 1988. Cloned CD3+ TCRalpha/beta- TigammaA- peripheral blood lymphocytes compared to the TigammaA+ counterparts : structural differences of the gamma/delta receptor and functional heterogeneity. Eur. J. Immunol. 18:1671.

Mami-Chouaib, F., S.Jitsukawa, F.Faure, B.Vasina, C.Genevee, T.Hercend and F.Triebel. 1989. cDNA cloning of funtional gamma and delta T cell receptor chains expressed in human peripheral blood lymphocytes. Eur. J. Immunol. 19:1545.

Miossec, C., F.Faure, L. Ferradini, S. Roman-Roman. S.Jitsukawa, S.Ferrini, A.Moretta, F.Triebel and T.Hercend. 1990. Further analysis of the TCR gamma/delta$^+$ peripheral lymphocyte subset : the "Vdelta1" gene segment is expressed with either Calpha or Cdelta. J. Exp. Med. 171:1171.

Moingeon, P., S.Jitsukawa, F.Faure, F.Troalen, F.Triebel, M.Graziani, F.Forestier, D.Bellet, C.Bohuon and T.Hercend. 1987. A gamma-chain complex forms a functional receptor on cloned human lymphocytes with natural killer-like activity. Nature. 325:723.

Rimm. I. J., S. F. Schlossman., E. L. Reinherz. 1981. Antiboy-dependent cellular cytotoxicity and Natural-Killer-Like activity are mediated by subsets of activated T cells. Clin. Immunol. Immunopathol. 21:134.

Seeley. J. K., G. Masuci., A. Poros., E. Klein., S. H. Golub 1989. Studies on cytotoxicity generated in human mixed lymphocyte cultures. II. Anti-K562 effectors are distinct from allospecific CTL and can be generated from NK-depleted T cells. J. Immunol. 123:1303.

Triebel. F., F. Faure. M. Graziani., S. Jitsukawa., M.P. Lefranc., T. Hercend. 1988. A unique V-J-C rearranged gene encodes a gamma protein expressed on the majority of CD3$^+$ TCRα/β^- circulating lymphocytes. J. Exp. Med. 167:694.

Triebel, F., F.Faure, F.Mami-Chouaib, S.Jitsukawa, A.L.Griscelli, C.Genevee, S.Roman-Roman and T.Hercend. 1988. A novel human Vδ gene expressed predominantly in the TigammaA+ fraction of τ/δ+ peripheral lymphocytes . Eur. J. Immunol. 18:2021.

Activation and Deactivation of Cloned γ/δ T Cells

D. Kabelitz, S. Wesselborg, K. Pechold, and O. Janssen

Institut für Immunologie, Universität Heidelberg, Im Neuenheimer Feld 305, D-6900 Heidelberg, FRG

INTRODUCTION

T cells are activated by the antigen-specific TCR/CD3 molecular complex following interaction with foreign antigenic peptides in association with self MHC proteins. In addition, T cells can be activated by an "alternative" pathway *via* the CD2 antigen (Meuer et al 1984). Soluble mAbs directed against T11.1, the sheep erythrocyte binding epitope of CD2, inhibit certain T cell functions such as Il-2 production, Il-2-dependent proliferation, and cytotoxic effector activity (Bolhuis et al 1986, Yssel et al 1987). In contrast, pairs of mAbs directed against the T11.2 and T11.3 epitope in combination activate resting TCR αß+ T cells. One of the two anti-CD2 mAbs required for T cell activation may be replaced by LFA-3 (the natural ligand for CD2), by sheep erythrocytes, or by PMA (Holter et al 1986, Hünig et al 1987). Here we report that, in contrast to αß+ T cells, cloned TCR γδ+ T cells can be activated by immobilized *single* mAbs to T11.1. In addition, we demonstrate that cloned γδ+ T cells undergo programmed cell death (apoptosis) when cultured with soluble anti-TCR or anti-CD3 mAb in the presence of exogenous Il-2.

MATERIALS AND METHODS

Il-2-dependent γδ+ clones were established by culturing CD4-CD8- peripheral blood T cells at 0.3 cells per well in the presence of PHA and irradiated feeder cells. The following mAbs were used to characterize established γδ+ clones: TCRδ-1 ($C_δ$), 7A5 ($V_γ9$), BB3 ($V_δ2$), δTCS-1 ($V_δ1$), A13 ($V_δ1$), BMA031 (TCR αß).
Anti-CD2 (T11.1) mAbs used include OKT11 (ATCC), 9E8 (this laboratory), BW0110 (Behringwerke, Marburg, FRG), and M-T910 (Brown et al 1987). 96-well plates were coated with 1 μg purified mAb per well. After being washed, clone cells were added at 2 - 5 x 10^4 cells per well in RPMI 1640 with 10 % FCS. After 24 h, 75 μl of culture supernatant were tested for Il-2 activity on Il-2-dependent CTLL cells by the colori-metric MTT assay (Jooss et al 1988). After additional 12 to 24 h, 1 μCi ^{3}H-TdR was added to the remaining cells for determination of cell proliferation. Cytotoxic activity of cloned γδ+ T cells was determined in a standard 4 h ^{51}Cr release assay against P815 target cells in the presence or absence of anti-CD2 or anti-CD3/TCR mAbs.
For determination of anti-CD3/TCR mAb-induced apoptosis, cloned γδ+ T cells were cultured in 24-well plates at 1.5 x 10^6 cells/ml in the absence or presence of 2 ng/ml rIl-2 and/or 1 μg/ml purified mAb. After 24 h, ^{3}H-TdR incorporation was

Current Topics in Microbiology and Immunology, Vol. 173

measured, and cell viability was determined by propidium iodide staining. Degradation of DNA was monitored by gel electrophoresis of total cellular DNA on 2 % agarose gels as described (Janssen et al 1990).

RESULTS

Single Anti-CD2 (T11.1) mAbs Activate γδ⁺ but not αβ⁺ T Cell Clones.

Various γδ⁺ clones expressing different TCR γδ phenotypes (see Table 1) were cultured in microtiter plates coated with different single anti-CD2 (T11.1) or anti-CD3 mAbs. As shown in Table 1, γδ⁺ but not αβ⁺ clones proliferated vigorously in response to single immobilized anti-CD2 mAbs. The proliferative response of γδ⁺ clones to anti-CD2 stimulation was frequently associated with the release of substantial amounts of Il-2 into the culture medium (Table 2). In some cases, however, anti-CD2 triggered the proliferation of a γδ⁺ clone in the absence of detectable Il-2 secretion (compare Tables 1 and 2). In contrast, neither CD4⁺ nor CD8⁺ TCR αβ⁺ clones released Il-2 in response to a single immobilized anti-CD2 mAb (not shown). In addition, anti-CD2 mAbs triggered the cytotoxic effector activity in cloned γδ⁺ T cells as tested against FcR-positive P815 target cells (results not shown). It should be stressed that not all tested anti-CD2 mAbs were effective in stimulating γδ⁺ T cell clones. Of 7 anti-CD2 (T11.1) mAbs tested, 4 were stimulatory (Tables 1 and 2), while 3 were not. It is clear, however, that the stimulatory capacity of a given anti-CD2 mAb did not depend on a particular IgG subclass, since stimulatory mAbs included both IgG1 (OKT11, 9E8, M-T910) and IgG2b (BW0110) antibodies.

Anti-CD3/TCR mAbs Induce Apoptosis in Cloned γδ⁺ T Cells.

During our studies on the activation requirements of cloned human γδ⁺ T cells, we noticed that a significant fraction of a given γδ⁺ clone died when exposed to soluble anti-CD3 or anti-TCR mAb and exogenous Il-2. Anti-CD3/TCR mAbs have been shown to induce programmed cell death (termed apoptosis) in immature murine thymocytes (Smith et al 1989), T cell hybridomas (Mercep et al 1988) and certain transformed leukemic T cell lines (Takahashi et al 1989). Apoptosis is characterized by DNA fragmentation into oligonucleosomal bands that are multiples of 200 bp (Appleby and Modak 1977). As shown in Table 3, $V_\gamma9^+$ clone A37DN17 proliferated in response to immobilized anti-CD3 (BMA030) or anti-TCR $V_\gamma9$ (7A5) mAbs in the *absence* of exogenous Il-2. In the *presence* of exogenous Il-2, the proliferation of A37DN17 cells was inhibited by both soluble and immobilized anti-CD3 or anti-TCR mAb. As shown by propidium iodide staining and FACS analysis (Table 4), a significant fraction of γδ⁺ clone cells died when cultured for 20 h with soluble anti-CD3/TCR mAb and exogenous Il-2. Electrophoresis of DNA on 2 % agarose gels revealed the characteristic "ladder" pattern of DNA fragmentation when γδ⁺ clone cells were exposed to Il-2 and soluble anti-CD3 or anti-TCR mAb but not when γδ⁺ clone cells were cultured with soluble W6/32 (anti-HLA class I) mAb (results not shown). Taken together, these results demonstrate that anti-CD3/TCR mAbs can induce programmed cell death in non-transformed γδ⁺ clones.

Table 1. Proliferation of $\gamma\delta^+$ clones stimulated by single anti-CD2 mAbs

	Clone (cpm x 10^3)				
Immobilized mAb	D768/4 ($V_\gamma 9V_\delta 2$)	D768/5 ($V_\delta 1$)	A92DN3 ($V_\gamma 9V_\delta 1$)	D798/11 ($\alpha\beta^+CD4^+$)	D798/18 ($\alpha\beta^+CD4^+$)
none	0.2	3.7	4.8	0.6	0.7
OKT11 (CD2)	2.7	18.4	44.7	1.3	0.8
9E8 (CD2)	4.9	17.4	39.6	0.8	0.6
BW0110 (CD2)	5.7	12.9	35.3	nd[a]	nd
M-T910 (CD2)	4.1	18.6	nd	nd	nd
OKT3 (CD3)	2.1	3.9	13.5	33.7	15.6
Il-2 (2 ng/ml)	3.7	12.1	20.4	44.2	25.8

2 - 5 x 10^4 clone cells were cultured for 48 h in 96-well plates coated with 1 µg purified mAb per well. ^{3}H-TdR was added during the last 6 h. Results are mean cpm x 10^3 of triplicate cultures. SD did not exceed 15 %.

[a] nd, not done

Table 2. Il-2 production by anti-CD2-stimulated $\gamma\delta^+$ clones

	Clone (OD x 10^{-3})			
Immobilized mAb	D768/4 ($V_\gamma 9V_\delta 2$)	D768/5 ($V_\delta 1$)	D768/6 ($V_\gamma 9$)	A36DN33 ($V_\gamma 9V_\delta 2$)
none	11	15	7	9
OKT11 (CD2)	325	33	441	478
9E8 (CD2)	307	112	419	313
BW0110 (CD2)	25	310	34	13
OKT3 (CD3)	118	25	244	116
Il-2 (2 ng/ml)	421	417	431	484

2 - 5 x 10^4 clone cells were cultured in 96-well plates previously coated with 1 µg purified mAb per well. After 24 h, 75 µl of supernatant were transferred to flat-bottom microtiter plates, and 15.000 Il-2-dependent CTLL cells were added per well. After additional incubation for 24 h at 37° C, proliferation of CTLL cells was measured in a colorimetric assay (cleavage of tetrazolium salt MTT) as described (Jooss et al 1988). Results are given as mean OD of triplicate cultures measured at 570 nm. SD were less than 15 % .

Table 3. Stimulatory and suppressive effects of anti-CD3/TCR mAbs on the proliferation of cloned $\gamma\delta^+$ T cells

	^{3}H-TdR incorporation (cpm x 10^3 ± SD)	
mAb added	without Il-2	with Il-2 (2 ng/ml)
-	0.5 ± 0.1	8.7 ± 1.5
soluble 7A5	0.2 ± 0.1	0.5 ± 0.1
soluble BMA030	0.4 ± 0.1	2.5 ± 0.1
immobilized 7A5[a]	5.1 ± 0.1	0.8 ± 0.1
immobilized BMA030[a]	6.3 ± 0.2	2.7 ± 0.3

$V_\gamma9^+$ clone (A37DN17) was cultured at 3 x 10^4 cells per well in the presence or absence of 2 ng/ml Il-2. Anti-TCR $V_\gamma9$ mAb 7A5 or anti-CD3 mAb BMA030 were added at 1 μg/ml. ^{3}H-TdR incorporation was determined after 24 h.

[a] Culture plates were coated with 1 μg purified mAb per well

Table 4. Correlation of propidium iodide staining and ^{3}H-TdR incorporation

mAb added	% growth inhibition (^{3}H-TdR uptake)	% dead cells (propidium iodide)
-	--	15.5
7A5 (anti-TCR)	94	67.5
BMA030 (anti-CD3)	81	51.0
w6/32 (anti-HLA I)	24	18.2
PHA-P (1 μg/ml)	86	71.8

$V_\gamma9^+$ clone B54 was cultured at 1.5 x 10^6 cells/ml in 24-well culture plates in the presence of Il-2 (2 ng/ml) and the indicated mAbs (1 μg/ml). After 24 h, ^{3}TdR incorporation was measured, and the percentage of dead cells was determined by propidium iodide staining and FACScan analysis.

DISCUSSION

Our experiments revealed a striking difference between cloned $\alpha\beta^+$ and $\gamma\delta^+$ T cells regarding the stimulation *via* the "alternative" CD2 pathway. Two anti-CD2 mAbs directed against different epitopes are required for stimulation of TCR $\alpha\beta^+$ T cells. In contrast, stimulation by a *single* immobilized anti-CD2 (T11.1) mAb is sufficient to trigger proliferation of and Il-2 secretion by cloned $\gamma\delta^+$ T cells. These results confirm and extend the findings of Pawelec et al (1990) who described the activation of cloned $\gamma\delta^+$ T cells by selected single anti-CD2 mAbs in the presence of EBV-transformed LCL feeder cells. In addition, our data are in line with results of Goedegebuure et al (1989) who were able to trigger cytotoxic effector function in $\gamma\delta^+$ T cells utilizing bispecific heteroconjugates containing a single anti-CD2 mAb. It appears that the threshold of signal(s) required for efficient CD2-dependent stimulation is lower for $\gamma\delta^+$ clones when compared to $\alpha\beta^+$ clones. It remains to be established whether the "alternative" activation pathway is also more easily triggered in resting (polyclonal) $\gamma\delta^+$ T cells than in $\alpha\beta^+$ T cells. In addition, further studies are aimed at analyzing whether LFA-3, the natural ligand for CD2, can replace anti-CD2 mAbs in activating $\gamma\delta^+$ T cell clones. As noted above, not every anti-CD2 (T11.1) mAb tested was capable of stimulating $\gamma\delta^+$ clone activation. Further studies are required to precisely delineate the CD2 epitope recognized by "stimulatory" *versus* "non-stimulatory" anti-CD2 mAbs. The stimulatory activity of a given anti-CD2 mAb does not seem to depend on a particular IgG subclass, since both IgG1 and IgG2b antibodies were among the stimulatory mAbs.

The present results also showed that cloned $\gamma\delta^+$ T cells are susceptible to anti-CD3 or anti-TCR mAb-induced apoptosis. CD3/TCR-dependent programmed cell death, characterized by oligonucleosomal DNA fragmentation, has been previously observed with immature thymocytes, T cell hybridomas, and leukemic T cells (Smith et al 1989, Mercep et al 1988, Takahashi et al 1989) but not with Il-2-dependent normal T cells. We observed that growth of cloned $\gamma\delta^+$ T cells was suppressed when cells were cultured in the presence of exogenous Il-2 and anti-CD3 or anti-TCR mAb. A significant fraction of clone cells died under these conditions, as shown by propidium iodide staining. In addition, characteristic patterns of DNA fragmentation were observed. These results raise the possibility that proliferation of $\gamma\delta^+$ T cells is negatively regulated by antigen. In this scenario, a *resting* $\gamma\delta^+$ T cell is activated when it encounters the "right" antigen/ligand *via* its T cell receptor, and progresses through the cell cycle if sufficient Il-2 is provided by bystander "helper" ($CD4^+$/TCR $\alpha\beta^+$) T cells. When, in the *presence* of Il-2, the *activated* $\gamma\delta^+$ T cell recognizes the same antigen again, death by apoptosis may ensue, thereby limiting the extent of $\gamma\delta$ T cell proliferation.

ACKNOWLEDGMENTS

We would like to thank Drs. K.H.Enssle, L.Moretta and E.P.Rieber for kindly providing monoclonal antibodies.

REFERENCES

Appleby DA, Modak SP (1977) DNA degradation in terminally differentiating lens fiber cells from chick embryos. Proc Natl Acad Sci (USA) 74:5579

Bolhuis RLH, Roozemond RC, van de Griend RJ (1986) Induction and blocking of cytolysis in CD2+, CD3- NK and CD2+, CD3+ cytotoxic T lymphocytes via CD2 50 kD sheep erythrocyte receptor. J Immunol 136:3939

Brown MH, Sewell WA, Monostori E, Crumpton MJ (1987) Characterization of CD2 epitopes by Western blotting. *In* Leukocyte Typing III. AJ McMichael, editor. Oxford University Press, Oxford, p 110

Goedegebuure PS, Segal DM, Braakman E, Vreugdenhil RJ, van Krimpen BA, van de Griend RJ, Bolhuis RLH (1989) Induction of lysis by T cell receptor γδ+/CD3+ T lymphocytes via CD2 requires triggering via the T11.1 epitope only. J Immunol 142:1797

Holter W, Fischer GF, Majdic O, Stockinger H, Knapp W (1986) T cell stimulation via the erythrocyte receptor. Synergism between monoclonal antibodies and phorbol myristate acetate without changes of free cytoplasmic Ca^{++} levels. J Exp Med 163:654

Hünig T, Tiefenthaler G, Meyer zum Büschenfelde KH, Meuer SC (1987) Alternative pathway activation of T cells by binding of CD2 to its cell-surface ligand. Nature 326:298

Janssen O, Wesselborg S, Heckl-Östreicher B, Pechhold K, Bender A, Schondelmaier S, Moldenhauer G, Kabelitz D (1990) T-cell receptor/CD3 signalling induces death by apoptosis in human T-cell receptor γδ-positive T-cells. J Immunol, in press

Jooss J, Zanker B, Wagner H, Kabelitz D (1988) Quantitative assessment of interleukin 2-producing alloreactive human T cells by limiting dilution analysis. J Immunol Meth 112:85

Mercep M, Bluestone JA, Noguchi PD, Ashwell JD (1988) Inhibition of transformd T cell growth in vitro by monoclonal antibodies directed against distinct activating molecules. J Immunol 140:324

Meuer SC, Hussey RE, Fabbi M, Fox D, Acuto O, Fitzgerald KA, Hodgdon JC, Protentis JP, Schlossman SF, Reinherz (1984) An alternative pathway of T-cell activation: A functional role for the 50 kd T11 sheep erythrocyte receptor protein. Cell 36:897

Pawelec G, Schaudt K, Rehbein A, Olive D, Bühring HJ (1990) Human T cell clones with γ/δ and α/β receptors are differently stimulated by monoclonal antibodies to CD2. Cell Immunol 129:385

Smith CA, Williams GT, Kingston R, Jenkinson EJ, Owen JJT (1989) Antibodies to CD3/T-cell receptor complex induce death by apoptosis in immature T cells in thymic cultures. Nature 337:181

Takahashi S, Maecker HT, Levy R (1989) DNA fragmentation and cell death mediated by T cell antigen receptor/CD3 complex on a leukemia T cell line. Eur J Immunol 19:1911

Yssel H, Aubry JP, de Waal Malefijt R, de Vries JE, Spits H (1987) Regulation by anti-CD2 monoclonal antibody of the activation of a human T cell clone induced by anti-CD3 or anti-T cell receptor antibodies. J Immunol 139:2850

γδ T Cells in Leprosy Lesions

K. Uyemura[1], H. Band[4], J. Ohmen[1], M. B. Brenner[4], T. H. Rea[3], and R. L. Modlin[1,2]

[1] Division of Dermatology [2] Department of Microbiology and Immunology, UCLA School of Medicine Los Angeles, CA 90024, USA
[3] Section of Dermatology, USC School of Medicine, Los Angeles, CA 90033, USA
[4] Laboratory of Immunochemistry, Dana Farber Cancer Institute, and Department of Rheumatology and Immunology, Harvard Medical School, Boston, MA 02115, USA

Leprosy

Leprosy provides a useful model for understanding immunoregulatory mechanisms in man, since the disease form a spectrum in which the immunologic response of the patient correlates with the clinical and histopathologic classification. Since leprosy is predominantly a disease of skin, it provides an opportunity to study the immune response to infectious pathogens at the site of disease activity. At one end of the spectrum, patients with tuberculoid leprosy have one or several skin lesions in which bacilli can rarely be identified. CD4+ T-lymphocytes predominate in these lesions (12) and respond to Mycobacterium leprae in vitro (11). At the other end of the spectrum, patients with lepromatous leprosy have diffuse infiltration of skin and nerves with bacilli-laden macrophages. CD8+ cells predominate in lepromatous lesions and function as antigen specific T-suppressor cells in vitro (13,14). The CD4+ lymphocytes derived from these lesions are unresponsive to M. leprae in vitro.

The benchmark measure of delayed-type hypersensitivity in leprosy is the lepromin or Mitsuda reaction, a 21 day reaction to intradermal injection of M. leprae which is characterized by granuloma formation. This test is positive in tuberculoid patients and negative in lepromatous patients. Imposed upon the leprosy spectrum are the so called "reactional states" which include the reversal reaction, thought to be a delayed-type hypersensitivity reaction against M. leprae antigens.

In understanding the role of T-cells in the outcome of infectious processes, we have used leprosy as a model. Although most studies have dealt with the role of T-cells bearing αß antigen receptors, we have recently focused on the potential role of γδ T-cell antigen receptors in the immune response to human pathogens. Specifically we asked: 1) Are γδ T-cells increased in leprosy lesions?; 2) Do γδ in leprosy lesions proliferate in response to M. leprae?; 3) Do γδ T-cells in leprosy lesions respond to granuloma formation?; and, 4) What is the diversity of the γδ repertoire in leprosy lesions?

Current Topics in Microbiology and Immunology, Vol. 173

$\gamma\delta$ T cells are increased in leprosy lesions.

We compared the occurrence of lymphocytes bearing TCR $\gamma\delta$ or TCR $\alpha\beta$ in each of the immunologic reactions to M. leprae using immunoperoxidase staining of skin biopsy specimens with specific monoclonal antibodies (15). In both lepromin skin tests and in reversal reactions, TCRδ^+ lymphocytes comprised 25 to 35% of CD3$^+$ cells in the lesions compared to approximately 5% of the CD3$^+$ cells in lesions of other forms of the disease. These data suggest that TCR $\gamma\delta$ lymphocytes may be involved in early active granulomatous responses (such as lepromin skin tests and reversal reactions), in contrast to the more chronic and/or immunologically unresponsive leprosy lesions. Similarly in localized American cutaneous leishmaniasis, another infectious disease with active granuloma formation, TCR $\gamma\delta^+$ cells were enriched in skin lesions relative to peripheral blood.

$\gamma\delta$ T cells in leprosy lesions respond to M. leprae.

To examine the antigen specificity of lymphocytes accumulated in leprosy lesions, we derived $\gamma\delta$ T cell lines from skin lesions and peripheral blood (15). One T cell line was obtained from the biopsy of a lepromin skin test of a patient with tuberculoid leprosy in reversal reaction. This line was expanded in vitro with M. leprae antigens and with IL-2, and then was subjected to cell sorting to enrich for TCR $\gamma\delta^+$ cells. Both $\gamma\delta$ T-cell lines proliferated specifically in response to M. leprae cell wall antigens and PPD, but not to the 65 kDa or 18 kDa recombinant heat-shock proteins of mycobacteria, or to tetanus toxoid which served as a control.

$\gamma\delta$ T cells in leprosy lesions contribute to granuloma formation.

Granuloma formation in leprosy is presumed to be important for restricting the spread or growth of the pathogens. The finding that antigen-specific TCR $\gamma\delta$ cells accumulate in large numbers in immunologic reactions characterized by active granuloma formation, but not established granulomas, suggested the possibility that these cells or their secreted products may play a role in the development of such lesions. Bone marrow-derived macrophages were cultured in presence of GM-CSF and supernatants obtained from anti-CD3 or antigen-stimulated TCR $\gamma\delta$ lymphocyte lines (15). Striking macrophage aggregation and cell division were seen in cultures containing GM-CSF with supernatants of activated TCR $\gamma\delta$ cells, but not those containing supernatants of non-stimulated TCR $\gamma\delta$ cells or either component alone. These studies suggest that $\gamma\delta$ T-cells derived from lesions appear to release a lymphokine(s) that synergizes with other cytokines to induce macrophage adherence, aggregation and proliferation. These cellular events are likely to be necessary for the granulomatous response.

Diversity of $\gamma\delta$ T-cells in leprosy lesions

An important unsolved questions about $\gamma\delta$ T-cell responses regards the breadth of the $\gamma\delta$ T-cell receptor repertoire. On the one hand, there is extensive junctional diversity, particular in the δ chain. On the other hand, the finding of a limited genetic diversity of the $\gamma\delta$ TCR resident populations at peripheral interfaces such as skin (1), gut (17) and lungs (2) is consistent with the recognition of a narrow antigen repertoire.

The microanatomic distribution of γδ T-cell subpopulations within lepromin skin tests was investigated by immunohistologic analysis of frozen sections with monoclonal antibodies directed against Vδ polypeptide chains using immunohistologic techniques. Within the dermal granulomas, Vδ1 and Vδ2 bearing cells accounted for the majority of infiltrating δ+ cells, with the Vδ2:Vδ1 ratio approximately 2:1 compared to 9:1 in the peripheral blood of these same individuals. TCRδ cells infiltrating the epidermis invariably were found to express the Vδ1 receptor, with Vδ2+ cells rarely present in the epidermal layer.

The determination of the genetic diversity of the TCR complex of these Vδ1 and Vδ2 populations in leprosy skin lesions is fundamental to understanding the nature of the γδ response to infectious pathogens. Since Vδ gene segments have been shown to rearrange with three known Jδ gene segments, we were able to determine the V-J gene recombinations of DNA extracted from lepromin skin tests by PCR amplification utilizing appropriate pairs of V and J oligonucleotide primers. PCR analysis confirmed that Vδ1 and Vδ2 were the predominant Vδ genes used by γδ T-cells in these lesions, and that both V regions were found to rearrange with the Jδ1 gene segment. Furthermore, Vδ2-Jδ3 gene rearrangements were detected in 5/5 blood samples but only 1/5 lesions. These data suggest a selective localization of a subpopulation of Vδ2 bearing cells to lesions.

To extend the above findings, the junctional diversity of γδ T-cells in lepromin skin tests was determined by cloning and sequencing of PCR amplified products. In three lepromin skin tests, the Vδ1-Jδ1 and Vδ2-Jδ1 junctions were found to lack diversity and differed in sequence from one another. This was in striking contrast to the peripheral blood of these individuals, which exhibited extensive junctional sequence diversity. The junctional sequences in lesions were not found to be expressed in any of the peripheral blood lymphocyte sequences.

The most likely explanation for the limited diversity of the V-J junctions of the δ TCR chain in lesions is that these cells are clonally expanded in lesions in recognition of a small number of antigens. It is noteworthy that there is greater amino acid sequence diversity at the V-J junction between individuals than between clones within an individual. While these results could be due to primary genetic difference in repertoire generation, it is tempting to speculate that it is due to polymorphic host-derived restriction elements that are involved in the clonal selection.

Discussion

T-cells bearing γδ antigen receptors may function as a first line of defense against infectious pathogens. This hypothesis is based on the finding of large numbers of γδ T-cells at peripheral interfaces including normal murine skin, gut and lungs (2,3,5,10). Furthermore, the murine primary, but not secondary immune response to challenge with M. tuberculosis is characterized by expansion of γδ T-cells (8). In fact, γδ T-cell hybridomas derived from antigen unselected murine neonatal thymocytes have been shown to respond to mycobacterial antigens (16). In humans, γδ T-cells from the peripheral blood have been shown to proliferate in response to mycobacteria (6,7,15) and γδ T-cells accumulate in early but not chronic granulomatous lesions of leprosy and leishmaniasis (15). The finding of a limited genetic diversity of γδ T-cells in leprosy lesions suggests the recognition of a limited antigen repertoire. Consistent with this hypothesis, γδ T-cells may recognize heat shock proteins (4,6,7,9,16) a highly conserved family of proteins that may be induced by a variety of cellular injury mechanisms. Limited receptor diversity would permit $\gamma\delta$ T-cells to recognize and respond to these antigens on mycobacteria, or homologous proteins present on autologous cells stressed by intracellular infection.

REFERENCES

1. Asarnow DM, Kuziel WA, Bonyhadi M, Tigelaar RE, Tucker PW, Allison JP (1988) Limited diversity of gamma delta antigen receptor genes of Thy-1+ dendritic epidermal cells. Cell 55:837-847

2. Augustin A, Kubo RT, Sim GK (1989) Resident pulmonary lymphocytes expressing the gamma/delta T-cell receptor. Nature 340:239-241

3. Bonneville M, Janeway CA Jr, Ito K, Haser W, Ishida I, Nakanishi N, Tonegawa S (1988) Intestinal intraepithelial lymphocytes are a distinct set of gamma delta T cells. Nature 336:479-481

4. Born W, Hall L, Dallas A, Boymel J, Shinnick T, Young D, Brennan P, OBrien R (1990) Recognition of a peptide antigen by heat shock--reactive gamma delta T lymphocytes. Science 249:67-69

5. Goodman T, Lefrancois L (1988) Expression of the gamma-delta T-cell receptor on intestinal CD8+ intraepithelial lymphocytes. Nature 333:855-858

6. Haregewoin A, Soman G, Hom RC, Finberg RW (1989) Human gamma delta+ T cells respond to mycobacterial heat-shock protein. Nature 340:309-312

7. Holoshitz J, Koning F, Coligan JE, De Bruyn J, Strober S (1989) Isolation of CD4- CD8- mycobacteria-reactive T lymphocyte clones from rheumatoid arthritis synovial fluid. Nature 339:226-229

8. Janis EM, Kaufmann SH, Schwartz RH, Pardoll DM (1989) Activation of gamma delta T cells in the primary immune response to Mycobacterium tuberculosis. Science 244:713-716

9. Kabelitz D, Bender A, Schondelmaier S, Schoel B, Kaufmann SHE (1990) A large fraction of human peripheral blood gamma/delta^{+} T cells is activated by Mycobacteium tuberculosis but not by its 65-kD heat shock protein. J Exp Med 171:667-679

10. Kuziel WA, Takashima A, Bonyhadi M, Bergstresser PR, Allison JP, Tigelaar RE, Tucker PW (1987) Regulation of T-cell receptor gamma-chain RNA expression in murine Thy-1+ dendritic epidermal cells. Nature 328:263-266

11. Modlin RL, Brenner MB, Krangel MS, Duby AD, Bloom BR (1987) T-cell receptors of human suppressor cells. Nature 329:541-545

12. Modlin RL, Hofman FM, Taylor CR, Rea TH (1983) T lymphocyte subsets in the skin lesions of patients with leprosy. J Am Acad Dermatol 8:182-189

13. Modlin RL, Kato H, Mehra V, Nelson EE, Xue-dong F, Rea TH, Pattengale PK, Bloom BR (1986) Genetically restricted suppressor T-cell clones derived from lepromatous leprosy lesions. Nature 322:459-461

14. Modlin RL, Mehra V, Wong L, Fujimiya Y, Chang W-C, Horwitz DA, Bloom BR, Rea TH, Pattengale PK (1986) Suppressor T lymphocytes from lepromatous leprosy skin lesions. J Immunol 137:2831-2834

15. Modlin RL, Pirmez C, Hofman FM, Torigian V, Uyemura K, Rea TH, Bloom BR, Brenner MB (1989) Lymphocytes bearing antigen-specific gamma/delta T-cell receptors in human infectious disease lesions. Nature 339:544-548

16. OBrien RL, Happ MP, Dallas A, Palmer E, Kubo R, Born WK (1989) Stimulation of a major subset of lymphocytes expressing T cell receptor gamma delta by an antigen derived from Mycobacterium tuberculosis. Cell 57:667-674

17. Takagaki Y, DeCloux A, Bonneville M, Tonegawa S (1989) Diversity of gamma delta T-cell receptors on murine intestinal intra-epithelial lymphocytes. Nature 339:712-714

ACKNOWLEDGMENTS

Supported by grants from the National Institutes of Health (AI 22553, AR 40312), the UNDP/World Bank/World Health Organization Special Programme for Research and Training in Tropical Diseases (IMMLEP and THELEP) and the Heiser Trust. H.B. is a Leukemia Society of America Special Fellow and an Arthritis Foundation Investigator.

Modifications of γδ T Lymphocytes in the Rheumatoid Arthritis Joint

T. Rème, I. Chaouni, Florence Frayssinoux, B. Combe and J. Sany

Inserm Unité 291 and Unité d'Immuno-Rhumatologie du CHR 99,
Rue Puech-Villa, 34090 Montpellier, France

Introduction

T lymphocytes recognize processed antigens through a specific cell surface receptor composed of an α/β heterodimeric unit associated with the CD3 protein complex. Another minor heterodimeric receptor has been more recently evidenced (Brenner et al. 1986), made up by γ and δ glycoprotein subunits encoded for by somatic gene rearrangements of V, D, J and C segments (review by Lefranc, this meeting). However, due to limited germline diversity, the TCRγ/δ variability largely depends on nucleotide modifications in the "N" region of the V-J or V-D-J junctions (Hata et al. 1988), and following the model of Davis and Bjorkman (1988) for $\alpha\beta$ T cells, $\gamma\delta$ T cells should recognize antigens in the context of poorly or non polymorphic antigen-presenting determinants (Strominger 1989). The in vivo development, functions, and antigens of TCRγ/δ-bearing cell populations, although still largely unknwon, are under intensive investigations and general discussions (Haas et al. 1990). Many attempts have been made to delineate functional $\gamma\delta$ T cell populations in pathological situations (De Maria et al. 1988, Holoshitz et al. 1989, Modlin et al. 1989) where self-mimicking mycobacterial antigens or heat shock proteins (HSP) could be involved (Raulet 1989). Following our recent data on the tissue distribution of $\gamma\delta$ T cell subsets within the rheumatoid synovium (Chaouni et al. in press), we overview here some of the modifications of $\gamma\delta$ T cell in patients with rheumatic diseases, in terms of phenotypic expression, gene usage and tissue distribution. Implications for understanding their eventual participation in the disease, and more generally the T cell involvement in the pathogenesis of autoimmunity are also discussed.

T $\gamma\delta$ DISTRIBUTION

1-Among species: A large portion of the lymphocytes infiltrating mouse and avian gut epithelium (Goodman and Lefrancois 1988, Bucy et al. 1988) and epidermis (Stingl et al. 1987, Kuziel et al. 1987) express a $\gamma\delta$ receptor. Elevated levels of $\gamma\delta$ T cells have been evidenced in other species like cattle and sheep (Mackay et al. 1989). In contrast, the human $\gamma\delta$ cells seem to be evenly distributed within T cell containing organs, in a rather constant and low proportion of 1 to 10% of $\gamma\delta$+ among $\alpha\beta$+ cells (Groh et al. 1989). However, an increasing number of histological studies using available antibodies against TCR chains analyze $\gamma\delta$ T cell distribution in human tissues, and may modulate speculations regarding epithelial surveillance (Janeway et al. 1988).

Current Topics in Microbiology and Immunology, Vol. 173

2-<u>In normal human</u>: The distribution of cd T cells in man has been extensively analyzed during the last two years, and if most of the published data from several groups are in agreement for peripheral blood (mean 5%, range 1-10%), spleen (10-20%), lymph nodes (2%), Thymus (1%) and tonsil (less than 1%) (Borst et al. 1988, Groh et al. 1989, Bucy et al. 1989, Falini et al. 1989), results for human gut epithelium largely diverge from one group to the other (Brandtzaeg et al. 1989). Actually, in contrast to lamina propria, human gut epithelium seems to contain more $\gamma\delta$ T cells than previously believed (Deusch and Pfeffer, this meeting). A most important feature in human is the differences in the V gene usage in the δ chain-encoding rearrangements between the thymus and the periphery (Lanier et al. 1988, Casorati et al. 1989). Two main subsets of $\gamma\delta$ T cells have been delineated, the "peripheral" type using a Vγ9JγPCγ1 rearranged gene product, preferential associated with a Vδ2-encoded δ chain (Triebel and Hercend 1989), and a "thymic" type using a Vδ1-encoded δ chain associated with various VγCγ-gene products.

3-<u>In pathology (especially RA)</u>: Analysis of the T cell receptor expressed by T lymphocytes infiltrating pathological sites has been performed first in the synovial fluid of juvenile arthritis patients (De Maria et al. 1987), where an elevated number of TCR $\alpha\beta$-negative T cell clones could be expanded. We and others (Brennan et al, 1988, Rème et al. 1990) have used the anti-Ti-γA monoclonal antibody (Jitsukawa et al. 1987) to shown that the TCR $\gamma\delta$ lymphocyte content is significantly increased in the synovial fluid of RA patients compared to normal blood. T $\gamma\delta$ clones with mycobacterial reactivity have been isolated from RA synovial fluid (Holoshitz et al. 1989) and lepromatous lesions (Modlin et al. 1989). To get further insight in the eventual involvement of $\gamma\delta$ T cells in the pathology of RA, we used a panel of monoclonal antibodies in immuno-histology to study the $\gamma\delta$ T cell infiltrate within the synovial membrane. Interestingly, we have shown that the overall number of uniformely distributed $\gamma\delta$ T cells is comparable to that of peripheral blood, while the ratio of Vγ9-negative, mostly Vδ1 cells to Vγ9-Vδ2 lymphocytes was much higher than that of paired blood (Chaouni et al. in press), and comparable to that of normal thymus. Table 1 summarizes our data on the modifications of $\gamma\delta$ T cells in the RA joint.

Table 1 Modifications of $\gamma\delta$ T cell subsets in the synovial joint from patients with rheumatoid arthritis

Quantitative modifications in the synovial fluid:
Increase in the absolute number of Vγ9Vδ2-using cells compared to blood.
Ratio Vγ9-using $\gamma\delta$ T / Total $\gamma\delta$ T cells > 75%
Qualitative modifications in the synovial membrane:
Uniform distribution of a number of $\gamma\delta$ cells comparable to that of blood
Relative dicrease in the Vγ9 subpopulation.
Ratio Vγ9-using $\gamma\delta$ T / Total $\gamma\delta$ T cells < 45%

THE T-$\gamma\delta$ - HSP CONNECTION

1-<u>What?</u>: In humans, $\gamma\delta$ cell lines reactive with mycobacterial HSP-65 have been generated from synovial fluid of a rheumatoid arthritis patient, leprosy skin lesions and a healthy individual immune to PPD (review by Born et al. 1990). Mycobacterial HSP-65-reactive T cell clones in rheumatoid arthritis synovial fluid recognize a NH2-terminal 15 amino-acid peptide (Gaston et al. 1990). The human "Vδ2" subset has been shown to proliferate in response to mycobacteria (review by Haas et al. 1990). Since we have shown that this subset is relatively decreased in RA synovial membrane, and since most of the studies in RA dealing with T cell responses to mycobacteria were achieved in the synovial fluid, one can speculate on the relatively lower concentration of stressed-cell products within the synovial membrane and/or their elimination through the synovial fluid.

2-How?: Due to a receptor structure similar to that of TCRαβ, complementary determinig regions encoded by V segments of TCRγδ could interact with the restricting element, and hypervariable junctional regions with the peptide. The limited diversity of CDR1 and CDR2 implies a restricted polymorphism of the restricting element, in agreement with studies showing a number of T γδ cells with specificity for non-polymorphic MHC-like molecules (Strominger 1989, Porcelli 1989).

CLONAL DOMINANCE

Several attempts to evidence in RA a dominant set of unknwon antigens have been made by looking for clonal rearrangements in the supposedly resulting dominant T cell clones in the rheumatoid joint. Conflicting results showing either rare or no TCRαβ oligoclonality (Stamenkovic et al. 1988, Keystone et al. 1988, Miltenburg et al. 1990) may depend on culture conditions and even vary within a synovial membrane (van Laar et al. 1990). Positive observations could result from "shared rearrangements" rather than antigen-specific TCR-bearing cell expansion (Kurnick 1990). Results from our group presented here suggest for TCRγδ a shift from the usual peripheral γδ cell type to the usually less frequent Vδ1 cell type. A similar preferential Vδ gene usage has recently been reported for synovial fluid cells (Sioud et al. 1990). Preliminary RFLP analysis in our laboratory reveals unusual VγI family gene usage in RA blood and numerous Vγ abnormal rearrangements to be elucidated in RA synovial membrane T cell clones (to be published).

IMPLICATIONS OF THE γδ T CELL MODIFICATIONS IN RA

The increase of γδ cells in RA synovial fluid and the shift in Vγ gene usage of γδ cells in RA synovial membrane could imply that these cells are involved in RA pathogenesis or development. Regarding the non-MHC restricted cytotocity and mycobacterial-HSP reactivity of γδ cells, and the preferential homing of mouse γδ T cells expressing a given Vγ gene product to specific target epithelia (see Haas et al. 1990 for a review), this population could be involved either positively in surveillance of synovial tissue growth, or negatively in bone and cartilage resorption.

However, the limited number of γδ T cells in the synovial membrane, added to the virtual absence of clonality in the rheumatoid joint, would argue against local expansion of antigen-specific T cells, and favor modifications of a preexisting T cell repertoire, established during ontogeny. Either by tolerizing T cells to self-MHC or deleting autoreactive T cells, intra-thymic selection could lead to a normally incomplete repertoire. Autoimmunity would then first manifest in filling the "blanks" in the ontogenic repertoire by expanding T (αβ or γδ?) cell populations, under particular inducing conditions (viral or bacterial infections, stress, ...), exerted on particularly susceptible (MHC?) individuals. Alternatively, "blank" creation in the repertoire could be modulated in individuals at ontogeny by the presence of susceptibility MHC alleles. Whatever the mechanism, such hypotheses should deserve an appropriate experimental approach.

REFERENCES

Born W et al. (1990) Recognition of heat shock proteins and $\gamma\delta$ cell function. Immunol Today 11:40-43

Borst J et al. (1988) Distinct molecular forms of human T cell receptor γ/δ detected on viable T cells by a monoclonal antibody. J Exp Med 167:1625-1644

Brandtzaeg P et al., correspondence from Bucy RP et al. and Janeway CA (1989) Epithelial homing of $\gamma\delta$ T cells? Nature 341:113-114

Brennan FM et al. (1988) T cells expressing $\gamma\delta$ chain receptors in rheumatoid arthritis. J Autoimmunity 1:319-326

Brenner MB et al. (1986) Identification of a putative second T-cell receptor. Nature 322:145-149

Bucy RP et al. (1988) Avian T cells expressing $\gamma\delta$ receptors localize in the splenic sinusoids and the intestinal epithelium. J Immunol 141:2200-2205

Bucy RP et al. (1989) Tissue localization and CD8 accessory molecule expression of $T\gamma\delta$ cells in humans. J Immunol 142:3045-3049

Casorati G et al. (1989) Molecular analysis of human γ/δ+ clones from thymus and peripheral blood. J Exp Med 170:1521-1535

Chaouni I et al. (in press) Distribution of T cell receptor-bearing lymphocytes in the synovial membrane from patients with rheumatoid arthritis. J Autoimmunity

Davis MM and Bjorkman PJ (1988) T-cell antigen receptor genes and T-cell recognition. Nature 334:395-402

De Maria A et al. (1987) CD3+4-8-WT31- (T cell receptor γ+) cells and other unusual phenotypes are frequently detected among spontaneously interleukin 2-responsive T lymphocytes present in the joint fluid in juvenile rheumatoid arthritis. A clonal analysis. Eur J Immunol 17:1815-1819

Falini B et al. (1989) Distribution of T cells bearing different forms of the T cell receptor γ/δ in normal and pathological human tissues. J Immunol 143:2480-2488

Gaston JSH et al. (1990) Recognition of a mycobacteria-specific epitope in the 65-kD heat-shock protein by synovial fluid-derived T cell clones. J Exp Med 171:831-841

Goodman T and Lefrancois L (1988) Expression of the γ-δ T-cell receptor on intestinal CD8+ intraepithelial lymphocytes. Nature 333:855-858

Groh V et al. (1989) Human lymphocytes bearing T cell receptor γ/δ are phenotypically diverse and evenly distributed throughout the lymphoid system. J Exp Med 169:1277-1294

Haas W et al. (1990) The development and function of $\gamma\delta$ T cells. Immunology Today 11:340-343

Hata S et al. (1988) Extensive junctional diversity of rearranged human T cell receptor δ genes. Science 240:1541-1546

Holoshitz J et al. (1989) Isolation of CD4- CD8- mycobacteria-reactive T lymphocyte clones from rheumatoid arthritis synovial fluid. Nature 339:226-229

Janeway CA et al. (1988) Specificity and function of T cells bearing $\gamma\delta$ receptors. Immunology Today 9:73-76

Jitsukawa S et al. (1987) A novel subset of human lymphocytes with a T cell receptor-γ complex. J Exp Med 166:1192-1197

Keystone EC et al. (1988) Structure of T cell antigen receptor β chain in synovial fluid cells from patients with rheumatoid arthritis. Arthritis Rheum 31:1555-1557

Kurnick J (1990) Clonal restriction and synovitis. 10th Rheumatic Immunopathology Seminar. Clermont-Ferrand, october 5, 1990.

Kuziel WA et al. (1987) Regulation of T-cell receptor γ-chain RNA expression in murine Thy-1+ dendritic epidermal cells. Nature 328:263-266

Lanier LL et al. (1988) Structural and serological heterogeneity of γ/δ T cell antigen receptor expression in thymus and peripheral blood. Eur J Immunol 18:1985-1992

Mackay CR et al. (1989) γ/δ T cells express a unique surface molecule appearing late during thymic development. Eur J Immunol 19:1477-1483

Miltenburg AMM et al. (1990) Dominant T-cell receptor β-chain gene rearrangements indicate clonal expansion in the rheumatoid joint. Scand J Immunol 31:121-125

Modlin RL et al. (1989) Lymphocytes bearing antigen-specific $\gamma\delta$ T-cell receptors accumulate in human infectious disease lesions. Nature 339:544-548

Porcelli S et al. (1989) Recognition of cluster of differentiation 1 antigens by human CD4-CD8- cytolytic T lymphocytes. Nature 341:447-450

Raulet DH (1898) Antigens for $\gamma\delta$ T cells. Nature 339:342-343

Rème T et al. (1990) T cell receptor expression and activation of rheumatoid arthritis synovial lymphocyte subsets. Phenotyping of multiple synovial sites. Arthritis Rheum, 33:485-492

Sioud M et al. (1990) The Vδ gene usage by freshly isolated T lymphocytes from synovial fluids in rheumatoid synovitis: a preliminary report. Scand J Immunol 31:415-421

Stamenkovic I et al. (1988) Clonal dominance among T-lymphocyte infiltrates in arthritis. Proc Natl Acad Sci USA 85:1179-1183

Stingl G et al. (1987) Thy-1+ dendritic epidermal cells express T3 antigen and the T-cell receptor γ chain. Proc. Natl. Acad. Sci. USA 84:4586-4590

Strominger JL (1989) The $\gamma\delta$ T cell receptor and class Ib MHC-related proteins: enigmatic molecules of immune recognition. Cell 57:895-898

Triebel F and Hercend T (1989) Subpopulations of human peripheral T gamma delta lymphocytes. Immunol Today 10:186-188.

van Laar JM et al. (1990) Analysis of T-cell receptor β-chain gene rearrangements in patients with rheumatoid artrhitis. Joint meeting Société Française d'Immunologie-Nederlandse Vereniging voor Immunologie. Paris, september 27-28, 1990. Abstract W1-9

Phenotypic and Functional Heterogeneity of Double Negative ($CD4^-CD8^-$) $\alpha\beta$ TcR^+ T Cell Clones

J. G. Murison, Sonia Quaratino, and M. Londei

Charing Cross Sunley Research Centre, 1 Lurgan Avenue, Hammersmith, London W68LW, England

INTRODUCTION

The CD4 and CD8 T cell markers are useful phenotypic landmarks which in general terms divide the T cell pool into those cells which aid B cell responses and T cells with cytotoxic capabilities. $CD4^+$ T cells respond to antigen in the context of class II MHC and are predominantly helpers of B cell responses whereas $CD8^+$ T cells are stimulated by antigen in the context of class I MHC and function as cytotoxic T cells. It has been assumed that all mature T cells express either of these markers, however, recently a T cell subpopulation has been described which lack both CD4 and CD8. The majority of these double negative T cells bear the alternative $\gamma\delta$ T cell receptor (TcR) with between 7% (Bender and Kabelitz, 1990) and 38% (Scott, 1990) being $\alpha\beta$ T cell receptor positive. T cells with this phenotype are known to exist in the thymus but are thought to be immature and destined to either die or differentiate into mature single-positive ($CD4^+CD8^-$ or $CD4^-CD8^+$) T cells. Therefore their existence in the periphery indicates either the existence of a novel mature T cell subset or leakage of immature cells from the thymus.

The potential relevance of $CD4^-CD8^-$ $\alpha\beta$ TcR^+ T cells in disease conditions has been highlighted by the findings of Shivakumar *et al* (1989) that the numbers of these cells in the blood of SLE patients increases proportionally with the severity of the disease. These workers were also able to show that double negative $\alpha\beta$ TcR^+ T cells were capable of helping B cells produce pathogenic anti-DNA antibodies. This study when taken in conjunction with those in mice showing an expansion of double negative $\alpha\beta$ TcR^+ T cells in mice predisposed to a variety of autoimmune conditions implicates

Current Topics in Microbiology and Immunology, Vol. 173

these cells as being important in the pathogenesis of autoimmunity (Datta *et al*, 1987; Davidson *et al*, 1986).

Despite their potentially important role in autoimmune conditions these cells have been poorly studied. In an attempt to establish the baseline characteristics of these cells we have generated a panel of T cell clones which are $CD4^-CD8^-$ αβ TcR^+ from the peripheral blood of 2 healthy individuals. These cells have been extensively phenotyped and various functions analysed in an attempt to provide some insight into their role *in vivo*.

Generation of T cell clones

To evaluate the basic biology of double negative TcR $\alpha\beta^+$ T cells we purified cells with the appropriate phenotype using a FACStar Plus (Becton Dickinson) and cloned by stimulation with allogeneic irradiated peripheral blood mononuclear cells (PBM) and PHA and expanded on IL-2. In this way we were able to generate a panel of clones from 2 healthy individuals which were used to assess their functional capabilities and phenotypic characteristics.

Phenotypic and Northern Analysis of T Cell Clones

The clones were derived from lines which had been selected for their lack of CD4 or CD8 cell surface markers but for the expression of the αβ TcR. To ensure that the clones were double-negative they were checked for the presence of CD4 or CD8 mRNA by northern analysis and were all found to be negative. When the clones were more extensively phenotyped it was found that they shared most of the characteristics of single-positive T cells. That is they expressed the 'pan' T cell markers CD2, CD3, and CD5 as well as HLA-DR, however only a proportion of the clones expressed the CD7 marker which is thought to be present on all T cells. The presence or absence of CD7 did not correlate with either of the individuals from which the clones were derived indicating that this is a common occurrence in double negative αβ TcR^+ T cells. CD7 is present on T cells from the early stages of T cell ontogeny in the thymus and therefore the absence of this moiety on some of the clones may indicate that the cells have either avoided maturation in the thymus or lost this marker during ontogeny. It seems unlikely that this sub-population of T cells are 'immature' as they do not express any of the thymocyte markers such as CD1, and they do not gain either CD4 or CD8 upon activation with PHA or immobilised anti-CD3 antibodies.

The other marker which was differentially expressed on the clones was CD28 which has been shown to be an important moiety in both T cell activation and the stabilisation of cytokine mRNA. The ligand for CD28 has been recently described as BB1 which is found almost exclusively on B cells (Linsley et al, 1990) and therefore the presence or absence of CD28 on the clones may point to the existance of at least two functionally distinct groups within the double negative $\alpha\beta^+$ T cell population.

Cytokine induced proliferation.

Cytokines are important intracellular messengers, and the cytokines which cells produce or respond to give an insight into the *in vivo* capabilities of a particular cell type. With this in mind we analysed which cytokines the $CD4^-CD8^-$ $\alpha\beta$ TcR^+ clones could respond to by a standard proliferation assay. Cytokines which are known stimulators of T cell growth such as IL-2, IL-4 and IL-7 were used as well as a number of others ie. IL-1, IL-3, IL-6, GM-CSF, and TNF. All of the clones tested proliferated in response to IL-2 stimulation whereas 3 clones were unable to respond to IL-4 and 1 did not proliferate when cultured with IL-7. This inability of some clones to proliferate when stimulated with IL-4 or IL-7 parallels that of T cell clones of other phenotypes which have been cultured from the peripheral blood. Most striking was that 70% of the clones were able to proliferate strongly to IL-3 which has been traditionally regarded as a colony stimulating factor for precursor cells and cells of myeloid lineage.

Cell mediated cytotoxicity

A study performed by Groh *et al* (1989) showed that 2 double negative T cell lines propagated from the skin of humans were able to kill target cells in an NK cell-like manner. This type of killing is consistent with that found for the majority of $\gamma\delta$ T cells which lack both CD4 and CD8 leading to the speculation that the double negative phenotype may be linked with an NK cell-like cytotoxicity.

To analyse the cytotoxic capabilities of the clones they were cultured with a variety of targets in a standard ^{51}Chromium release assay. Unlike the cell lines propagated by Groh *et al* (1989) none of the 5 clones tested killed the MOLT4, Raji, P815, or K562 cell lines whereas 4 of the 5 clones tested were able to lyse the anti-CD3 secreting hybridoma UCHT1 in a titrable fashion. This discrepancy in the results of our group with that of Groh and co-workers may be due to the cells having been obtained from different tissues. This data however indicates that not all double negative αβ TcR$^+$ cells function in an NK cell-like manner.

Analysis of T Cell Receptor

The T cell receptor for antigen encodes the ability of the cell to respond to a specific epitope in conjunction with an MHC molecule and as such the study of this complex is vital to the understanding of the action of these cells. We have used PCR to analyse the spectrum of Vβ regions used by these clones. Thus far the Vβ regions of 8 clones have been analysed. Only two of the clones tested used the same Vβ region indicating that the double negative cells like the majority of T cells in the periphery can potentially respond to a wide variety of antigens.

DISCUSSION

Double negative αβ TcR$^+$ T cells have been shown to be important in autoimmune diseases in both mice and men (Seman *et al*, 1990; Shivakumar *et al*, 1989), but the role these cells play in these conditions is unclear. MRL/lpr and C3H/gld mice spontaneously develop autoimmune diseases which resemble SLE or Rheumatoid arthritis in humans (Andrews *et al*, 1978; Hang *et al*, 1982) and are characterised by the proliferation of CD4$^-$CD8$^-$ αβ TcR$^+$ T cells. Both of these mouse strains produce a large array of autoantibodies one of which has been found to bind to the murine IL-3 receptor in an agonistic manner (Sugawara *et al*,1988). This finding taken in conjunction with our own that human CD4$^-$CD8$^-$ αβ TcR$^+$ T cells respond to IL-3 may provide an explanation for the expansion of these cells in their respective autoimmune conditions.

REFERENCES

Andrews, B.S., Eisenberg, R.A., Theofilopoulos, A.N., Izui, S., Wilson, C.B., McConahey, P.J., Murphey, E.D., Roths, J.B., and Dixon, F.J. (1978) Spontaneous lupus-like syndromes. J. Exp. Med. 148: 1198

Bender, A., and D. Kabelitz. (1990) $CD4^-CD8^-$ human T cell: Phenotypic heterogeneity and activation requirements of freshly isolated "double-negative" T cells. Cell. Immunol. 128: 542-554.

Datta, S.K., Patel, H., and Berry, D. (1987) Induction of a cationic shift in IgG anti-DNA autoantibodies. Role of T helper cells with classical and novel phenotypes in three models of murine models of lupus nephritis. J. Exp. Med. 165: 1252-1268.

Davidson, W.F. Dumont, F.J., Bedigian, H.G., Fowlkes, B.J., and Morse, H.C. (1986) Phenotypic, functional and molecular genetic comparisons of the abnormal lymphoid cells of C3H-lpr/lpr and C3H gld/gld mice. J. Immunol. 136: 4075-4084.

Groh, V., M. Fabbi, F. Hochstenbach, R.T. Maziarz, and J.L. Strominger. (1989) Double negative ($CD4^-CD8^-$) lymphocytes bearing T cell receptor α and β chains in normal human skin. Proc. Natl. Acad. Sci. USA 86: 5059-5063.

Hang, L., Theofilopoulos, A.N., anf Dixon, F.J. (1982) A spontaneopus rheumatoid arthritis-like disease in MRL/l mice. J. Exp. Med. 155: 1690.

Linsley, P.S., Clark, E.A., and Ledbetter, J.A. (1990) T-cell antigen CD28 mediates adhesion with B cells by interacting with activation antigen B7/BB-1. Proc. Natl. Acad. Sci. USA. 87: 5031-5035.

Porcelli, S., M.B. Brenner, J.L. Greenstein, S.P. Balk, C. Terhorst, and P.A. Bleicher. (1989) Recognition of cluster of differentiation 1 antigens by human $CD4^-8^-$ cytolytic T lymphocytes. Nature, 341: 447-450.

Scott, C.S., S.J. Richards, and B.E. Roberts. (1990) Patterns of membrane $TcR\alpha\beta$ and $TcR\gamma\delta$ chain expression by normal blood $CD4^+CD8^-$, $CD4^-CD8^+$, $CD4^-CD8^{dim+}$ and $CD4^-CD8^-$ lymphocytes. Immunol. 70: 351-356.

Seman, M., Boudaly, S., Roger, T., Morisset, J., and Pham, G. Autoreactive T cells in normal mice: unrestricted recognition of self peptides on dendritic cell I-A molecules by $CD4^-CD8^-$ T cell receptor α/β^+ T cell clones expressing $V\beta8.1$ gene segments. Eur. J. Immunol. 20: 1265-1272.

Shivakumar, S., G.C. Tsokos, and S.K. Datta. (1989) T cell receptor α/β expressing double-negative ($CD4^-CD8^-$) and CD4 helper cells in humans augment the production of pathogenic anti-DNA autoantibodies associated with lupus nephritis. J. Immunol. 143: 103-112.

Sugawara, M., Hattori, C., Tezuka, E., Tamura, S., and Ohta, Y. (1988) Monoclonal autoantibodies wih interleukin 3-like activity derived from a MRL/lpr mouse. J. Immunol. 146: 526-530.

Restriction

Specificity of Human T Lymphocytes Expressing a γ/δ T Cell Antigen Receptor. Recognition of a Polymorphic Determinant of HLA Class I Molecules by a γ/δ^+ Clone

E. Ciccone[1], O. Viale[1], Daniela Pende[1], M. Malnati[1], A. Moretta[2], and L. Moretta[1]

[1] Istituto Nazionale per la Ricerca sul Cancro, Genova, Italy
[2] Istituto di Istologia ed Embriologia Generale, Universitá di Genova, Italy

In addition to the major T lymphocyte population, expressing surface receptors (TCR) for antigen/MHC composed of a disulphide-linked heterodimer (alpha/beta chains), a minor subset has recently been identified which expresses a different form (gamma/delta) of CD3-associated molecules (Brenner 1986, 1987). Different molecular forms of TCR gamma/delta have been identified by the use of anti-TCR gamma/delta mAbs (Bottino 1988, Hochstenbach 1988, Moretta A. 1988). Thus, BB3 (Ciccone 1988a) and A13 (Ferrini 1989) mAbs (or analogous delta-TCS1) have been shown to recognize two non-overlapping lymphocyte subsets (Bottino 1988) of peripheral blood-derived TCR gamma-delta+ cells. BB3 mAb have been shown to specifically recognize Vdelta2 expressing cells, on the other hand A13 recognizes Vdelta1+ lymphocytes. More than 95% of TCR gamma/delta+ lymphocytes in the peripheral blood belong to the BB3+ or A13+ cells. In the peripheral blood molecules reactive with BB3 mAb correlate with the disulphide-linked ($C\gamma1$ encoded) form of the receptor, whereas A13 reacts with two distinct non disulphide-linked ($C\gamma2$ encoded) forms of TCR gamma/delta. Although in the first functional studies the MHC-unrestricted cytolytic activity of these cells was emphasized (Borst 1987), recently evidence has been provided both in man and in mouse that TCR gamma/delta+ cells are able to specifically recognize and kill allogeneic cells (Matis 1987, Ciccone 1988b). In our laboratory it has been demonstrated that TCR gamma/delta+ cells were able to recognize allogeneic cells in conventional MLC. Both proliferation and lysis of target cells bearing the stimulating alloantigens could be detected (Ciccone 1988b). This phenomenon was clearly specific since neither autologous nor allogeneic unrelated blasts were lysed. The same responding population cultured with autologous irradiated cells did not lyse either autologous or

Current Topics in Microbiology and Immunology, Vol. 173

allogeneic target blast cells. An additional demonstration that TCR gamma/delta+ cells can recognize allogeneic cells was obtained by the analysis of MLC-derived T-cell clones (Ciccone 1988a). Clones that specifically lysed allogeneic target cells did not lyse autologous or unrelated blasts. In addition they were either BB3+ or A13+ (see Table I), thus, indicating that both types of TCR gamma/delta can mediate alloantigen recognition. The analysis of allospecific cell clones also provided direct evidence that the specific cytolytic activity is not necessarily correlated with the expression of MHC-unrestricted cytotoxicity.

SURFACE PHENOTYPE			CYTOLYTIC ACTIVITY		
Clone	BB3	A13	Specific allogeneic PHA blasts	Autologous PHA blasts	Nonspecific allogeneic PHA blasts
E2	+	-	68	6	0
E17	+	-	18	4	0
E33	+	-	62	5	0
E68	+	-	18	0	ND
E39	+	-	16	1	0
E41	-	+	37	9	8
L33	-	+	29	2	0
LM1	-	+	45	0	0
LM12	-	+	33	0	2

Results are expressed as % specific 51Cr release at an E/T cell ratio of 3:1

Since alloreactive TCR gamma/delta+ clones are well known susceptible to inhibition by mAb specific for CD3 and TCR molecular complex, we investigated whether also anti-CD3 or anti-TCR gamma/delta mAbs could inhibit the specific cytolysis mediated by alloreactive gamma/delta+ clones. This experiment was performed on 3 representative TCR gamma/delta+ clones.

We found that anti-CD3 and anti-TCR gamma/delta mAbs strongly inhibited the cytolytic activity, of all 3 clones, of the specific allogeneic cell bearing the stimulating alloantigen. It is noteworthy that 2 clones were A13+ and were inhibited by the corresponding antibodies, whereas the another (BB3+) was inhibited by BB3 mAb. Since all 3 clones displayed strong cytolytic activity against K562 target cells, we also analyzed whether this type of cytolytic activity was susceptible to mAb-mediated inhibition. We found that neither anti-CD3 nor anti-TCR gamma/delta mAbs had inhibitory effect. In contrast anti-LFA1 mAb used, as control, inhibited the lysis of K562 target cells. It should be noted that, in this experiments, the mAbs used as inhibitors were added at the onset of the test. Recognition

of MHC class I molecules by the TCR gamma/delta+ clone LM12. Clone LM12 was A13+ and did not react with BB3 or WT31 mAbs. As shown by immunoprecipitation experiments the TCR of the clone LM12 was compared by a non disulphide linked heterodimer, as suggested by A13 surface phenotype.
Clone LM12 had been derived from donor LM after stimulation in MLC against allogeneic cells derived from donor MM (HLA typing: Aw68, 24; B35, w55; DR1, 7). Since preliminary experiments suggested that the A24 allele could represent the antigen recognized by this clone, we analyzed a panel of PHA-induced target cells derived from 9 HLA-A24 positive and 9 HLA-A24 negative donors. As shown in table II, clone LM12 selectively lysed only those target cells which expressed the A24 allele. Therefore these data suggest that the A24 allele represents the restriction element of the lysis mediated by clone LM12.

TABLE II

Specific cytolytic activity of clone LM 12 (TCRγ/δ+,A13+) against HLA-A24+ target cells.

Donor*	HLA-Typing A	HLA-Typing B	HLA-Typing DR	Cytolytic activity **
L.M.	3 ,29	35,44	5,7	0
G.M.	1 ,29	44,62	7,10	0
G.B.	2 ,29	44,50	2,7	0
M.B.	3 ,28	18,44	7,14	1%
O.V.	3 ,11	7,35	2,5	2%
M.P.	3 ,30	14,40	1,7	0
M.E.	3 ,30	14,40	3,7	0
Ki.	3 ,30	14,13	5,7	3%
M.L.	3 ,68	14,35	5,7	3%
R.M.	3,24	18,55	1,5	26%
M.M	68,24	35,55	1,7	45%
S.P.	30,24	49,60	4,5	43%
A.S.	2,24	49,51	5,6	37%
G.A	2,24	39,50	3,8	48%
V.O.	2,24	35,29	2,3	23%
A.C.	2,24	27,35	4,5	37%
E.R.	1,24	35,49	2,6	62%
C.M.	2,24	13,51	7,8	40%

*The target cells are Cr-labelled PHA-derived blasts from PBL of different donors.
**Results are expressed as % of specific Cr-release at an effector:target cell ratio of 10:1.

In order to directly verify this possibility, we next analysed the ability of clone LM12 to lyse murine P815 cells transfected with the human DNA fragment coding for HLA-A24 and expressing this antigen at the cell surface (Maryanski

1985); as control we used untransfected P815 cells, or cells transfected with Cw3. The A24-positive P815 cells were efficiently lysed by clone LM12, whereas virtually no lysis was observed against untransfected or Cw3-transfected P815 target cells.
Another A13+ clone (derived from the same donor) with no specific cytolytic activity against A24+ allogeneic cells, did not lyse P815 transfected with A24. These data provide direct evidence that the A24 molecule is specifically recognized by clone LM12.

CONCLUSION

Our data indicated that the MHC class I molecules may represent a possible molecular target for TCR gamma/delta+ lymphocytes and that polymorphic determinants of this molecules are recognized (Ciccone 1989). Another reports indicated that TCR gamma/delta+ cells can recognize various surface molecules including MHC class I (Spits 1990) or MHC class II (Kozbor 1989), CD1 (Porcelli 1989) and in the mouse TL antigens. In addition cells specific for a given molecules (e.g. CD1) may represent a small percentages of all TCR gamma/delta+ cells. Taken together these data may suggest that different surface molecules may be recognized by TCR gamma/delta+ cells.
These structures could represent restriction element which allow antigen recognition by this T cell subset.

This work was supported in part by grants awarded by the Consiglio Nazionale delle Ricerche (CNR) "Piano Finalizzato Oncologia" to L. Moretta and A. Moretta and by Associazione Italiana per la Ricerca sul Cancro (AIRC). O. Viale and D. Pende are recipients of a fellowship awarded by AIRC.

REFERENCES

1. Borst, J., R.J. van de Griend, J.W. van Oostveen, S. Ang, C.J. Melief, J.G. Seidman, and R.L.H. Bolhuis. Nature (Lond.) 1987. 325:683.

2. Bottino, C., G. Tambussi, S. Ferrini, E. Ciccone, P. Varese, M.C. Mingari, L. Moretta, and A. Moretta. J. Exp. Med. 1988. 168:491.

3. Brenner, M.B., J. McLean, D.P.B. Dialynas, J.L. Strominger, J.A. Smith, F.L. Owen, J.G. Seidman, S.Ip, F. Rosen, and M.S. Krangel. Nature (Lond.) 1986. 322:145.

4. Brenner, M.B., J. McLean, H. Scheft, J. Riberdy, S. Ang, J.G. Seidman, P. Devlin, and M.S. Krangel. Nature (Lond). 1987. 325:689.

5. Ciccone, E., S. Ferrini, C. Bottino, O. Viale, I. Prigione, G. Pantaleo, G. Tambussi, A. Moretta, L. Moretta. J. Exp. Med. 1988a. 168:1.

6. Ciccone, E., O. Viale, C. Bottino, D. Pende, N. Migone, G. Casorati, G. Tambussi, A. Moretta, and L. Moretta. J. Exp. Med. 1988b. 167:1517.

7. Ciccone, E., O. Viale, D. Pende, M. Malnati, G.B. Ferrara, S. Barocci, A. Moretta, L. Moretta. Eur. J. Immunol. 1989. 19:1267.

8. Ferrini, S., I. Prigione, C. Bottino, E. Ciccone. G. Tambussi, S. Mammoliti, L. Moretta, A. Moretta. Eur. J. Immunol. 1989. 19:57.

9. Hochstenbach, F., C. Parker, J. McLean, V. Greselmann, H. Band, I. Bank, L. Chess, H. Spits, J.L. Strominger, J.G. Seidman, and M.B. Brenner. J. Exp. Med. 1988. 168:761.

10. Kozbor, D., G. Trinchieri, D. Monos, M. Isobe, G. Russo, J.A. Haney, C. Zmijewski, C.M. Croce. J. Exp. Med. 1989. 169:1847.

11. Maryanski,J.L., A. Moretta, B. Jordan, E. De Plaen, A. Van Pel, T. Boon, and J.C. Cerottini. Eur. J. Immunol. 1985. 15:1111.

12. Matis, L.A., R. Cron, and J.A. Bluestone. Nature (Lond). 1987. 330:262.

13. Moretta, A., C. Bottino, E. Ciccone, G. Tambussi, M.C. Mingari, S. Ferrini, G. Casorati, P. Varese, O. Viale, N. Migone, and L. Moretta. J.Exp. Med. 1988. 168:2349.

14. Porcelli, S., M.B. Brenner, J.L. Greenstein, S.P. Balk, C. Terhorst, P.A. Bleicher. Nature (Lond.) 1989. 341:447.

15. Spits, H., X. Paliard, V.H., Engelhard, J.E. De Vries. J. Immunol. 1990. 144:4156.

Antigens and Antigen-Presenting Molecules for γδ T Cells

H. BAND[1,2], ST. A. PORCELLI[1,2], G. PANCHAMOORTHY[1], J. MCLEAN[1], C. T. MORITA[1,2], S. ISHIKAWA[3], R. L. MODLIN[4], and M. B. BRENNER[1,2]

[1] Laboratory of Immunochemistry, Dana Farber Cancer Institute
[2] Department of Rheumatology and Immunology, Harvard Medical School, Boston, MA, USA
[3] Department of Medicine, Tufts University School of Medicine and Unit of Pulmonary Medicine, St. Elizabeth's Hospital, Boston, MA, USA
[4] Division of Dermatology, Department of Microbiology and Immunology, U.C.L.A. School of Medicine, Los Angeles, CA, USA

The T cell receptor (TCR) γδ is expressed on a distinct subset of T cells (Brenner et al., 1986). Unlike the TCR αβ, little is known about γδ T cell recognition. TCR α and β and TCR γ and δ genes are assembled during development by recombination of distinct gene segments (V, D, J). While the germline gene segment repertoire for both TCR γ and δ chains is small compared to the TCR αβ, extensive junctional diversity is observed particularly in the TCR δ chain (reviewed in Brenner et al. 1988). The diversity resides in a region presumed to form the CDR3 homolog of the immunoglobulin molecule considered important in antigen recognition (Davis and Bjorkman, 1988). These structural features of the γδ TCR suggest a role for T cells bearing this receptor in specific antigen recognition.

PREFERENTIAL RECOGNITION OF MYCOBACTERIAL ANTIGENS BY γδ T CELLS: Studies on the receptor repertoire and anatomical localization of γδ T cells have provided important clues in the search to identify relevant antigens and antigen-presenting molecules for the γδ TCR. In both mouse and man, γδ T cells show evidence of localization to epithelia, such as in the gut (Goodman and Lafrancois, 1988; Bucy et al., 1989), the epidermis at least in mice (Koning et al., 1987), and the lung (Augustin et al., 1989). Analysis of the V-gene repertoire of γδ T cells in different locations shows the usage of TCRs encoded by particular V-gene segments or pairs of V-gene segments, which may arise at specific periods in thymic ontogeny (Asarnow et al., 1988; Itohara et al., 1990; Havran and Allison, 1990). The limited V-gene repertoire, tissue localization in the proximity of epithelial cell sheets, and lack of junctional diversity on certain murine γδ T cells suggested that γδ T cells may recognize phylogenetically conserved antigens, such as the heat shock proteins (HSPs). As such, γδ T cells may represent a first line of defence against invading pathogens and cellular stress (Janeway, 1988). To date, a number of studies have shown that γδ T cells respond specifically and perhaps predominantly to mycobacterial antigens including their heat shock proteins. For example, murine thymocyte-derived hybridomas (O'Brien et al. 1989), murine lymph node and lung cells (Janis et al. 1989, Augustin et al. 1989), human cutaneous lesions of mycobacterial infection (Modlin et al. 1989), rheumatoid arthritis synovial T cells (Holoshitz et al. 1989) and human peripheral blood γδ T cells (Kabelitz et al 1990) all respond strikingly to mycobacterial antigens. This reactivity may not be unique to mycobacteria, as several other bacteria, yeast and a human B cell line (Daudi) are able to induce a similar expansion of γδ T cells (Fisch et al. 1990; Abo et al., 1990; our unpublished results). Reactivity to mycobacteria supports the suggestion that γδ T cells may specialize in host defence against intracellular microbes and perhaps

Current Topics in Microbiology and Immunology, Vol. 173

<u>THE γδ T CELL EXPANSION BY MYCOBACTERIA DOES NOT REQUIRE PRIOR ANTIGENIC EXPOSURE</u>: The marked expansion of γδ T cells by mycobacteria may result from repeated prior exposures to conventional antigens present in microorganisms. Alternatively, certain mycobacterial molecules may stimulate whole populations of γδ T cells characterized by their sharing receptors encoded by common germline V or J gene segments, analogous to the recognition of enterotoxin superantigens by TCR Vß regions of αß TCRs (Marrack and Kappler, 1990). To distinguish between conventional antigen and superantigen types of recognition, we characterized the repertoire of the human γδ T cell response to <u>Mycobacterium</u> <u>tuberculosis</u> (H37Ra) in PBMC derived from: individuals with active tuberculosis; healthy individuals that had recently converted to PPD-positive status as a result of household contact with active cases of tuberculosis; PPD-negative healthy individuals; and newborns (umbilical cord blood) representing antigenically naive individuals. The αß and γδ T cells in <u>in vitro</u> antigen-stimulated cultures were analyzed by flow cytometry using mAb anti-TCRδ1 (pan γδ, Band et al. 1987) and BMA-031 (pan αß, gift of Dr. I.V. Kurrle). A dramatic expansion of the γδ T cells occurs in the seven to ten day <u>in vitro</u> response to H37Ra in the peripheral blood of all individuals tested, irrespective of prior mycobacterial infection. Importantly, γδ T cell expansion is observed also in PBMC derived from fresh umbilical cord blood, further demonstrating that previous antigenic stimulation is not required for the marked γδ T cell response to H37Ra (unpublished results).

<u>MYCOBACTERIA-INDUCED γδ T CELL EXPANSION IS V-GENE RELATED</u>: The above experiments which show innate reactivity of γδ T cells to H37Ra, suggested that this recognition might be determined by germline gene sequences. Thus, we used mAb against Vδ1 (δTCS1 and A13, gifts from T Cell Sciences and Dr. S. Ferrini, respectively), Vδ2 (BB3, and in some experiments G1; gifts from Drs A and L. Moretta, E. Ciccone and S. Ferrini), and Vγ2 (anti-TiγA, a gift from Dr. T. Hercend) to assess the V-gene repertoire of γδ T cells that expand in response to H37Ra. In adult peripheral blood, the majority of γδ T cells express Vγ2 paired to Vδ2, while Vδ1 in association with any of several Vγ gene segments accounts for most of the remaining γδ T cells at this site (Faure et al., 1988; Bottino et al. 1988; Parker et al., 1990). The V-gene repertoire shows nearly equal usage of Vδ1 and Vδ2 in peripheral blood at birth, after which the Vδ2 subset expands to account for the majority of peripheral blood γδ T cells in man (Parker et al., 1990). Interestingly, the γδ T cell expansion in response to H37Ra occurs almost exclusively in the Vγ2/Vδ2 expressing subset. That expansion of Vγ2/Vδ2 bearing cells is not solely due to their predominance in most adults or due to their activation state (these T cells are CD45RO^{hi}) is demonstrated by the results on umbilical cord blood derived PBMC where Vδ1 and other non-Vγ2 bearing cells represent a significant fraction of the γδ T cells. Yet, as in adult individuals, the Vγ2/Vδ2 bearing cells expand in response to H37Ra (unpublished results).

<u>TCR REPERTOIRE OF MYCOBACTERIA-STIMULATED γδ T CELLS IS DIVERSE</u>: To examine the repertoire of H37Ra-responsive γδ T cells, we determined the nucleotide sequences of the TCR δ genes by polymerase chain reaction (PCR). Both the Vδ2-Jδ1 and Vδ2-Jδ3 bearing T cells expanded in response to H37Ra show a high degree of diversity characterized by the use of 2 and sometimes 3 Dδ gene segments in tandem and by the

a few sequences were found to be repeated in these analyses, the number of identically repeated sequences in the H37Ra-stimulated cultures was similar to that found in the unstimulated PBL. Thus, the responding cells are characterized by a high degree of junctional diversity, similar to that of freshly isolated γδ T cells, but with a striking preference for utilization of Vγ2 and Vδ2 germline gene segments. This, together with a high precursor frequency for mycobacteria among γδ T cells as reported by Kaufmann and colleagues (Kabelitz et al., 1990), suggests a mode of recognition analogous to superantigen recognition by αβ T cells. Recently, reactivity to a classic superantigen, SEA, was observed in this γδ T cell subset (Rust et al., 1990). We have earlier demonstrated an age-related post-thymic expansion of γδ T cells in the peripheral blood in man (Parker et al., 1990). Interestingly, this expansion exclusively involves γδ T cells that bear Vγ2/Vδ2 encoded receptors, which is the same subset that undergoes <u>in vitro</u> expansion in response to mycobacteria. The post-thymic expansion of γδ T cells may, therefore, follow similar stimulations by antigen <u>in vivo</u>.

<u>PRESENTING MOLECULES FOR γδ TCR RECOGNITION OF MYCOBACTERIAL ANTIGENS</u>: The mechanism by which T cells expressing γδ TCRs recognize antigens is incompletely defined. It is noteworthy that nearly all γδ TCR bearing cells lack expression of CD4, and most either lack CD8 or express it at a relatively low level (Groh, et al, 1988). This suggests that γδ T cells may be less frequently restricted than αβ T cells to recognition of antigens presented by classical MHC class I or II molecules, and may more often use other cell surface molecules for antigen presentation. Obvious candidates for these "other" antigen presenting molecules" are the products of so-called "non-classical" MHC class I genes. Southern blot analyses have demonstrated the existence of at least 14 other class I related genes in humans (Orr, et al, 1987) and approximately 30 such loci in mice (Weiss, et al, 1984). These include in humans the HLA-E, -F and -G loci, and in mice the Q and TL regions. At least some of these loci are known to be expressed at the protein level in both species. In addition, other molecules with structural similarity to MHC products but encoded by genes located outside of the MHC may also function as target structures for γδ TCRs. The best known molecules in this category are the members of the CD1 family, which are distantly related to MHC class I and II molecules (Calabi, F & Milstein, C, 1986) and are present in both human and mouse.

Currently available data, although limited, support the idea that T cells bearing γδ TCRs are restricted by a broader array of MHC and MHC-like molecules than are their TCR αβ bearing counterparts. To date, approximately equal numbers of γδ T cell clones recognizing classical MHC encoded molecules and nonclassical MHC or MHC-like molecules have been reported. Two laboratories have described murine γδ T cells specific for products of genes mapping to the TL locus (Bluestone, et al, 1988; Ito, et al, 1990). In one case, the molecule recognized has been shown to be encoded by a previously undescribed TL region gene encoding a broadly expressed class I molecule (Ito, et al, 1990). One example has also been described of a murine γδ T cell hybridoma specific for a synthetic polymer of glutamic acid and tyrosine (GT) and restricted by the widely expressed Qa-1 product of the Q locus (Vidovic, et al, 1989). A limited number of reports in both mouse and man have also described γδ T cells that recognize classical MHC class I and II molecules, in a manner similar to that

1990). Another study demonstrated recognition of MHC class I related molecules by human γδ T cell clones, but did not define the nature (i.e, classical vs. nonclassical class I) of the molecules recognized (Rivas, et al, 1989).

Recognition of non-MHC encoded molecules has also been found for human γδ T cells. In 1987 we proposed that a γδ T cell line (IDP2) possessed specific MHC unrestricted recognition based on the pattern of cytolysis of a panel of target cell lines (Brenner et al, 1987). Subsequently, this was shown by monoclonal antibody blocking and DNA transfection experiments to be due to specific recognition of a CD1 molecule by the γδ receptor (Porcelli, et al, 1989). More recently, another laboratory has also isolated a γδ T cell clone specific for CD1 (Faure, et al, 1990). In both cases, these γδ T cell clones were specific for CD1c, one of the five potential glycoproteins encoded by the human CD1 gene locus (Calabi F & Milstein C, 1986). CD1 reactive γδ T cells have not yet been described in the mouse, perhaps because reagents necessary to define murine CD1 glycoproteins have not yet become generally available. Interestingly, we have also isolated several examples of human αß T cells specific for CD1 molecules. All of these display the $CD4^-8^-$ surface phenotype typical of γδ T cells (Porcelli et al, 1989). This finding is consistent with our hypothesis that the absence of CD4 or CD8, which is generally the rule for γδ T cells and the exception for αß T cells, may promote self-restriction by a broader array of MHC and MHC-like molecules.

In man, analyses of the presentation of mycobacterial antigens to γδ T cells have yielded conflicting results. Haregewoin et al described a HSP65 reactive γδ T cell line that required autologous plastic-adherent peripheral blood mononuclear cells for optimal responses <u>in vitro</u> (Haregewoin, et al, 1989). Another study demonstrated a requirement for autologous or MHC matched antigen presenting cells to stimulate proliferation of PPD specific γδ T cells isolated from a lepromin skin test specimen, although the relevant MHC or MHC linked loci could not be precisely defined (Modlin et al 1989). These studies were interpreted as evidence that polymorphic MHC linked antigen presenting molecules restrict the response of γδ T cells to mycobacterial antigens. In contrast, a HSP65 reactive human γδ clone isolated from synovial fluid of a patient with rheumatoid arthritis showed equal responsiveness to antigen presented by PBMC of autologous origin or from eight different randomly selected allogeneic individuals (Holoshitz et al, 1989). In this case, it is possible that presentation of HSP65 may not be restricted by classical MHC molecules, but by relatively nonpolymorphic putative antigen presenting molecules, such as the HLA-E, -F, -G or CD1 loci products.

As for γδ T cells responsive to PPD or purified HSP65, studies of γδ T cell clones reactive to crude lysates of <u>M. tuberculosis</u> demonstrate that an accessory cell is required for optimal responses (Kabelitz, et al , 1990; our unpublished results). However, PBMC from a panel of MHC nonidentical donors all work equally well, suggesting that if a specific antigen presenting molecule is involved it is likely to have limited or no polymorphism. Alternatively, the stimulatory moiety of mycobacteria sonicates may act in a manner more analogous to the Staphylococcal enterotoxin superantigens, which involves binding to and presentation by MHC class II molecules but with minimal or no restriction by polymorphic residues (Marrack, P & Kappler J., 1990). Resolution of the mechanism involved will probably first require the for the dramatic stimulatory effect of M. tuberculosis preparations for $\gamma\delta$ T cells.

In conclusion, responses elicited by antigens such as those of mycobacteria appear to be superantigen-like responses for $\gamma\delta$ T cells. The requirement for and the nature of self antigen-presenting molecules recognized by the $\gamma\delta$ TCR are still poorly characterized; however a role for nonclassical MHC or non-MHC molecules, such as CD1, appears likely. Together, these reactivities provide a means to study the recognition by the $\gamma\delta$ TCR, such studies should move us closer to understanding the functions of this unique subset of T cells.

ACKNOWLEDGEMENTS

This work was supported by grants from the NIH and received financial support from the UNDP/World Bank/WHO Special Program for Research and Training in Tropical Diseases. H.B. is a Leukemia Society of America Special Fellow and an Arthritis Foundation Investigator. S.P. is a fellow of the Medical Foundation.

REFERENCES

Abo, T., Sugawara, S., Seki, S., Fujii, M., Rikiishi, H., Takeda, K., and Kumagai, K. (1990). International Immunology 2:775-785.

Asarnow, D.m., Kuziel, W.A., Bonyhadi, M., Tigelaar, R.E., Tucker, P.W., and Allison, J.P. (1988). Cell 55:837-847.

Augustin, A., Kubo, R.T., and Sim, G-K. (1989). Nature 340:239-241.

Band, H., Hochstenbach, F., McLean, J., Hata, S., Krangel, M.S., Brenner, M.B. (1987). Science 238:682.

Bluestone, J.A., Cron, R.Q., Cotterman, M., Houlden, B.A., and Matis, L.A. (1988). J. Exp. Med. 168:1899-1916.

Bosnes, V., Qvigstad, E., Lundin, K.E.A., and Thorsby, E. (1990). Eur. J. Immunol. 20:1429-1433.

Bottino, C., Tambussi, G., Ferrini, S., Ciccone, E., Varese, P., Mingari, M.C., Moretta, L., and Moretta, A. (1988). J. Exp. Med. 168:491-505.

Brenner, M.B., Mclean, J., Dialynas, D., Strominger, J.L., Smith, J.A., Owen, F.L., Seidman, J.G., Ip, S., Rosen, F. and Krangel, M.S. (1986). Nature 322:145-149.

Brenner, M.B., Mclean, J., Scheft, H., Riberdy, J., Ang, S.L., Seidman, J.G., Devlin, P., Krangel, M.S. (1987). Nature 325:689-694.

Brenner, M.B., Strominger, J.L. and Krangel, M.S. (1988). Adv. Immunol. 43:133-192.

Bucy, R. P., Chen, C-L.H., and Cooper, M.D. (1989). J. Immunol. 142:3045-3049.

Calabi, F & Milstein C. (1986). Nature. 323, 540-543.

Ciccone, E., Viale, O., Pende, D., Malnati, M., Ferrara, G.B., Barocci, S., Moretta, A., Moretta, L. (1989). Eur. J. Immunol. 19:1267-1271.

Davis, M.M. and Bjorkman, P.J. (1988). Nature 334:395-402.

Faure, F., Jitsukawa, S., Triebel, F., and Hercend, T. (1988). J. Immunol. 141:3357-3360.

Faure, F, Jitsukawa, S, Miossec, C & Hercend, T. (1990). Eur. J. Immunol. 20: 703-6.

Fisch, P., Malkovsky, M., Braakman, E., Strum, E., Bolhuis, R.L.H., Prieve, A., Sosman, J.A., Lam, V.A., and Sondel, P.M. (1990). J. Exp. Med. 171:1567-1579.

Goodman, T and Lefrancois, L. (1988). Nature 333:855-858. (1989) J Exp Med. 169:1277.
Haregewoin, A., Soman, G., Hom, R.C., and Finberg, R.W. (1989). Nature 340:309-312.
Havran, W., and Allison, J.P. (1990). Nature 344:68-70.
Holoshitz, J., Koning, F., Coligan, J.E., De Bruyn, J., and Strober, S. (1989). Nature 339:226-229.
Ito, K., Van Kaer, L., Bonneville, M., Hsu, S., Murphy, D.B., and Tonegawa, S. (1990). Cell 62:549-561.
Itohara, S., Farr, A.G., LaFaille, J.J., Bonneville, M., Takagaki, Y., Haas, W., and Tonegawa, S. (1990). Nature 343:754-757.
Janeway, C.A. (1988). Nature 333:804-806.
Janis, E.M., Kaufmann, S.H.E., Schwartz, R.H., Pardoll, D.M. (1989). Science 244:713-716.
Jitsukawa, S., Triebel, F., Faure, F., Miossec, C., and Hercend, T. (1988). Eur. J. Immunol. 18:1671-1679.
Kabelitz, D., Bender, A., Schondelmaier, S., Schoel, B., and Kaufmann, S.H.E. (1990). J. Exp. Med. 171:667-679.
Koning, F., Stingl, G., Yokoyama, W.M., Yamada, H., Maloy, W.L., Tscachler, E., Shevach, E.M., and Coligan, J.E. (1987). Science 236:834-837.
Kozbor, D., Trinchieri, G., Monos, D.S., Isobe, M., Russo, G., Haney., J.A., Zmijewski, C., and Croce, C. (1989). J. Exp. Med. 169:1847-1851.
Marrack, P. and Kappler, J. (1990). Science 248:705-711.
Matis, L.A., Fry, A.M., Cron, R.Q., Cotterman, M.M., Dick, R.F., and Bluestone, J.A. (1989). Science 245:746-749.
Modlin, R.L., Pirmez, C., Hofman, F.M., Torigian, V., Uyemura, K., Rea, T.H., Bloom, B.R., and Brenner, M.B. (1989). Nature 339:544-548.
O'Brien, R. L., Happ, M.P., Dallas, A., Palmer, E., Kubo, R., Born, W.K. (1989). Cell 57:667-674.
Orr, HT, Koller, BH, Geraghty, D and DeMars, R. (1987). In Kelsoe, G and Schulze, DH, eds., Evolution and Vertebrate Immunity, p. 349. University of Texas Press, Austin, Tx.
Parker, C.M., Groh, V., Band, H., Porcelli, S.A., Morita, C.T., Fabbi, M., Glass, D., Strominger, J.L., and Brenner, M.B. (1990). J. Exp. Med. 171:1597-1612.
Porcelli, S.A., Brenner, M.B., Greenstein, J.L., Balk, S.P., Terhorst, C., and Bleicher, P.A. (1989). Nature 341:447-450.
Rivas, A., Koide, J., Cleary, M., and Engleman, E.G. (1989). J. Immunol. 142:1840-1846.
Rust, C.J.J., Verreck, F., Vietor, H., and Koning, F. (1990). Nature 346:572-574.
Spits, H., Paliard, X., Engelhard, V.H., and de Vries, J.E. (1990). J. Immunol. 144:4156-4162.
Vidovic, D., Roglic, M., McKune, K., Guerder, S., MacKay, C., Dembic, Z. (1989). Nature 340:646-650.
Weiss, EH, Golden, L, Fahrner, K., Mellor, AL, Devlin, JJ, Bullman, H, Tiddens, H, Bud, H & Flavell RA. (1984). Nature 310: 650.

Correlation Between TCR V Gene Usage and Antigen Specificities in Human γδ T Cells

G. De Libero[1,2], G. Casorati[1], N. Migone[3], and A. Lanzavecchia[1]

[1] Basel Institute for Immunology, Grenzacherstraße 487, CH-4005 Basel
[2] Experimental Immunology, Department of Research, Basel University, CH-4031 Basel
[3] Dipartimento di Genetica, Biologia e Chimica Medica, Universitá di Torino and CNR, Immunogenetica e Istocompatibilitá, I-10126 Torino, Italy

INTRODUCTION

In peripheral blood of normal donors a major fraction of γδ cells express a disulphide-linked (Cγ1) receptor with Vγ9 and Vδ2 gene products (Casorati 1989, Groh 1989). Since cells bearing Vγ9/Vδ2 TCR constitute only a minor fraction of γδ cells in the postnatal thymus (Casorati 1989) these cells might be selectively expanded in the periphery. The expansion of Vγ9/Vδ2 cells may occur after stimulation with antigen (Parker 1990), or it may be dependent on encountering ligands that bind to the Vγ or Vδ, irrespective of the specificities encoded by the V-(D)-J junctional sequences. To discriminate between these possibilities we investigated the antigen reactivity of Vγ9/Vδ2 bearing cells. Here we describe two frequent reactivities which correlate with the V gene usage and might explain expansion in peripheral blood.

MATERIALS AND METHODS

T cell cloning and assays.

T cell clones were established from peripheral blood of normal donors as reported (De Libero 1989). 72 hr proliferation assays were performed using $5x10^4$ responder cells/well and irradiated EBV-B ($3x10^4$/well) or PBMC (10^5/well) as APC. IL-2 was used at 5 U/ml. ^{3}H-Thymidine (Amersham, U.K.) was added in the last 18 hr. FACS analysis was performed with mAbs previously described (Casorati 1989).

Southern and Northern analyses.

Clones were studied for Vγ and Vδ gene rearrangements and expression by Southern and Northern analyses as already reported (Casorati 1989).

---

* This project was in part supported by the "CNR Progetto Finalizzato Biotecnologie e Biostrumentazioni". The Basel Institute for Immunology is founded and supported by Hoffmann La Roche, Ltd. Basel.

Current Topics in Microbiology and Immunology, Vol. 173

RESULTS AND DISCUSSION

A large fraction of peripheral blood γδ cells recognize Molt-4 lymphoma cells and M. tuberculosis-pulsed APC.

In order to find antigen reactivities of γδ cells, we generated 189 γδ T cell clones from peripheral blood using PHA and tested their ability to lyse different tumor target cells. Most of the clones were capable to lyse Molt-4 cells. Antibody-inhibition experiments with δ1 and TiγA mAbs, which recognize Cδ and Vγ9-encoded determinants respectively, completely abolished Molt-4 killing, thus suggesting that TCR is involved in Molt-4 recognition. Analysis with mAbs specific for TCR variable regions indicated that Molt-4 recognition is restricted to cells that express Vγ9 chain and that almost all Vγ9$^+$ cells from peripheral blood recognize Molt-4.

We next investigated the relative role of Vγ and Vδ chains in Molt 4 recognition using 31 clones which express "unusual" pairings of Vγ and Vδ chains (Casorati 1989). Analysis of these clones showed that, with only one exception, the common feature of the Molt-4 reactive cells is expression of Vγ9Cγ1. The Vδ chain seems to be less critical, since at least two different Vδ chains can pair with Vγ9Cγ1 and confer Molt-4 reactivity. Killing is not inhibited by anti-MHC class I and II or anti-CD1 mAbs.

The recognition of Molt-4 by human Vγ9$^+$ cells resembles the recognition of genetically determined Mls ligands by murine αβ cells expressing particular Vβ gene products (Abe 1989). In both cases the partner chain and junctional regions seems to be largely irrelevant. The molecular nature of the Molt-4 associated and Mls gene-controlled superantigens remains to be elucidated. Recently, Vγ9-related ligands have also been found to be expressed by Daudi cells (Fisch, 1990) and by staphilococcal enterotoxin A (SEA) (Rust, 1990). It is possible that cellular superantigens identical or related to the Molt-4 and Daudi superantigens, or microbial superantigens like SEA might play a role in the expansion of a subset of Vγ9$^+$ cells in the periphery. However, this mechanism cannot explain the preferential use of Vδ2 by circulating Vγ9$^+$ cells.

A second common specificity of Vγ9/Vδ2 cells was identified when PBMC were stimulated with heat-killed M. tuberculosis. Analysis with Vγ and Vδ specific mAbs showed that virtually all the proliferating γδ cells carried a Vγ9/Vδ2 receptor. A preferential expansion of γδ cells was not detected when PPD or tetanus toxin were used as stimulating antigens and γδ clones established from these cultures proliferated in response to APC pulsed with M. tuberculosis, but not to APC pulsed with PPD or tetanus toxoid. These cells also killed mycobacteria-pulsed APC (data not shown).

Molt4 and M. tuberculosis specificities are partially overlapping.

The TCR requirements for recognition of *M. tuberculosis*-pulsed APC were investigated with two sets of clones: one derived from *M. tuberculosis*-specific cell lines, and one from PHA-stimulated PBMC. All clones derived from *M. tuberculosis*-reactive cell lines are positive with TiγA and BB3 mAbs and therefore express a Vγ9/Vδ2 TCR. Furthermore, 13 clones were molecularly analyzed and all were characterized by a Vγ9-JP-Cγ1/Vδ2-J1 receptor. The molecular analysis thus confirmed that *M. tuberculosis*-reactive clones bear a Vγ9/Vδ2 receptor.

We next asked whether random (PHA-derived) Vγ9/Vδ2 clones would also recognize APC pulsed with *M. tuberculosis*. About 2/3 (11/16) of the Vγ9/Vδ2 tested clones proliferated to *M. tuberculosis*-pulsed APC, while none of the $V\gamma9^+/V\delta2^-$, or $V\gamma9^-/V\delta2^+$, or $V\gamma9^-/V\delta2^-$ clones showed proliferative responses to *M. tuberculosis* (table I).

These results indicate that expression of both Vγ9 and Vδ2 is required for recognition of *M. tuberculosis*-pulsed APC, but it is not sufficient, since 1/3 of the $V\gamma9^+/V\delta2^+$ clones do not proliferate to this stimulus. Involvement of both TCR chains resembles αβ TCR requirements for antigen recognition, and inappropriate junctional regions could account for lack of *M. tuberculosis* reactivity. In addition, these results also suggest that Molt-4 and *M. tuberculosis* reactivities are different and only partially overlapping. This latter inference was confirmed by the finding that *M. tuberculosis*-non reactive $V\gamma9^+/V\delta2^+$ clones recognize and kill Molt-4 (table I).

In conclusion, we have detected two specificities among Vγ9/Vδ2 cells in peripheral blood. The first is defined by the recognition of a tumor cell line, Molt-4, which appears to be a property of γδ cells expressing a TCR

Table I

Vγ9/Vδ2 clones recognize both Molt4 lymphoma and *M. tuberculosis*-pulsed APC

Number of clones	Vγ9	Vδ2	Recognition of Molt4	Recognition of APC+M.tub.
19	+	+	+	+
5	+	+	+	-
2	+	-	+	-
3	+	-	-	-
2	-	+	-	-
11	-	-	-	-

containing a Vγ9-Cγ1 chain. The second specificity is directed against antigen presenting cells pulsed with M. tuberculosis, and is characteristic of the majority, but not all, Vγ9/Vδ2 cells. While the Molt-4 specificity resembles that of cellular superantigens, recognition of mycobacteria-pulsed APC is more similar to that of a peptide associated with a restriction molecule. These findings show a correlation between V gene usage and antigen recognition in γδ cells and may help in understanding the mechanisms of specific expansion of Vγ9/Vδ2 cells in periphery.

REFERENCES

Abe, R., and R.J. Hodes. (1989) T-cell recognition of minor lymphocyte stimulating (Mls) gene products. Ann Rev Immunol 7:683.

Casorati, G., G. De Libero, A. Lanzavecchia, and N. Migone. (1989) Molecular analysis of human γ/δ^+ clones from thymus and peripheral blood. J Exp Med 170:1521.

De Libero, G. and A. Lanzavecchia. (1989) Establishment of human double-positive thymocyte clones. J Exp Med 170:303.

Fisch P., M. Malkovsky, E. Braakman, E. Sturm, R.L.H. Bolhuis, A. Prieve, J.A., Sosman, V.A. Lam, and P.M. Sondel. (1990) γδ T cell clones mediate distinct patterns of non-major histocompatibility complex-restricted cytolysis. J Exp Med 171:1567.

Groh, V., S. Porcelli, M. Fabbi, L.L. Lanier, L.J. Picker, T. Anderson, R.A. Warnke, A.K. Bhan, J.L. Strominger, and M.B. Brenner. (1989) Human lymphocytes bearing T cell receptor γδ are phenotypically diverse and evenly distributed thoroughout the lymphoid system. J Exp Med 169:1277.

Parker, C.M., V. Groh, H. Band, S.A. Porcelli, C. Morita, M. Fabbi, D. Glass, J.L. Strominger, and M.B. Brenner. (1990) Evidence for extrathymic changes in the T cell receptor γ/δ repertoire. J Exp Med 171:1612.

Rust, C.J.J., F. Verreck, H. Victor, and F. Koning (1990) Specific recognition of staphylococcal enterotoxin by human T cells bearing receptors with the Vγ9 region. Nature. 346:572.

Qa-1 Restricted γδ T Cells Can Help B Cells

D. VIDOVIC'[1] and Z. DEMBIC'[2]

[1] Basel Institute for Immunology, Grenzacherstraße 487, CH-4005 Basel
[2] Central Research Units, F. Hoffmann-La Roche Ltd., CH-4002 Basel

1. Introduction

T lymphocytes recognize antigens presented in the context of self MHC molecules via specific heterodimeric protein receptors (TCR) (Saito and Germain 1988). Two types of TCR have been described in vertebrates, e.g. αβ and γδ. In contrast to αβ T cells that recognize Ag in conjunction with classical class I and II MHC molecules and are responsible for Ag-specific cytotoxicity and help, respectively, the question regarding functions of γδ T cells remains basically unsolved.

It is well documented that relatively invariant antigens structurally similar to classical class I MHC molecules (Qa-1, Tla, CD1) serve as ligands for the γδ TCR, providing a restriction context for the presentation of foreign antigens (Bluestone et al. 1988, Vidovic' et al. 1989, Boneville et al. 1989, Porcelli et al. 1989). Although the majority of the γδ cell lines examined exhibit cytolytic activity, production of lymphokines has also been demonstrated (Raulet 1989). Our aim was to find out if Ag-specific γδ T cells were capable of helping B cells to produce specific antibodies.

2. Results

The proliferative T cell response of $H\text{-}2^d$ $Qa\text{-}1^b$ inbred mouse strains to the random copolymer poly (Glu^{50} Tyr^{50}) (GT) was found to fall into two categories: BALB/c mice respond only marginally, whereas DBA/2 mice mount a substantial proliferative reaction (Vidovic' et al. 1985). Two subsets of cells are present in GT-reactive DBA/2 populations. Cells expressing αβ TCR recognize processed GT in the context of class II MHC molecule A^d, while γδ cells react to unprocessed GT in association with $Qa\text{-}1^b$ (Vidovic' and Matzinger 1988, Vidovic' et al. 1989). BALB/c mice harbor a non-MHC gene Im^{gt}, the product of which in association with A^d mimics the GT/A^d complex. Cross-reactivity of DBA/2-derived GT/A^d-specific T cells to Im^{gt}/A^d-expressing BALB/c B cell blasts clearly demonstrates that the low BALB/c anti-GT responsiveness is due to lack of A^d restricted αβ T cells as a consequence of self-tolerance (Table 1).

Since GT/$Qa\text{-}1^b$-specific γδ T cells from DBA/2 (clone DGT3) did not cross-react with Im^{gt}/$Qa\text{-}1^b$ complex (Vidovic' 1989), we expected to detect such cells in low-responder BALB/c strain. Indeed, a marginal GT-specific proliferative response of BALB/c mice is selectively inhibited by $Qa\text{-}1^b$ specific antibodies (Figure 1). Moreover, from GT-primed BALB/c mice we obtained an IL-2 producing T cell hybridoma clone, CGT3, with a phenotype virtually indistinguishable from the DBA/2-originated DGT3 clone (e.g. it is GT/$Qa\text{-}1^b$-specific, $Thy1^+$, Ig^-, $CD4^-8^-$, MHC class I^+, class II^-, $TCR\beta^-$ $\gamma^+\delta^+$) (Figures 2 and 3). Therefore, the low proliferative response of BALB/c mice could be attributed to the presence of minor subpopulations of Qa-1-restricted γδ T cells. The discrepancy concerning CD8 expression on CGT3 (which is $CD8^-$) and bulk BALB/c GT-reactive populations (which are inhibited by CD8-specific antibodies) might be explained, either by heterogeneity of Qa-1 restricted populations (containing $CD8^+$ and $CD8^-$ cells), or by downregulation of CD8 molecules on fusion with the BW5147 thymoma (Carbone et al. 1988).

Although Qa-1 restricted γδ T cells did produce IL-2, we wanted to determine if they could help B cells to release antibodies. As shown in Figure 4, BALB/c mice that lack A^d-restricted αβ subpopulations did produce significant amounts of GT-specific antibodies after in vivo immunization

Current Topics in Microbiology and Immunology, Vol. 173

with GT. The titer was lower than in DBA/2 strains, which can be explained by the combined αβ and γδ helper T cell response of DBA/2 mice. The helping capabilities of Qa-1 restricted γδ T cells (hybridoma clones DGT3 and CGT3, as well as polyclonal BALB/c T cell populations) were confirmed by direct in vitro PFC assay. As shown in Table 2, their functional potential was indistinguishable from the help provided by A^d-restricted αβ T cells (clone DGT1).

3. Conclusion

The nature of the ligands recognized by γδ T cells as well as the function of these cells are not known. Structural similarities between αβ and γδ TCRs and the association of γδ T cell response to non-classical MHC molecules indicate that αβ and γδ ligands can be similar. Because of the (i) frequent cytolytic activity, (ii) smaller V segment receptor repertoire, (iii) association with epithelia, (iv) low polymorphism of the potential restriction elements (non-classical MHC), and (v) frequent reactivity to heat shock proteins, it has been suggested that γδ T cells might act as a first line of defense and respond to stress signals of damaged tissues, rather than recognizing a variety of foreign antigens (Janeway et al. 1988, Raulet 1989, Born et al. 1990). Recently it has been shown that γδ T cells that responded to a syngeneic B lymphoma could induce its differentiation and Ig secretion in the absence of the antigen (Sperling and Wortis 1990). In this report we describe γδ T cells that recognize foreign antigen GT complexed with a non-classical MHC molecule Qa-1, and are perfectly capable of helping B cells to produce GT-specific antibodies, supporting the notion that the possible physiological role of Qa-1-restricted recognition by gd subsets is not too distinct from classical MHC restricted recognition by αβ T cells.

4. Acknowledgements

We thank Dr. Leo Lefrancois for GL3 antibody, Ms. Chantal Guiet for excellent technical help, Drs. Antonio de la Hera and Charles Mackay for critical reading of the manuscript, Ms. Janette Millar for secretarial help and Mr. Hanspeter Stahlberger for the artwork. The Basel Institute for Immunology was founded and is supported by F. Hoffmann-La Roche Ltd., CH-4002 Basel, Switzerland.

5. References

Babu UM, Maurer PH (1981) The expression of anti-poly (l Glu^{60} l Phe^{40}) idiotypic determinants dictated by the gene products in the major histocompatibility complex. J Exp Med 154:649-658

Bluestone JA, Cron RQ, Cotterman M, Houlden BA, Matis LA (1988) Structure and specificity of T cell receptor γδ on major histocompatibility complex antigen-specific $CD3^+$, $CD4^-$, $CD8^-$ T lymphocytes. J Exp Med 168:1899-1916

Boneville M, Ito K, Krecko EG, Itohara S, Kapper D, Ishida I, Kanagawa O, Janeway CA Jr, Murphy DB, Tonegawa S (1989) Recognition of a self major histocompatibility complex TL region product by γδ T-cell receptors. Proc Natl Acad Sci USA 86:5929-5932

Born W, Happ MP, Dallas A, Reardon C, Kubo R, Shinnick T, Brennan P, O'Brien R (1990) Recognition of heat stock proteins and γδ cell function. Immunol Today 11:40-43

Carbone A, Marrack P, Kappler J (1988) Remethylation at sites 5' of the murine Lyt-2 gene in association with shutdown of Lyt-2 expression. J Immunol 141:1369-1375

Goodman T, Lefrancois L (1989) Intraepithelial lymphocytes. Anatomical site, not T cell receptor form dictates phenotype and function. J Exp Med 170:1569-1581

Janeway CA Jr, Jones A, Hayday A (1988) Specificity and function of T cells bearing γδ receptors. Immunol Today 9:73-76

Kappler JW (1974) A micro-technique for hemolytic plaque assay. J Immunol 112:1271-1274

Perucca PJ, Faulk WP, Fustenberg HH (1969) Passive immune lysis with chromic chloride-treated erythrocytes. J Immunol 102:812-820

Porcelli S, Brenner MB, Greenstein JL, Balk SP, Terhorst C, Bleicher PA (1989) Recognition of cluster of differentiation 1 antigens by human CD4- CD8- cytolytic T lymphocytes. Nature 341:447-450

Raulet DH (1989) The structure, function and molecular genetics of the γδ T cell receptor. Ann Rev Immunol 7:175-207

Saito T, Germain RN (1988) The generation and selection of the T cell repertoire: Insights from studies of the molecular basis of T cell recognition. Immunol Rev 101:81-113

Sperling AL, Wortis HH (1989) $CD4^-$, $CD8^-$ γδ cells from normal mice respond to a syngeneic B cell lymphoma and can induce its differentiation. Int Immunol 1: 434-442

Vidovic' D (1989) Elimination of self-tolerance turns nonresponder mice into responders. Immunogenetics 30: 194-199

Vidovic' D, Klein J, Nagy ZA (1985) Recessive T cell response to poly (Glu^{50} Tyr^{50}) possibly caused by self tolerance. J Immunol 134:3563-3568

Vidovic' D, Matzinger P (1988) Unresponsiveness to a foreign antigen can be caused by self-tolerance. Nature 336:222-225

Vidovic' D, Roglic' M, McKune K, Guerder S, Mackay C, Dembic' Z (1989) Qa-1 restricted recognition of foreign antigen by γδ T cell hybridoma. Nature 340:646-650

Voller A, Bidwell DE, Bartlett A (1979) The enzyme linked immunosorbent assay (ELISA). Dynatec, Guernsey

Table 1: Characteristics of different GT-reactive T cell subsets in H-2[d] Qa-1[b] mice

TCR	Specificity	Cross-reactivity	Presence in strain	
			DBA/2 (High responder)	BALB/c (Low responder)
$\alpha\beta$	GT/A[d]	Im[gt]/A[d] (BALB/c B cells)	+	-
$\gamma\delta$	GT/Qa-1[b]	-	+	+

Table 2: Frequency of positive responses in cultures containing B cells, antigens, SRBC and different sources of T cell help

Source of GT primed B cells	T cell help	Ag	SRBC normal	SRBC GT-coated
DBA/2	-	-	0/32	0/32
		GT	1/32	3/32
		OVA	1/32	1/32
	GT primed T cells	-	0/32	3/32
		GT	2/32	32/32
		OVA	1/32	2/32
	DGT1	-	1/32	2/32
		GT	0/32	32/32
		OVA	0/32	2/32
	DGT3	-	0/32	1/32
		GT	2/32	32/32
		OVA	1/32	1/32
BALB/c	-	-	0/32	0/32
		GT	0/32	0/32
		OVA	0/32	0/32
	GT primed T cells	-	0/32	1/32
		GT	1/32	30/32
		OVA	0/32	2/32
	CGT3	-	0/32	2/32
		GT	0/32	31/32
		OVA	0/32	2/32

Hemolytic plaque assay was performed in microwells as described by Kappler (1974). SRBC were coated with GT (20mg/ml) by treatment with chromic chloride (Perucca et al. 1969). OVA, ovalbumin.

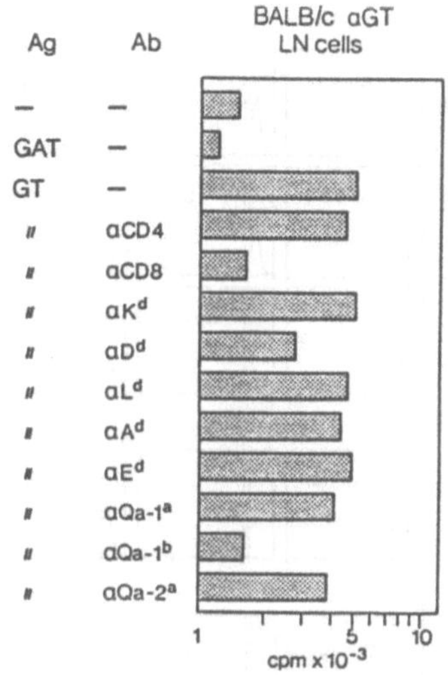

Figure 1: Antibody blocking of BALB/c proliferative response to GT.
Data represent proliferation of pooled lymph node cells of five GT-immunized mice. Purified polyclonal (Qa-1-specific) and monoclonal (all others) Ab were used at a final concentration of 5μg/ml. Materials and methods (including antibody sources) have been described previously (Vidovic' et al. 1985, Vidovic' et al. 1989). GAT, poly (Glu^{60} Ala^{30} Tyr^{10})

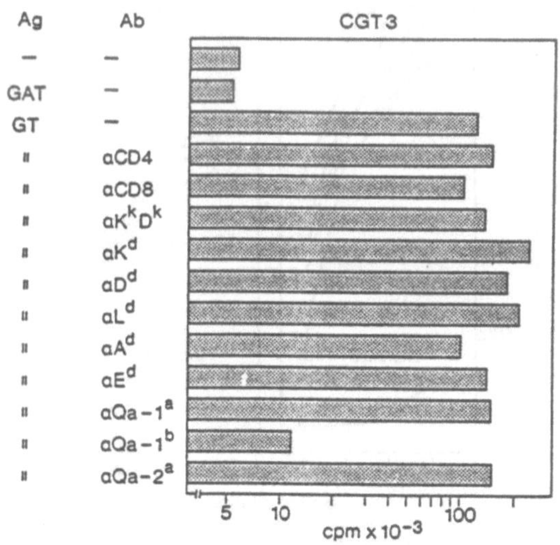

Figure 2: Antibody blocking of CGT3 response to GT.
Stimulaton of CGT3 was assayed by the production of IL-2 (Vidovic' et al. 1979).

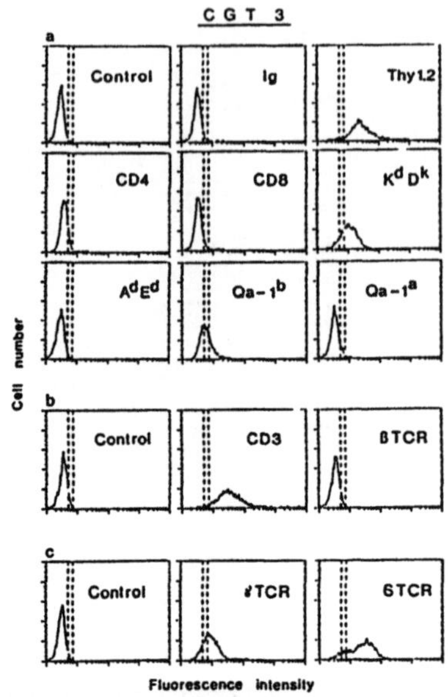

Figure 3: Cell surface phenotype of clone CGT3.
Immunofluorescence staining and antibody sources have been described previously (Vidovic' et al. 1989). Pan-γ and pan-δ TCR specific mAbs were obtained from B cell hybridomas GL3 (Goodman and Lefrancois 1989) and Dorotea (our laboratory, unpublished data), respectively.

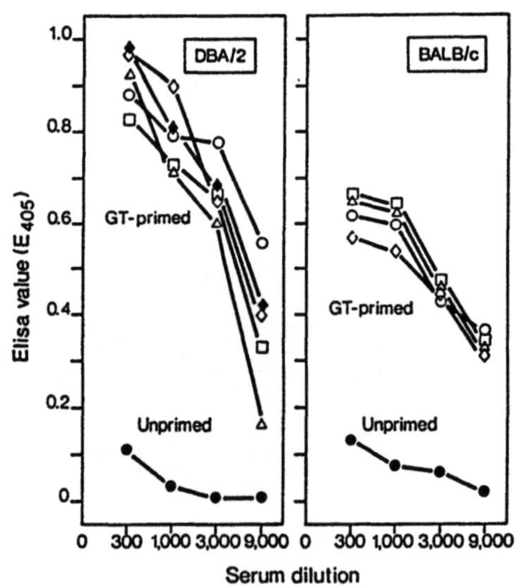

Figure 4: Humoral GT-specific responses.
Mice were immunized with GT as described (Babu and Maurer 1981). Individual sera samples were tested for GT reactivity by ELISA (Voller et al. 1979)). Background values (binding to third party antigen ovalbumin) have been subtracted.

CD5 B Cells

CD5 B Cells in the Mouse

I. FÖRSTER, H. GU, W. MÜLLER, M. SCHMITT, D. TARLINTON, and K. RAJEWSKY

Institute for Genetics, University of Cologne, Weyertal 121, D-5000 Cologne 41, FRG

Similar to γ/δ T cells in the T cell compartment, CD5 (Ly1) B cells represent a small subset of peripheral B cells which are thought to be involved in the maintenance of natural immunity. In this paper we discuss the main characteristics of CD5 B cells to compare them to those of γ/δ T cells. For this purpose, we focus mainly on the timepoint of CD5 B cell generation in ontogeny and on the antibody repertoire expressed by these cells.

1. Timing of CD5 B cell generation

CD5 B cells are among the first B cells generated in early ontogeny. While the absolute number of these cells increases with age, their frequency in the total B cell pool is highest at the time of birth and steadily decreases thereafter. Thus, in newborn mice about 30% of splenic B cells express low levels of the CD5 marker on the surface compared to less than 2% in the adult (Table 1). Similarly, in the peritoneal cavity (the predominant location of CD5 B cells in the adult) the frequency of CD5 B cells decreases from about 100% in neonatal mice to about 50% of all B cells in the adult. However, due to clonal expansion and development of chronic leukemia, the frequency of CD5 B cells often increases again in aged mice (see also Förster et al., 1988).

Table 1: Occurrence of CD5 B cells in normal mice

	Spleen		Peritoneal cells	
	% of* B220+ cells	total no.	% of B220+ cells	total no.
newborn	30	0.3×10^6	nd	
4 week-old	< 5	$< 2 \times 10^6$	85	0.3×10^6
adult	< 2	$< 2 \times 10^6$	50	1.6×10^6

* Values were obtained by fluorescence staining with anti-CD5 and anti-B220 antibodies in CB.20 or BALB/c mice.

Current Topics in Microbiology and Immunology, Vol. 173

Over the last years it has been argued that CD5 B cells represent a distinct B cell lineage because progenitors of CD5 B cells appeared to be different from progenitors of conventional B cells (reviewed in Herzenberg et al., 1986). This conclusion was mainly drawn from the experimental finding that - in contrast to conventional B cells - CD5 B cells could not be reconstituted from bone marrow of adult mice upon transfer into allotype-congenic irradiated recipients. A recent experiment by Vakil and Kearney indicates, however, that the capacity of adult bone marrow to reconstitute CD5 B cells is dependent on the environment into which the cells are transferred. Thus, transfer of bone marrow cells from adult BALB/c donors into newborn CB.17 scid mice led to reconstitution of CD5 B cells whereas transfer into adult scid mice did not (M. Vakil and J.F. Kearney, personal communication). Experiments by Lalor et al. (1989) indicated, however, that in intact mice de novo generation of CD5 B cells ceases by the age of 4-6 weeks due to some kind of feedback inhibition. In these experiments immunoglobulin (Ig) allotype heterozygous mice (IgH^axIgH^b) were neonatally treated with anti-IgM^b. After recovery from the treatment, the mice were able to produce conventional but not CD5 B cells of the b allotype if the anti-IgM^b antibodies were present for at least the first 4-6 weeks of life. In contrast, both B cell subsets were present after recovery when IgM^b homozygous mice were treated in the same way. It could be shown that normal mature CD5 B cells or even monoclonal populations such as a slowly growing Ly1 B tumor cell line present in the suppressed mice during recovery would inhibit the de novo generation of other CD5 B cells.

In agreement with the data described above, we have determined the timepoint of generation of CD5 B cells in normal mice by studying the frequency of N-sequence insertion in V_HDJ_H rearrangements of CD5 B cells (Gu et al., 1990). For this purpose, V_HDJ_H regions of CD5 B cells from various stages of ontogeny were either isolated from amplified cDNA libraries of sorted CD5 B cells or from CD5 B cell-hybridomas or -lymphomas (Förster et al., 1988; Pennell et al., 1988). In Table 2 a summary of these data is given. It is apparent

Table 2: Incidence of N-sequence insertion

cells	age	mean length of N regions (nucleotides) VD	DJ
pre B cells	2 days	0.9	0.7
	4 months	5.1	3.8
conventional B cells	4 weeks	4.6	2.8
	4 months	4.7	2.4
CD5 B cells	4 days	0.6	0.7
	4 weeks	2.2	1.9
	6-10 months	4.2	0.8
CD5 lymphomas	> 1 year	1.8	0.8

that CD5 B cells of 4 week-old mice possess shorter N-regions than conventional B cells of the same age. However, CD5 B cells are still generated at this time since in aged mice CD5 B cells with relatively long N-sequences can also be detected. Interestingly, the CD5 lymphomas contain very little N sequence and, for this and other reasons, are likely to be generated early in ontogeny (see Gu et al., 1990).

In summary, the data available so far show that CD5 B cells are generated around the time of birth and during the first weeks of life from stem cells which are or are not identical to those of conventional B cells. Thereafter the de novo generation of these cells stops and they are propagated as mature B cells over the lifetime of the animal.

2. Antibody repertoire

As far as immune responses of CD5 B cells have been analyzed, mostly specificities against bacterial antigens and self-antigens were detected. Studied in most detail are antibodies which bind to bromelain-treated mouse red blood cells (brmRBC), in particular to phosphatidylcholine (Ptc) (Reininger et al., 1988; Pennell et al., 1989; Carmack et al., 1990). Such antibodies arise spontaneously in mice and are almost exclusively produced by CD5 B cells. The only antigens against which specifically induced immune responses of CD5 B cells have so far been obtained in vivo are the bacterial antigens $\alpha(1,3)$dextran (dex) (Förster and Rajewsky, 1987) and phosphorylcholine (PC) (Masmoudi et al., 1990). In the latter case, the dominant idiotype in the immune response against PC (T15) was shown to be exclusively produced by CD5 B cells. It should be noted that the immune responses to dex and PC represent the classical examples of idiotypically interconnected immune responses (for review see Kearney and Vakil, 1986). It is therefore tempting to speculate that the antibody repertoire of CD5 B cells is selected by environmental antigens as well as autoantigens such as anti-idiotypic antibodies.

Concerning the Ig variable (V) gene repertoire expressed by CD5 B cells, we have previously shown that CD5 B cells which had been propagated in vivo over long periods of time expressed a highly restricted set of germline V_H and V_L genes (Förster et al., 1988). This selection was only visible looking at the representation of individual V_H genes but not at the level of V_H gene family utilization. We were then interested to see whether a similar restriction could also be observed in CD5 B cells at earlier stages of ontogeny and whether the V gene repertoire expressed by these cells was different from that of conventional B cells. In this analysis we focussed on the representation of V_H genes belonging to the J558 V_H gene family which comprises about half of all V_H genes of the murine IgH^b haplotype. Amplified cDNA libraries were prepared from the various B cell subsets at different stages of ontogeny and 10-20 cDNA clones containing J558 V_H genes were randomly picked from each library and were sequenced. The results showed that CD5 B cells isolated from 4-day- and 4-week-old mice expressed a similarly restricted set of V_H genes as CD5 B cells from aged animals, indicating that the selection of the CD5 B cell repertoire occurs already early in ontogeny (Förster et al., 1989). Surprisingly, conventional $\mu^+\delta^{high}$ B cells of adult mice, most of which appear to represent long-lived B cells (see Förster and

Rajewsky, 1990), also expressed a highly restricted V_H gene repertoire which was at least in part similar to that of CD5 B cells. Together with data obtained by other groups on the V_H gene repertoire of CD5 B cells (Pennell et al., 1988; Reininger et al., 1988; Tarlinton et al., 1988; Andrade et al., 1989; Carmack et al., 1990) these data indicate that CD5 B cells and long-lived conventional B cells are both selected on the basis of their antibody specificity but that this selection may vary with different organ location, activation requirements and growth properties of the cells (H. Gu, D. Tarlinton, W. Müller, K. Rajewsky and I. Förster, manuscript in preparation).

3. T cell dependency of CD5 B cells

The spectrum of antigen specificities in the CD5 B cell population and the absence of somatic hypermutation in CD5 B cells following clonal expansion suggest that these cells are activated in a T cell independent fashion. On the other hand, lymphokine responsiveness and production of T cell dependent Ig isotypes have also been described (Herzenberg et al., 1986; Wetzel, 1989). We have started to examine the T cell dependency of CD5 B cells by transferring the cells into scid mice in the presence or absence of T cells. In this system the high spontaneous Ig production by CD5 B cells which had been noticed before in other systems (Herzenberg et al., 1986; Förster and Rajewsky, 1987) was found to be largely dependent on the presence of T cells. Thus, 4 weeks following transfer of $3x10^5$ BALB/c CD5 B cells together with $6\text{-}8x10^4$ T cells we were able to detect in the order of 1.5 mg/ml IgM and about 10fold lower levels of IgA and IgG1 of the donor allotype in the serum of the reconstituted scid mice. However, in the presence of only trace amounts of T cells (<0.2% of the transferred CD5 B cells) the levels of serum IgM and IgA were reduced by a factor of 10 and the concentration of IgG1 was below 1 μg/ml. We are planning to use this system to test whether CD5 B cells can, in principle, be activated to respond to antigens in a T cell dependent manner.

4. Common features of CD5 B cells and γ/δ T cells

An obvious similarity between CD5 B cells and γ/δ T cells (at least those of the epithelial type) is that both lymphocyte subsets are mainly generated during early ontogeny and subsequently persist in certain anatomical locations like the gut, skin or peritoneal and pleural cavities over long periods of time. Since the antigen receptors of both CD5 B cells and γ/δ T cells may be mainly directed against bacterial antigens and self antigens and are encoded by a restricted set of germline V region genes, it is possible that these cells are responsible for the maintenance of evolutionarily selected natural immunity in the mouse.

Acknowledgements:

This work was supported by the Deutsche Forschungsgemeinschaft through SFB 243 and the FAZIT Foundation.

References:

Andrade L, Freitas AA, Huetz F, Poncet P, Coutinho A (1989) Eur. J. Immunol. 19: 1117-1122
Carmack CE, Shinton SA, Hayakawa K and Hardy RR (1990) J. Exp. Med. 172: 371-374
Förster I and Rajewsky K (1987) Eur. J. Immunol. 17: 521-528
Förster I, Gu H, Rajewsky K (1988) EMBO J. 7: 3693-3703
Förster I, Gu H, Rajewsky K (1989) Progress in Immunology VII, Springer-Verlag, Heidelberg, pp.389-393
Förster I and Rajewsky K (1990) Proc. Natl. Acad. Sci. USA 87: 4781-4784
Gu H, Förster I and Rajewsky K (1990) EMBO J. 9: 2133-2140
Herzenberg LA, Stall AM, Lalor PA, Sidman C, Moore WA, Parks DR, Herzenberg LA (1986) Immunol. Rev. 93: 81-102
Kearney JF, Vakil M (1986) Immunol. Rev. 94: 39-50
Lalor PA, Herzenberg LA, Adams S, Stall AM (1989) Eur. J. Immunol. 19: 507-513
Masmoudi H, Mota-Santos T, Huetz F, Coutinho A, Cazenave P-A (1990) Intern. Immunol. 2: 515-520
Pennell CA, Arnold LW, Haughton G, Clarke SH (1988) J. Immunol. 141: 2788-2796
Pennell CA, Mercolino TJ, Grdina TA, Arnold LW, Haughton G, Clarke SH (1989) Eur. J. Immunol. 19: 1289-1295
Reininger L, Kaushik A, Izui S, Jaton JC (1988) Eur. J. Immunol. 18: 1521-1526
Tarlinton D, Stall AM, Herzenberg LA (1988) EMBO J. 7: 3705-3710
Wetzel GD (1989) Eur. J. Immunol. 19: 1701-1707

Mucosal and Dermal Immunity

Selection of Vδ^+ T Cell Receptors of Intestinal Intraepithelial Lymphocytes is Dependent on Class II Histocompatibility Antigen Expression

L. LEFRANCOIS[1], R. LECORRE[1], JUDY MAYO[1], J. A. BLUESTONE[2], and T. GOODMAN[1]

[1] Dept. of Cell Biology, The Upjohn Company, Kalamazoo, MI 49001, USA
[2] Ben May Institute, University of Chicago, Chicago, Illinois 60637, USA

INTRODUCTION

The in vivo specificity of T cell antigen receptors (Tcr) composed of γ and δ chains remains largely unknown. In the mouse, a major population of γ-δ T cells is found in the CD8$^+$ intraepithelial lymphocyte (IEL) subset of the small intestine with 30-80% of the cells being Tcr γ-δ^+ (Goodman and Lefrancois, 1988; Bonneville et al., 1988). IEL are readily isolated and thus provide an optimal population for analyzing interactions with antigen and MHC in an unmanipulated system. We have demonstrated that freshly isolated IEL exhibit constitutive lytic activity directed towards unknown antigens (Goodman and Lefrancois, 1988). However, IEL from germ-free mice are not cytolytic suggesting that environmental antigens are responsible for their activation (Lefrancois and Goodman, 1989). Indeed, when germ-free mice are adapted to non-sterile conditions, IEL eventually become cytolytic. This result agrees well with the recent work indicating recognition of bacterial antigens by γ-δ Tcr$^+$ T cells (Janis et al., 1989; O'Brien et al., 1989; Holoshitz et al., 1989)

Many studies have clearly demonstrated positive and negative thymic selection by MHC or Mls of α-β Tcrs that use particular V-regions (Kappler et al.,1988; MacDonald et al.,1988; Kisielow et al.,1988; Kisielow et al., 1988; Sha et al.,1988; Bill et al., 1989; Liao et al., 1989). In normal animals, positive or negative selection of γ-δ Tcrs and any effects on γ-δ Tcrs by determinants encoded within the MHC have not been reported. Two groups have recently produced transgenic mice expressing γ-δ Tcrs with specificity for TL region products. In strains where the autoantigen was expressed thymic T cells expressing the transgenic Tcr were either deleted (Dent et al., 1990) or were made tolerant without deletion (Bonneville et al., 1990). Thus, γ-δ Tcrs can function in repertoire selection but whether the transgenic systems parallel normal in vivo differentiation patterns and Tcr specificities is unclear.

Current Topics in Microbiology and Immunology, Vol. 173

In order to analyze Tcr selection in IEL we have generated monoclonal antibodies (MAb) specific for γ-δ Tcrs (Goodman and Lefrancois, 1989). One of these MAb recognizes γ-δ Tcr expressing the δ chain variable region four (Vδ4). Using this MAb in conjunction with a MAb that is reactive with all γ-δ T cells we have analyzed the influence of MHC antigens on Vδ4 expression. Surprisingly, we find that certain class II MHC molecules have a positive effect on the usage of Vδ4 by $CD8^+$ IEL and that this effect occurred independently of the thymus. Thus, in analogy to selection of α-β Tcrs in the thymus, some γ-δ Tcrs also appear to be selected by MHC molecules, but in an anatomically distinct site. The consequences of this selection in the intestinal mucosa are discussed in light of CD8 expression by IEL and the proposed extrathymic maturation of IEL.

MATERIALS AND METHODS

Mice. Mice were obtained from the Jackson Laboratory, Bar Harbor, Maine. B10.A(5R) mice were provided by Dr. Louis Matis, FDA, and B10.HTT mice were provided by Dr. Philippa Marrack, NJH, Denver.

Production of radiation bone marrow chimeras. Bone marrow cells from tibias and femurs of $B6C3F_1$ mice were treated with anti-Thy1, anti-CD8, and anti-CD4 and complement prior to i.v injection of $5x10^6$ cells into C57BL/6J(Lyt2.2) or C3H/OuJ(Lyt2.1) mice that had received 1100rad from a ^{137}Cs source. Two days before use bone marrow donors were given a single intraperitoneal injection of the anti-Thy1 MAb, T24 to aid in removal of contaminating T cells. The recipient mice had either been thymectomized (ATXBM) or not. At the time of sacrifice, all thymectomized mice were examined and found to be free of thymus tissue.

Monoclonal antibodies. The anti-γ-δ Tcr MAb have been described in detail elsewhere (Goodman and Lefrancois, 1989). GL3 is a hamster MAb specific for all γ-δ Tcr. GL2 is a hamster MAb that reacts with a subset of $GL3^+$ IEL. Other MAb used were: 3.168, anti-CD8 (Sarmiento et al., 1982); H57.597, anti-α-β Tcr (Kubo et al., 1989), generously provided by Dr. Ralph Kubo, National Jewish Hospital, Denver, CO; 116-13.1 (anti-Lyt2.1; Shen, 1981); 2.43 (anti-Lyt2.2, Sarmiento et al., 1980).

Purification of IEL. IEL were isolated essentially as described previously (Goodman and Lefrancois, 1988). Briefly, the small intestines of 2-6 mice were cut into 5mm pieces that were then stirred at 37°C in Hank's balanced salt solution with the addition of 1mM dithioerythritol. The resulting supernatants containing a mixture of lymphocytes and epithelial cells were centrifuged through a 44%/67.5% Percoll gradient. Cells at the interface were panned on petri dishes coated with an anti-CD8 MAb. The final populations were 88-95% $CD8^+$.

Immunofluorescence analysis. IEL were resuspended in phosphate buffered saline (PBS)-0.2% bovine serum albumin-0.1% NaN_3 at a concentration of $1x10^7$ cells/ml followed by incubation at 4°C for 30min with 100ul of properly diluted MAb. The Mab were either directly labeled with fluorescein isothiocyanate (FITC) or were biotinylated. For the latter, avidin-phycoerythrin (Biomeda, Foster City, CA) was used for detection. After staining, cells were washed twice and resuspended at $1x10^6$ cells/ml for cytofluorimetric analysis. Relative fluorescent intensities of individual cells were measured with an Ortho Cytofluorograph model 50H. Forward and right angle light scatter was used to exclude dead and aggregated cells.

Production of IEL hybridomas. Fresh C57Bl/6J IEL or IEL that had been cultured after activation with an anti-Tcr γ-δ specific MAb ,were fused with the Tcr α gene negative variant of BW5147 with 50% polyethylene glycol at a IEL:BW5147 ratio of 1:1. After selection in medium containing hypoxanthine, aminopterin and thymidine (HAT) the resulting hybrids were analyzed for TCR γ-δ expression by fluorescence analysis using the GL3 and GL2 MAbs.

Northern blot analysis. RNA was isolated from fresh IEL or from IEL hybridomas by cell lysis with guanidine isothiocyanate followed by centrifugation over a $CsCl_2$ cushion. Northern blots were performed according to the method of Church and Gilbert (1984). 5µg of RNA/lane was loaded on agarose gels after glyoxal denaturation. The following cDNA fragments were used as probes: C-δ, 900bp EcoR1; Vδ1, 410bp Xbal/EcoRV; Vδ2, 410bp BglII; Vδ3, EcoR1/NcI1 270bp; Vδ4, 300bp EcoR1/EcoRV, Vδ5, 800bp HindIII; Vδ6, RI/AvaI 380bp as described by Elliot et al.(1988). (cDNAs were generously provided by Y. Chien, Stanford, CA). The probes were labelled with ^{32}P-dATP by random priming. Autoradiography was performed with Kodak XAR-5 film and Cronex intensifiers at -70°C.

RESULTS

Vδ4 is recognized by the GL2 MAb

We have previously shown that the GL3 MAb reacts with all non-α-β Tcr-expressing IEL (Goodman and Lefrancois, 1990). In addition GL3 reacted with all Tcr γ-δ^+ T cell hybrids derived from non-IEL sources. GL3 also immunoprecipitated disulfide-linked dimers that were shown to be composed of γ and δ chains. Therefore this MAb is pan-reactive with γ-δ Tcrs. The GL2 MAb immunoprecipitates from IEL only those γ-δ Tcrs that contain the higher relative mass δ chain of 46Kd. Two T cell hybrids derived from non-IEL cell types that utilize Vδ4 are $GL2^+$, tentatively assigning the GL2 determinant to Vδ4 (Goodman and Lefrancois, 1990). We have now produced a panel of IEL hybridomas of $GL2^+$ and $GL2^-$ phenotype. Northern blot analysis with Vδ probes revealed that all of the $GL2^+$ hybrids expressed Vδ4 while hybrids not containing Vδ4 mRNA were $GL2^-$ (Fig. 1). γ chain usage did not correlate with GL2 expression (data not shown). We conclude that this MAb is specific for a determinant in the Vδ4 region of the δ chain.

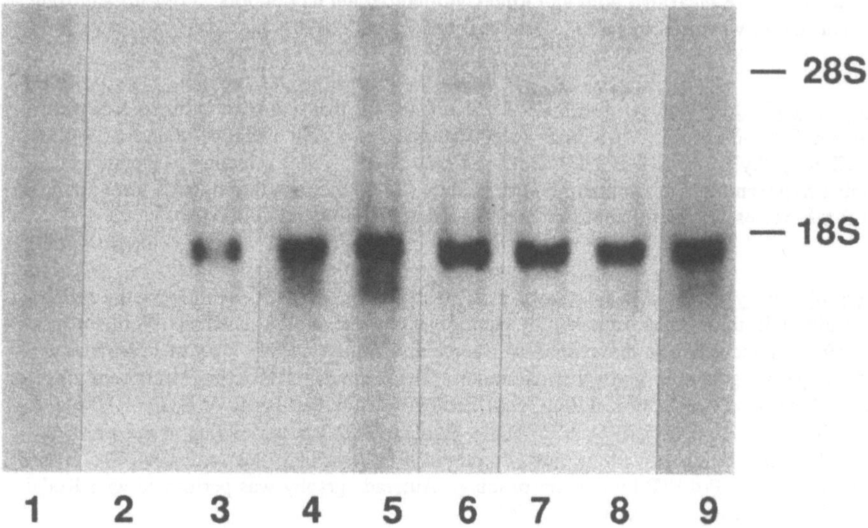

Fig. 1. GL2 reacts with Tcrs using Vδ4. RNA from IEL hybrids was subjected to Northern blot analysis using ^{32}P-labelled probe specific for Vδ4. Lane 1, BW5147 fusion partner, Lane 2, GL2^{-} IEL hybrid and lanes 3-9, GL2^{+} IEL hybrids.

IEL utilize Vδ4 to a high degree

We tested mRNA from C57Bl/6J IEL for usage of the various Vδ genes (Fig. 2). Little usage of Vδ1,2 or 3 genes was detected (lanes 1-3). However, relatively large quantities of Vδ4 mRNA were present (lane 4). The Vδ regions encoded by V5 and V6 were also utilized but to a lesser extent than V4 (lanes 5 & 6). Thus, Vδ4 appears to be the predominant Vδ region utilized by γ/δ IEL.

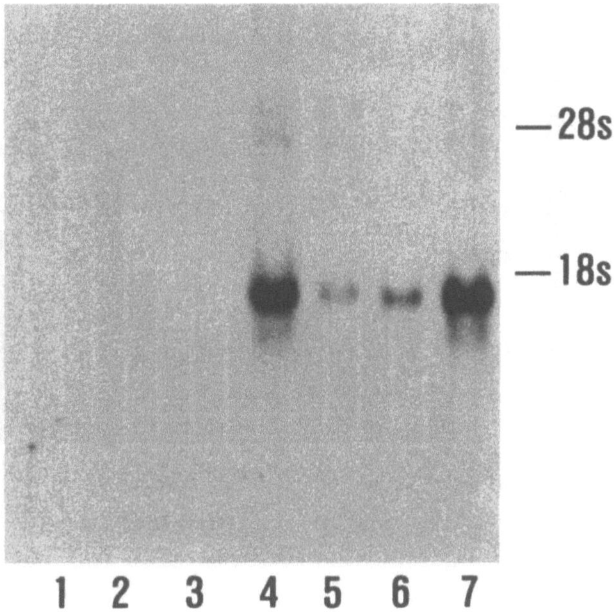

Fig. 2. Northern blot analysis of Vδ usage in IEL. RNA from C57B1/6J IEL was subjected to Northern blot analysis using ^{32}-P labelled Vδ region-specific cDNA probes. Lane 1, Vδ1; lane 2, Vδ2; lane 3, Vδ3; lane 4, Vδ4; lane 5, Vδ5; lane 6, Vδ6; lane 7, Cδ.

Comparison of Vδ4 usage in IEL from various MHC haplotypes

Since GL2 was Vδ4 specific we wished to determine if MHC expression affected usage of this V-region. Our previous results from C57Bl/6J IEL indicated that approximately 30% of Tcr γ-δ^+ IEL were $GL2^+$. In order to determine if GL2 expression was influenced by MHC we analyzed Tcr expression of IEL isolated from a large panel of inbred mice of various haplotypes (Fig. 3). The percentage of $GL2^+$ IEL is expressed as the proportion of $GL3^+$ cells. IEL from mice of $H\text{-}2^{b,d,u,v}$ haplotypes contained an average of 29+/-8% $GL2^+$ cells although a high degree of variability even among strains of identical MHC was evident, suggesting that other non-MHC genes may effect Vδ4 usage in some situations. In contrast, the $H\text{-}2^k$ strains analyzed exhibited markedly higher numbers of $GL2^+$ IEL (Fig. 3). The average percentage of $GL2^+$ IEL of the $H\text{-}2^k$ strains shown in Fig. 2 was 57+/-5%. Among $H\text{-}2^k$ strains CBA/J exhibited a relatively high degree of variability, with 51+/-9%, n=7, of IEL being $GL2^+$ with values ranging from 35-62%. Thus, as with the non-$H\text{-}2^k$ strains, non-MHC genes may influence Vδ4 use.

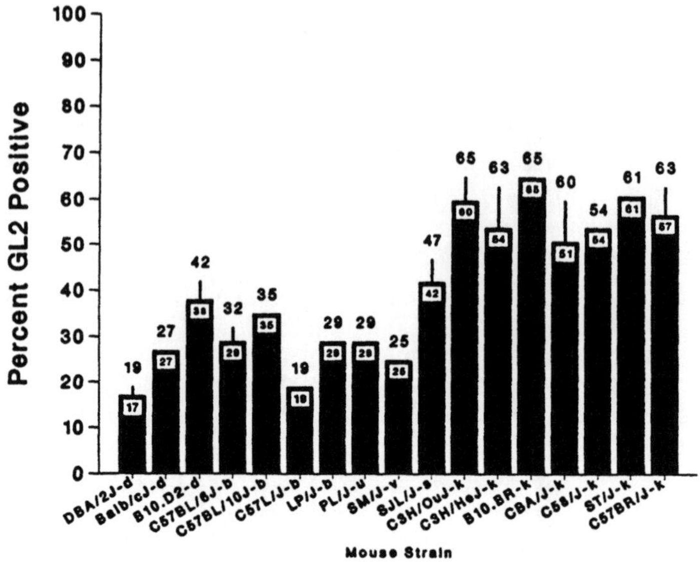

Fig. 3. Analysis of Vδ4 usage in inbred mice. IEL were isolated from the mouse strains indicated and were analyzed for GL2 and GL3 expression by fluorescence flow cytometry. The values for GL2 are presented as the percent of GL3$^+$ cells. Where standard deviations are shown at least three determinations were performed, otherwise an average of two experiments is given. Two to six mice per determination were used.

Analysis of IEL from recombinant inbred mice

Considering that positive selection of Vδ4 appeared to be linked to H-2^k expression we wished to determine if in fact the effect mapped to the MHC and whether genes outside the MHC could alter the GL2 percentage. Thus, IEL from a set of H-2^k x H-2^b recombinant inbred strains (BXH) were tested (Fig. 4). Of 12 strains available six were H-2^k and a high percentage of IEL (66+/-5%) from all of these utilized Vδ4. In contrast, IEL from the six H-2^b strains were of the GL2 low phenotype (20+/-4%). Thus, the selection mapped to the MHC. Furthermore, the fact that no intermediate Vδ4 users were found suggested that non-MHC genes in these strains were not influencing Vδ4 usage in this limited analysis.

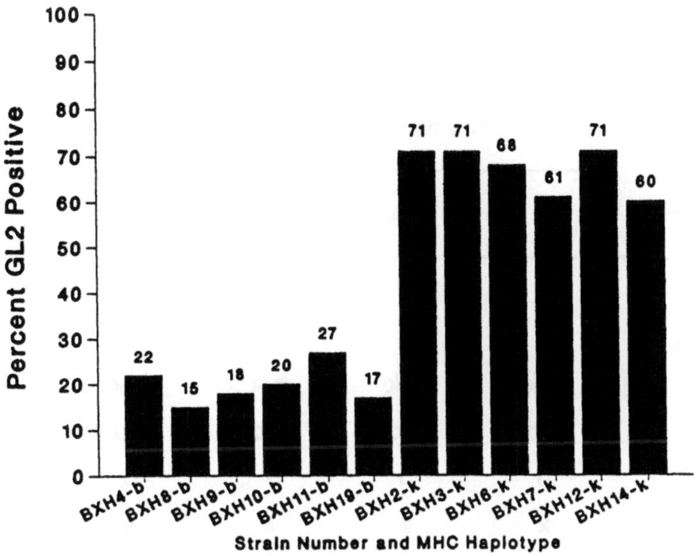

Fig. 4. Increased Vδ4 usage is linked to H-2^k expression. IEL were isolated from the BXH strains indicated and were analyzed for GL2 and GL3 expression by fluorescence flow cytometry. The values for GL2 are presented as the percent of $GL3^+$ cells. The average value of two determinations per strain using 1-3 mice per experiment is given.

Selection of Vδ4 TCRs is dependent on I-E expression

A more precise mapping of the MHC effect was performed using IEL from various strains recombinant at the MHC. When IEL from C3H.SW (H-2^b) mice were analyzed the $GL2^{lo}$ phenotype was observed indicating that non-MHC genes from a $GL2^{hi}$ strain (C3H) did not cause increased Vδ4 expression (Table 1). Conversely, IEL from B6.AKR (H-2^k) mice contained high numbers of $GL2^+$ cells indicating that H-2^k expression was required to generate the high phenotype. These results corroborate our findings with IEL from RI strains. We also tested IEL from B6.TLA^a mice and found low Vδ4 usage. Since this strain expresses the TL products found in at least some of the H-2^k strains tested, TL by itself does not appear to be involved in the selection, but could be involved if the selection requires more than one selecting element.

Further analysis of IEL from B10 congenic strains was performed to define the MHC region(s) required for selection. IEL from B10.A and B10.A(2R) mice had high numbers of Vδ4^+ cells (63%,n=2 and 57%, n=2, respectively) while IEL from B10.A(4R) mice contained lower numbers of Vδ4^+ cells (32+/-5%, n=3; Table 1). Since H-2D does not appear to contribute to the selection the only other difference between these strains is at the I-E locus. B10.A and B10.A(2R) express I-E^k and B10.A(4R) does not express I-E suggesting that expression of I-E^k was responsible for the effect. Further proof that increased Vδ4 usage mapped to I-E was obtained by the demonstration that B10.A(5R) IEL exhibited the $GL2^{hi}$ phenotype (58+/-3%, n=3). B10.A(3R) whose MHC is identical to B10.A(5R) also had high numbers of Vδ4^+ IEL (data not shown). In these strains the I-E molecule is composed of a β chain derived from H-2^b and an α chain derived from H-2k. IEL from B10.HTT mice whose I-E molecules are composed of EBs and EαK chains also had high levels of $GL2^+$ cells. The results indicated that the expression of I-E from H-2 b,k or s haplotypes was required for selection of γ-δ TCRs utilizing Vδ4. Since Eα chains are nonpolymorphic amongst mouse strains, (McNicholas et al, 1982; Mathis et al, 1983) then any Tcr variations due to I-E expression must be linked to the polymorphic Eβ chains.

Table 1. Increased expression of Vδ4 is controlled by I-E

Strain	H-2 Molecules Expressed				Vδ4 High Phenotype
	K	I-A	I-E	D	
C3H	k	k	k	k	+
C57Bl/6	b	b	-	b	-
B6C3F$_1$	b/k	b/k	b/k	b/k	+
C3D2F$_1$	k/d	k/d	k/d	k/d	+
C3H.SW	b	b	-	b	-
B6.AKR	k	k	k	k	+
B6.TLA[a]	b	b	-	b	-
B10.A	k	k	k	d	+
B10.A (2R)	k	k	k	b	+
B10.A (4R)	k	k	-	b	-
B10.A (5R)	b	b	$E_\beta^b:E_\alpha^k$	d	+
B10.HTT	b	s	$E_\beta^s:E_\alpha^k$	d	+

IEL were isolated from the mouse strains listed and analyzed with the GL3 and GL2 MAb by fluorescence flow cytometry.

Increased Vδ4 usage in F_1 hybrid mice suggests positive selection

Thus far the results are consistent with either positive selection of Tcrs using Vδ4 in the H-2^k strains or negative selection in H-2^b and other strains. However, Vδ4$^+$ cells comprised 47+/-0.6% (n=4) of IEL obtained from B6C3F_1 and 53% of IEL from C3D2F_1 mice (n=2) (Table 1). These hybrids are derived from a Vδ4 high (C3H) and two Vδ4 low strains (B6 & DBA\2). This result was consistent with positive selection since deletion would be expected to occur if the deleting element was present and in an F_1 the positively selecting component, if expressed codominantly, may be effectively "diluted" by 50%.

Selection of γ-δ TCRs can occur extrathymically

Previous findings had suggested that some IEL may mature extrathymically (Ropke and Everett, 1975; Mayrhofer, 1980; Mayrhofer and Whately, 1983). In addition, two recent reports demonstrate that in nude mice IEL are present that are expressing γ-δ Tcrs (De Geus et al. 1990, Bonneville et al. 1990). We wished to determine whether γ-δ Tcr$^+$ IEL could mature and be selected via an extrathymic mechanism. Bone marrow from (B6 x C3H)F^1 mice was T cell depleted and used to reconstitute lethally irradiated B6 or C3H mice that had either been thymectomized or not. TCR γ-δ$^+$ IEL could be isolated from all groups of mice, regardless of thymectomy, although in some cases thymectomized mice yielded significantly fewer γ-δ$^+$ IEL than controls (range 20-80% of controls). IEL from the B6 and the B6-thymectomized mice contained 15% and 26% GL2$^+$ cells, respectively (Fig. 5). In contrast, IEL from the C3H and C3H-thymectomized mice contained 54% and 65% GL2$^+$ cells, respectively. Thus, positive selection of γ-δ Tcrs can occur extrathymically. Moreover, these results demonstrate that the Tcr γ-δ$^+$ IEL isolated from these mice were not derived from mature T cells contaminating our bone marrow inoculum, since those cells would be expected to display the Vδ4 usage of the F_1 donor (47% of F_1 IEL were GL2$^+$) in either recipient.

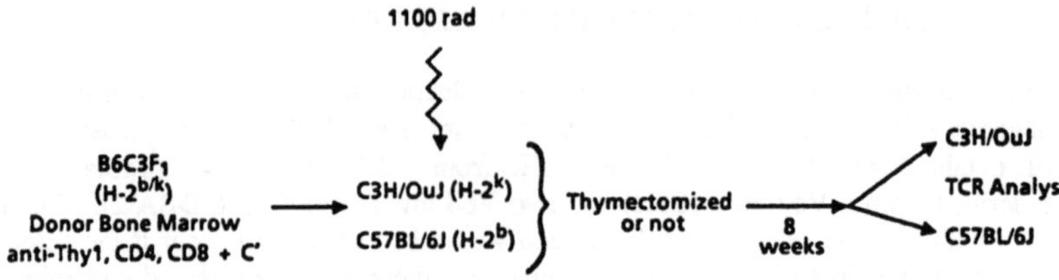

	% of TCR γ,δ^+ IEL using Vδ4
$B6C3F_1$	47
$F_1 \rightarrow B6$	15
$F_1 \rightarrow B6\text{-}Tx$	23
$F_1 \rightarrow C3H$	54
$F_1 \rightarrow C3H\text{-}Tx$	65

Fig. 5. Positive selection of Vδ4 Tcrs can occur in the absence of a thymus. Bone marrow from $B6C3F_1$ mice was injected into irradiated C3H/OuJ or C57BL/6J mice as depicted. The values for Vδ4 are presented as the percent of $GL3^+$ cells. The values represent the average of two determinations using 2-3 mice per group.

DISCUSSION

Although evidence exists for selection during T cell development of α-β Tcrs, negative or positive selection of γ-δ Tcrs by MHC had not been reported. In the case of α-β Tcrs the thymus has been clearly demonstrated to be the site of α-β Tcr selection and the thymic epithelium is likely to be involved in this process (Kappler et al., 1988; Berg et al., 1989; Benoist and Mathis, 1989; Bill and Palmer, 1989). In addition, the α-β Tcr repertoire may be influenced by clonal anergy induced in the the thymus or in the periphery (Morahan et al., 1989; Ramsdell et al., 1989; Burkly et al., 1989). For γ-δ Tcrs selection has not been readily demonstrated perhaps due to limited V-region usage and in some cases, lack of reagents. In mice lacking class I MHC due to disruption of the B-2 microglobulin gene, thymic γ-δ T cells were normal suggesting that class I molecules are not necessary for their development (Ziljstra et al., 1990). Furthermore, $CD8^+$ Tcr $\gamma\text{-}\delta^+$ IEL also appear to be normal in these mice (Raulet, personal communication). Our results demonstrating selection of γ-δ Tcrs of IEL by class II MHC agree well with these findings. It will be interesting to learn if $\gamma\text{-}\delta^+$ T cells in other anatomic locations exhibit similar restriction or are MHC non-restricted.

IEL had been previously shown to mature extrathymically in irradiated thymectomized bone marrow reconstituted (ATXBM) rats (Mayrhofer et al. 1980; Mayrhofer and Whately, 1983) and are present in nude mice (Ropke and Everett, 1975; Mayrhofer, 1980). IEL Tcr expression was recently analyzed in nude mice and the results indicate that at least some γ-δ Tcr⁺ IEL mature in these animals (De Geus et al. 1990, Bonneville et al, 1990). Our results with ATXBM mice extend these findings and further demonstrate that γ-δ Tcr selection can occur extrathymically. Does this selection occur in response to epithelial cell I-E expression in analogy to thymic Tcr selection? Constitutive class II protein expression in non-hematopoietic tissues is primarily, but not exclusively, restricted to epithelium. Class II MHC expression of intestinal epithelium can be up-regulated by IEL derived factors (Cerf-Bensussan et al., 1984) and intestinal epithelial cells are able to present antigens to class II-restricted T cells (Bland and Warren, 1986; Kaiserlian et al., 1989). Thus, in the absence of thymic trafficking by IEL it seems likely that the positive selection observed occurred within the intestinal epithelium. Whether selection occurs in a localized site, for example the villous crypts, or in diffuse sites throughout the villi, or in a site other than the intestine, remains to be learned. The use of tissue specific promoters in production of class II MHC transgenics should aid in answering these questions.

The apparent restriction to class II MHC of a large population of $CD8^+$ IEL is unique among the Tcr γ-δ^+ subsets described thus far. Whether CD8 of IEL is utilized for antigen/MHC recognition in the conventional sense is not known. Interestingly, $CD8^+$ class II restricted peripheral T cells, that presumably utilize the α-β Tcr, have been reported (Shinohara and Kojima, 1984; Golding et al., 1987). However, these cells do not require interaction with class II molecules to mature (Mizuochi et al., 1988) but it is not known if engagement of CD8 by class I MHC is required.

In sum, the γ-δ Tcr repertoire of IEL was influenced by extrathymic MHC protein expression, perhaps in the intestinal epithelium, in analogy to α-β Tcr selection on thymic epithelium. Although IEL were $CD8^+$, the expression of certain I-E molecules resulted in positive selection of a subset of Vδ4-containing Tcrs. The consequences of the possible selection of the IEL Tcr repertoire within the intestinal epithelium by class II MHC, rather than in the thymus, are presently under investigation.

REFERENCES

Asarnow, DM, Goodman, T, Lefrancois, L, Allison, JP (1989) Distinct antigen receptor repertoires of two classes of murine epithelium-associated T cells. Nature 341:60-62.

Asarnow, DM, Kuziel, WA, Bonyhadi, M, Tigelaar, RE, Tucker, PW, Allison, JP (1988) Limited diversity of γ,δ antigen receptor genes of $Thy1^+$ dendritic epidermal cells. Cell 55:837-847.

Benoist, C, Mathis, D (1989) Positive selection of the T cell repertoire: Where and when does it occur? Cell 58:1027-1033.

Berg, LJ, Pullen, AM, Fazekas de St. Groth, B, Mathis, D, Benoist, C, Davis, MM (1989) Antigen/MHC-specific T cells are preferentially exported from the thymus in the presence of their MHC ligand. Cell 58:1035-1046.

Bill, J, Palmer, E (1989) Positive selection of $CD4^+$ T cells mediated by MHC class II-bearing stromal cell in the thymic cortex. Nature 341:649-651.

Bland, PW, Warren, LG (1986) Antigen presentation by epithelial cells of the rat small intestine I. Kinetics, antigen specificity and blocking by anti-Iα antisera. Immunology 58:1-7.

Bonneville, M, Itoharah, S, Krecko, EG, Mombaerts, P, Ishida, I, Katsuki, M, Berns, A, Farr, AG, Janeway, CA, Tonegawa, S (1990) Transgenic mice demonstrate that epithelial homing of γ,δ T cells is determined by cell lineages independent of T cell receptor specificity. J Exp Med 171:1015-1026.

Bonneville, M, Janeway, CA, Ito, K, Haser, W, Ishida, I, Nakanishi, N, Tonegawa, S (1988) Intestinal intraepithelial lymphocytes are a distinct set of γ,δ T cells. Nature (London) 336:479-481.

Burkly, LC, Lo, D, Kanagawa, O, Brinster, RL, Flavell, R (1989) T-cell tolerance by clonal anergy in transgenic mice with non-lymphoid expression of MHC class II I-E. Nature 342:564-566.

Cerf-Bensussan, N, Quaroni, A, Kurnick, JT, Bhan, AK (1984) Intraepithelial lymphocytes modulate Ia expression by intestinal epithelial cells. J Immunol 132: 2244-2252.

Church, GM, Gilbert, W (1984) Genomic sequencing. Proc Natl Acad Sci 81:1991-1995.

Cron, RQ, Gajewski, TF, Sharrow, SO, Fitch, FW, Matis, LA, and Bluestone, JA (1989) Phenotypic and functional analysis of murine $CD3^+$, $CD4^-$, $CD8^-$ TCR-γ,δ-expressing peripheral T cells. J Immunol 142:3754-3762.

Elliot, JF, Rock, EP, Patten, PA, Davis, MM, Chien, Y (1988) The adult T-cell receptor D-chain is diverse and distinct from that of fetal thymocytes. Nature 331: 627-631.

Goodman, T, Lefrancois, L (1988) Expression of the γ,δ T cell receptor on intestinal $CD8^+$ intraepithelial lymphocytes. Nature (London) 333:855-858.

Goodman, T, Lefrancois, L (1989) Intraepithelial lymphocytes. Anatomical site, not T cell receptor form, dictates phenotype and function. J Exp Med 170:1569-1581.

Holoshitz, J, Koning, F, Coligan, JE, De Bruyn, J, Strober, S (1989) Isolation of $CD4^-8^-$ mycobacteria-reactive T lymphocyte clones from rheumatoid arthritis synovial fluid. Nature 339:226-229.

Ito, K, Bonneville, M, Takagaki, Y, Nakanishi, N, Kanagawa, O, Grecko, EG, Tonegawa, S (1989) Different γ,δ T-cell receptors are expressed on thymocytes at different stages of development. Proc Natl Acad Sci USA 86:631-635.

Janis, EM, Kaufmann, SHE, Schwartz, RH, Pardoll, DM (1989) Activation of γ,δ T cells in the primary immune response to Mycobacterium tuberculosis. Science 244: 713-716.

Kaiserlian, D, Vidal, K, Revillard, J-P (1989) Murine enterocytes can present soluble antigen to specific class II-restricted $CD4^+$ T cells. Eur J Immunol 19:1513-1516.

Kappler, JW, Roehm, N, Marrack, P (1988) T cell tolerance by clonal elimination in the thymus. Cell 49:273-280.

Kappler, JW, Staerz, U, White, J, Marrack, PC (1988) Self-tolerance eliminates T cells specific for Mls-modified products of the major histo-compatibility complex. Nature 332:35-40.

Kisielow, P, Bluthmann, H, Staerz, UD, Steinmetz, M, von Boehmer, H (1988) Tolerance in T-cell-receptor transgenic mice involves deletion of nonmature $CD4^+8^+$ thymocytes. Nature 333:742-746.

Kisielow, P, Teh, HS, Bluthmann, H, von Boehmer, H (1988) Positive selection of antigen-specific T cells in thymus by restricting MHC molecules. Nature 335:730-733.

Kubo, R, Born, W, Kappler, JW, Marrack, P, Pigeon, M (1989) Characterization of a monoclonal antibody which detects all murine a,B T cell receptors. J Immunol 142:2736-2742.

Lafaille, JJ, Haas, W, Coutinho, A, Tonegawa, S (1990) Positive selection of gd T cells. Immunology Today 11:75-78.

Lefrancois, L, Goodman, T (1989) <u>In vivo</u> modulation of cytolytic activity and Thy-1 expression in TCR-γ,δ^+ intraepithelial lymphocytes. Science (Washington, DC) 243: 1716-1718.

Liao, N, Maltzman, J, Raulet, DH (1989) Positive selection determines T cell receptor Vβ14 gene usage by $CD8^+$ T cells. J Exp Med 170:135-143.

Macdonald, HR, Schneider, R, Lees, RK, Howe, RC, Acha-Orbea, H, Festenstein, H, Zinkernagel, RM, Hengartner, H (1988) T-cell receptor V_B use predicts reactivity and tolerance to Mls^a-encoded antigens. Nature 332:40-45.

MacDonald, HR, Schreyer, M, Howe, RC, Bron, C (1990) Selective expression of CD8a(Lyt-2) subunit on activated thymic γ,δ cells. Eur J Immunol, in press.

Mayrhofer, G (1980) Thymus-dependent and thymus-independent subpopulations of intestinal intraepithelial lymphocytes: A granular subpopulation of probable bone marrow origin and relationship to mucosal mast cells. Blood 55:532-535.

Morahan, G, Allison, J, Miller, JFAP (1989) Tolerance of class I histocompatibility antigens expressed extrathymically. Nature 339:622-624.

O'Brien, RL, Happ, MH, Dallas, A, Palmer, E, Kubo, R, Born, WK (1989) Stimulation of a major subset of lymphocytes expressing T cell receptor γ,δ by an antigen derived from Mycobacterium tuberculosis. Cell 57:667-674.

Pardoll, DM, Fowlkes, BJ, Bluestone, JA, Kruisbeek, A, Maloy, WL, Coligan, JE, Schwartz, R (1987) Differential expression of two distinct T-cell receptors during thymocyte development. Nature 326:79-81.

Parrott, DMV, Tait, C, MacKenzie, S, Mowat, AM, Davies, MDJ, Micklem, HS (1983) Analysis of the effector functions of different populations of mucosal lymphocytes. Ann NY Acad Sci 77:307-319.

Ramsdell, F, Lantz, T, Fowlkes, BJ (1989) A nondeletional mechanism of thymic self tolerance. Science 246:1038-1041.

Ropke, C, Everett, NB (1975) Kinetics of intraepthelial lymphocytes in the small intestine of thymus-deprived and antigen-deprived mice. Anat Rec 185:101-108.

Sarmiento, M, Glasebrook, AL, Fitch, FW (1982) IgG or IgM monoclonal antibodies reactive with the different determinants on the molecular complex bearing Lyt-2 antigen block T-cell mediated cytolysis in the absence of complement. J Immunol 125:2665-2672.

Sha, WC, Nelson, CA, Newberry, RD, Kranz, DM, Russell, JH, Loh, DY (1988) Positive and negative selection of an antigen receptor on T cells in transgenic mice. Nature 336:73-76.

Shen, FW (eds Hammerling, GJ, Hammerling, U, Kearney, JF) (1981) In: Monoclonal antibodies and T cell hybridomas, (Elsevier, Amsterdam) pp. 25-31.

Takagaki, Y, DeCloux, A, Bonneville, M, Tonegawa, S (1989) Diversity of γ,δ T-cell receptors on murine intestinal intraepithelial lymphocytes. Nature 339:712-714.

Ziljstra, M, Bix, M, Simister, NE, Loring, JM, Raulet, DH, Jaenisch, R (1990) B2-Microglobulin deficient mice lack $CD4^-8^+$ cytolytic T cells. Nature 344:742-746.

The Role of Fetal Epithelial Tissues in the Maturation/Differentiation of Bone Marrow-Derived Precursors into Dendritic Epidermal T Cells (DETC) of the Mouse

G. STINGL, ADELHEID ELBE, ELISABETH PAER, O. KILGUS, R. STROHAL, and SUSANNE SCHREIBER

Division of Cutaneous Immunology, Department of Dermatology I, University of Vienna Medical School, VIRCC, Brunner-Straße 59, A-1235 Vienna

INTRODUCTION

In addition to Langerhans cells, the epidermis of all strains of normal mice harbors another population of bone marrow-derived dendritic cells. Because of their abundant expression of surface-bound Thy-1 antigens these $CD45^+$, asialo-$GM1^+$, Ia^-, $CD4^-$, $CD5^-$, $CD8^-$ cells have originally been referred to as Thy-1^+ dendritic epidermal cells (Thy-1^+ DEC; Romani et al., 1985). The original assumption that Thy-1^+ DEC belong to the T cell lineage was proven conclusively with the subsequent demonstration that the overwhelming majority of Thy-1^+ DEC in normal adult epidermis expressed CD3-associated TCR γ/δ heterodimers (Stingl et al., 1987; Steiner et al., 1988). In view of this finding, we have proposed that Thy-1^+ DEC should be renamed dendritic epidermal T cells (DETC; Steiner et al., 1988).

When investigators from several laboratories started to analyze DETC-derived clones for their profiles of γ and δ gene utilization, they made the surprising observations that - with a few notable exceptions (McConnell et al., 1989) - the clones expressed similar γ ($V\gamma3/J\gamma1/C\gamma1$) and δ ($V\delta1/D\delta2/J\delta2/C\delta$) genes, and, furthermore, exhibited no evidence for junctional diversity (Asarnow et al., 1988). More recently, evidence has been presented that (1) most, if not all, CD3-bearing cells in uncultured murine epidermal cell suspensions coexpressed $V\gamma3$ (Havran et al., 1989) and that (2)

Current Topics in Microbiology and Immunology, Vol. 173

productively rearranged TCR cDNA clones amplified by PCR from mRNA obtained from uncultured epidermal cell suspensions preferentially used $V\gamma3/J\gamma1/C\gamma1$-$V\delta1/D\delta2/J\delta2/C\delta$ genes and exhibited almost no N additions between the various coding segments (Asarnow et al., 1989).

Ever since their discovery, the ontogeny of DETC has been a widely debated issue. Several observations have indicated that the perinatal period is of critical importance for both maturation and expansion of the DETC population. While DETC first appear during the last days of gestation and reach their mature phenotype and their full numerical density within the first few weeks of the postnatal period (Romani et al., 1986, Elbe et al., 1989), depletion of the DETC population of adult animals by physicochemical agents is not followed by its reappearance (Aberer et al., 1986).

Furthermore, we found that reconstitution of adult γ-irradiated thymus-containing AKR/Ola or B6Pl-Thy-1.a (both expressing the Thy-1.1 allele) mice with either T-cell depleted C3H/He/Han or C57Bl/6 (both Thy-1.2) bone marrow cells or fetal liver cells resulted in the appearance of round Thy-1.2$^+$ cells in the epidermis which, over the course of several months, remained CD3$^-$. In contrast, lymph nodes and spleens of chimeric mice were colonized by large numbers of Thy-1.2$^+$/CD3$^+$ lymphocytes (Honjo et al., 1990).

The mere possibility that the phenomenon observed was due to an adverse effect of γ-irradiation on the induction of CD3/TCR expression in DETC-precursors was ruled out in experiments in which newborn skin was transplanted onto chimeric animals. Again, Thy-1.2$^+$ cells immigrating into the transplant did not acquire CD3 antigens (E. Payer, unpublished observation).

While all these findings support the contention that DETC maturation happens early in life rather than in the adult animal, the important question remains whether this event occurs in a thymus-dependent or in a thymus-independent fashion. The former theory maintains that DETC are derived from thymocytes which migrate into the epidermis during the neonatal period. According to the latter hypothesis, DETC are derived from bone marrow precursors which enter the skin without having passed through the thymus and differentiate in situ. At the first glance, the finding of predominantly round Thy-1$^+$/CD3$^-$, but not of Thy-1$^+$/CD3$^+$, dendritic cells in the epidermis of athymic mice (Nixon-Fulton et al., 1988) could be taken as a strong argument for the thymus-dependence of DETC. On the other hand, the skin of athymic mice is grossly abnormal (nu/nu) and may not be able to provide the appropriate microenvironment for DETC maturation.

In the past two years, we have designed experimental strategies to test these converse hypotheses and will now briefly summarize the results obtained in these studies.

THE EPIDERMIS - A HOMING SITE FOR Vγ3$^+$ FETAL THYMOCYTES

The hypothesis that DETC originate from fetal thymocytes was originally put forward by Havran & Allison (1988) when they found that a mAb against the Vγ3 determinant reacts with both the first CD3-bearing fetal thymocytes and with the DETC population of adult animals. The similarity between these two cell systems was furthered by the observation that the entire TCR configuration of these early fetal thymocytes was essentially indistinguishable from that expressed by DETC (Lafaille et al., 1988, Asarnow et al., 1989). To test the assumption that these Vγ3$^+$ fetal thymocytes are the actual DETC precursors, we injected day 16 and day 19 fetal, as well as adult thymocytes into either syngeneic (C3H, Balb/c) or Thy-1-disparate (B6Pl-Thy-1.a → C57Bl/6 nu/nu) athymic nude mice.

Results obtained showed that the injection of day 16 fetal thymocytes resulted regularly, the injection of day 19 fetal thymocytes in most instances, in the appearance of distinct clusters of donor-type Vγ3$^+$/CD3$^+$ dendritic epidermal cells. In contrast, the injection of adult thymocytes was not followed by the emergence of the DETC population (Payer et al., submitted for publication). The presence of CD3$^+$/TCR Vγ3$^+$ cells in day 16 and day 19 fetal, but not in adult thymocyte populations together with the failure to detect DETC after transfer of Thy-1$^+$/CD3$^-$ fetal thymocytes strongly suggest that, under the experimental conditions chosen, CD3$^+$/TCR Vγ3$^+$ thymocytes are the DETC precursors.

The first donor-type cells entering the epidermis were identified 13/14 days after i.v. injection of day 16 fetal thymocytes. Some of these cells were round, others were dendritic in shape, and all expressed the CD3/TCR Vγ3 complex. When analyzed 4 to 12 weeks after cell transfer, most donor-type Thy-1$^+$ cells displayed a dendritic configuration. The enumeration of DETC-clusters found and of cell numbers within the clusters from 2 to 12 weeks after cell transfer revealed a dramatic increase in the cell density within one cluster which was not accompanied by an increase in the number of clusters. Thus, it appears that, upon arrival in the epidermis, Vγ3$^+$ day 16 fetal thymocytes undergo vigorous proliferative activity (Payer et al., submitted for publication).

Most recently, Havran & Allison (1990) reported that implantation of day 14 fetal thymic lobes, but not of day 2 newborn thymic lobes from euthymic mice into either syngeneic athymic nude mice or Thy-1-disparate euthymic newborn mice resulted in the appearance of Vγ3$^+$ DETC. These data together with the results of our cell transfer experiments (Payer et al., submitted for publication) lend strong support to the concept that CD3$^+$/TCR Vγ3$^+$ fetal thymocytes are the actual precursors of DETC.

THE FETAL SKIN - A SITE OF MATURATION OF CD3$^-$ LYMPHOCYTES INTO DETC

The above findings do not exclude the possibility that the microenvironment of the fetal skin/epidermis can provide T cell educating stimuli similar to those of the fetal thymic epithelium. Following this reasoning, we have considered the possibility that the CD45$^+$/Thy-1$^+$/CD3$^-$ cells present within day 17 fetal epidermis (Elbe et al., 1989) could represent the target for these putative stimuli. As a consequence, these cells would mature into DETC and, thus, should not be detectable in adult murine epidermis. In fact, this appears to be the case as we have failed to detect appreciable numbers of such Thy-1$^+$/CD3$^-$ epidermal lymphocytes in the adult animal (Elbe & Payer, unpublished observations).

Recent flow cytometric studies have shown that the above CD45$^+$/Thy-1$^+$/CD3$^-$ cells are the only lymphocytes present in day 16 fetal murine skin (epidermis and dermis) (Elbe et al., in preparation). Our contention that day 16 fetal skin is indeed devoid of mature T lymphocytes was further substantiated when we examined cell suspensions prepared from day 16 fetal skin for the presence of Thy-1 and CD3/TCR messages. While we detected high-abundant Thy-1 transcripts, our search for CD3-γ, CD3-ϵ, Cα, Cβ and Cδ mRNA yielded negative results (Elbe et al., in preparation). In further studies, we were able to propagate these CD45$^+$/Thy-1$^+$/CD3$^-$ day 16 fetal skin cells by culturing them in IL-2. All the resulting cell lines again displayed the CD45$^+$/Thy-1$^+$/CD3$^-$/CD4$^-$/CD8$^-$ phenotype, but contained CD3-γ mRNA and incomplete TCR Cγ1, Cγ4 and Cβ transcripts (Elbe et al., in preparation). We conclude from these findings that CD45$^+$/Thy-1$^+$/CD3$^-$ day 16 fetal skin cells are of T cell lineage and could qualify as potential DETC precursors. In order to test this hypothesis, we transplanted full-thickness grafts from body wall skin of day 16 C57Bl/6 (Thy-1.2) mice onto full-thickness wound beds of B6Pl-Thy-1.a (Thy-1.1) animals. Analysis of the grafts at various time points after transplantaton revealed the presence of steadily increasing numbers of donor-type Thy-1$^+$/CD3$^+$ dendritic cells which uniformly reacted with an anti-Vγ3 mAb (Elbe et al., in preparation).

These results demonstrate that fetal skin harbors the DETC precursors and imply that the microenvironment of the fetal epidermis can provide stimuli promoting the expression of CD3/TCR genes in these cells.

SUMMARY

Our attempts to clarify the contribution of the thymic vs. the cutaneous microenvironment in the maturation of dendritic epidermal T cell (DETC) precursors into DETC gave diverse results. In one series of experiments, we found that i.v. injection of fetal thymocytes (containing a TCR Vγ3-expressing subpopulation), but not of adult thymocytes (containing no TCR $V\gamma3^+$ cells) results in the appearance of CD3/TCR $V\gamma3^+$ dendritic epidermal cells (=DETC). In other experiments, we have obtained evidence that transplantation of day 16 fetal skin onto a Thy-1-disparate recipient results in the appearance of donor-type DETC. Our further observation that the transplanted skin contains $CD45^+/Thy\text{-}1^+/CD3^-$ lymphocytes, but no mature T cells, therefore implies that fetal skin can provide stimuli promoting the expression of CD3/TCR genes in immature ($CD3^-$) DETC precursors.
It remains to be seen whether both or only one of these maturational pathways are (is) followed under physiological conditions.

ACKNOWLEDGEMENT

We thank Mrs. Sabine Seizov for her excellent help in preparing this manuscript.

REFERENCES

Aberer W, Romani N, Elbe A, Stingl G (1986) Effects of physicochemical agents on murine epidermal Langerhans cells and Thy-1-positive dendritic epidermal cells. J Immunol 136: 1210-1216.

Asarnow DM, Kuziel WA, Bonyhadi M, Tigelaar RE, Tucker PW, Allison JP (1988). Limited diversity of γ/δ antigen receptor genes of Thy-1$^+$ dendritic epidermal cells. Cell 55: 837-847.

Asarnow DM, Goodman T, LeFrancois L, Allison JP (1989) Distinct antigen receptor repertoires of two classes of murine epithelium-associated T cells. Nature 341: 60-62.

Elbe A, Tschachler E, Steiner G, Binder A, Wolff K, Stingl G (1989) Maturational steps of bone marrow-derived dendritic murine epidermal cells. Phenotypic and functional studies on Langerhans cells and Thy-1$^+$ dendritic epidermal cells in the perinatal period. J Immunol 143: 2431-2438.

Havran WL, Allison JP (1988) Developmentally ordered appearance of thymocytes expressing different T-cell antigen receptors. Nature 335: 443-445.

Havran WL, Grell S, Duwe G, Kimura J, Wilson A, Kruisbeek AM, O'Brien RL, Born W, Tigelaar RE, Allison JP (1989) Limited diversity of T-cell receptor γ-chain expression of murine Thy-1$^+$ dendritic epidermal cells revealed by Vγ3-specific monoclonal antibody. Proc Natl Acad Sci USA 86: 4185-4189.

Havran WL, Allison JP (1990) Origin of Thy-1$^+$ dendritic epidermal cells of adult mice from fetal thymic precursors. Nature 344: 68-70.

Honjo M, Elbe A, Steiner G, Assmann I, Wolff K, Stingl G (1990) Thymus-independent generation of Thy-1^{+} epidermal cells from a pool of Thy-1^{-} bone marrow precursors. J Invest Dermatol (in press).

Lafaille JJ, DeCloux A, Bonneville M, Takagaki Y, Tonegawa S (1989) Junctional sequences of T cell receptor $\gamma\delta$ genes: implications for $\gamma\delta$ T cell lineages and for a novel intermediate of V-(D)-J joining. Cell 59: 859-870.

McConnell TJ, Yokoyama WM, Kikuchi GE, Einhorn GP, Stingl G, Shevach EM, Coligan JE (1989) δ-chains of dendritic epidermal T cell receptors are diverse but pair with γ-chains in a restricted manner. J Immunol 142: 2924-2931.

Nixon-Fulton JL, Kuziel WA, Santerse B, Bergstresser PR, Tucker PW, Tigelaar RE (1988) Thy-1^{+} epidermal cells in nude mice are distinct from their counterparts in thymus-bearing mice. A study of morphology, function, and T cell receptor expression. J Immunol 141: 1897-1903.

Romani N, Stingl G, Tschachler E, Witmer MD, Steinman RM, Shevach EM, Schuler G (1985) The Thy-1-bearing cell of murine epidermis. A distinctive leukocyte perhaps related to natural killer cells. J Exp Med 161: 1368-1383.

Romani N, Schuler G, Fritsch P (1986) Ontogeny of Ia-positive and Thy-1-positive leukocytes of murine epidermis. J Invest Dermatol 86: 129-133.

Steiner G, Koning F, Elbe A, Tschachler E, Yokoyama WM, Shevach EM, Stingl G, Coligan JE (1988) Characterization of T cell receptors on resident murine dendritic epidermal T cells. Eur J Immunol 18: 1323-1328.

Stingl G, Koning F, Yamada H, Yokoyama WM, Tschachler E, Bluestone JA, Steiner G, Samelson LE, Lew AM, Coligan JE, Shevach EM (1987) Thy-1$^+$ dendritic epidermal cells express T3 antigen and T-cell receptor γ chain. Proc Natl Acad Sci USA 84: 4586-4590.

Phenotypic and Functional Characterization of Human TCRγδ+ Intestinal Intraepithelial Lymphocytes

K. Deusch[2], K. Pfeffer[2], K. Reich[1], M. Gstettenbauer[1], S. Daum[1], F. Luling[1], and M. Classen[1]

[1] II. Department of Medicine and Institute of Microbiology
[2] Technical University, Ismaninger Straße 22, D-8000 Munich 80, FRG

ABSTRACT
Intestinal intraepithelial lymphocytes (IEL) appear to represent a peculiar set of immune cells compartmentalized at the interface between the organism and the external environment. In previous studies we observed that within human IEL TCR-τ/δ T cells represent a major fraction that predominantly express the CD8 molecule and preferentially uses the V-δ-1 gene segment. Thus these data suggested a preferential accumulation/homing of CD8+ V-δ-1$^+$ IEL within the human intestinal epithelium. However, to date the functional role of these cells with regard to immune regulation at this most critical immunological site is poorly understood. In this study, the cytotoxic potential and proliferative capacity of human IEL in response to mitogenic stimuli has been characterized with respect to IEL T cell receptor type and TCR-τ/δ variable gene segment usage as determined by flowcytometry. The frequeny of TCR-1+ IEL expressing both CD56 and CD16 which are considered to be NK-cell markers was found to be much higher (38.9±12.4%) than within intestinal lamina propria lympmocytes (LPL) (9.1±4.8%) or peripheral blood lymphocytes (PBL) (6.4±3.3%). In contrast, the fractions of CD16$^-$CD56$^+$ cells within IEL, LPL and PBL were comparable. Surprisingly, IEL mediated NK-cell activity (K562 lysis) was virtually absent whereas within PBL it was within the normal range. Furthermore, in cytotoxicity assays employing ^{51}Cr-labeled OKT3 hybridoma cells and P815 cells as targets, the cytotoxic potential of IEL was much lower than that of PBL. Finally, only a minority (41.9±4.4%) of TCR-τ/δ IEL were CD5+ whereas the fraction of CD5$^+$ TCR-τ/δ LPL and TCR-τ/δ PBL were 69.8±12.8% and 82.8±8.4%, respectively. The latter finding suggests an extrathymic pathway of differentiation for TCR-τ/δ+ IEL. Taken together, our data suport the notion that human IEL represent a functionally, phenotypically and ontogenetically distinct population of cells that may regulate the local mucosal response by mechanisms different from those described for other lymphoid tissues such as peripheral lymph nodes or the peripheral blood.

INTRODUCTION

Evidence has been accumulating that intestinal intraepithelial lymphocytes (IEL) represent a discrete lymphoid compartment of the immune system lymphoid exhibiting phenotypical, functional and morphological peculiarities compared to lymphocytes residing in the intestinal lamina propria or circulating in peripheral blood (McI. Mowat 1990). In normal small intestinal epithelium approximately 15%-20% of cells are lymphocytes. Similarly, in the stomach, colon and rectum large numbers of IEL can be found.Recently, it was found that in man intraepithelial T lymphocytes segregate into two almost equally sized subpopulations expressing either a TCR-α/ß or a TCR-τ/δ. Moreover, TCR-τ/δ$^+$ IEL were shown to preferentially use the V-δ-1 gene segment to build their TCR (Deusch et al., in press). Thus, with regard to both the distribution of TCR-τ/δ (TCR-1) and TCR-α/ß (TCR-2) cells and limited TCR-τ/δ V-gene segment usage, these findings resemble those observed previously for murine and avian IEL (Goodman and Lefrancois 1987, Asarnow et al. 1989, Tagakaki et al. 1989, Itohara et al. 1990, Bucy et al. 1988). However, the expression of phenotypic markers on TCR-τ/δ+ IEL indicating cell function, activational status or ontogentic origin have not been well characterized. Therefore, employing an isolation procedure that yields highly pure IEL and LPL, we characterized by flow cytometry the phenotype of such markers on human IEL.

MATERIAL AND METHODS

Tissue specimens. Intestinal samples from colon, 5 to 10 cm long were obtained from 14 adult patients. All underwent large bowel resection for for colorectal carcinoma. The samples were taken at least 10 cm away from the lesion and were found to be histologically normal. Peripheral blood was taken from the same patients prior to operation. **Isolation of human IEL, LPL and PBL.** IEL and LPL were isolated as desribed previously (2). Briefly, the resected tissue was washed with PBS and the mucosa was dissected from the underlying muscularis mucosae with scissors. To remove mucus, the mucosa was incubated with Ca^{++} and Mg^{++} free HBSS containing 100U/ml penicillin, 100 ug/ml streptomycin, 50 ug/ml gentamycin , 2,5 ug/ml amphotericin B (Seromed Biochrom KG, Berlin), 25 mM HEPES buffer, 2 mM

Current Topics in Microbiology and Immunology, Vol. 173

EDTA and 1 mM DL-Dithiothreitol in a shaker at 37°C for 10 min. Following incubation, the supernatant was discarded and the mucosal tissue was incubated in the same HBSS medium without DL-Dithiothreitol 4 times for 30 min each. The supernatants from each of these incubations were collected and washed in a 40% isotonic Percoll solution (Pharmacia, Uppsala, Sweden) and finally centrifuged over Ficoll-Hypaque. LPL were obtained by incubating the remaining mucosal tissue overnight in a shaking waterbath at 37°C in RPMI 1640 medium containing 0.01% collagenase(Sigma, Deisenhofen, FRG) and 0,01% deoxyribonuclease I (Sigma, Deisenhofen, FRG). The resulting cell suspension was passed though a stainless steel sieve (0.1mm mesh), washed twice in 40% isotonic Percoll (Pharmacia, Uppsala, Sweden) and centrifuged over Ficoll-Hypaque. Viability of both IEL and LPL as determined by Trypan blue exclusion was always >90% Viability of these cells. Peripheral blood lymphocytes were isolated by centrifugation over Ficoll-Hypaque. **Phenotypic analysis.** Single and two colour fluorescence measurements of antibody labeled isolated IEL, LPL and PBL were performed on a FACSSCAN (Becton Dickinson). Monoclonal antibodies used were anti-CD3 (Leu 4-PE) CD4 (Leu 3), CD8 (Leu 2a), anti-CD16 (Leu11), anti-CD56 (Leu 19), anti-CD5-PE, anti-CD7-PE anti-MHC Class II, anti-CD25 (all from Becton Dickinson), TCR-delta-1-FITC (gamma/delta-TCR) TCS-1 (anti-V-δ-1,JP; both purchased from T Cell Sciences), anti-V-δ-2,J1-3 (4G6), anti-V-δ-2,J1 (7A8), anti-V-τ-9 (4D7) (all kindly provided by Dr. Pfeffer)) and BMA 031 (anti-TCR-α/ß) (kindly provided by Behringwerke, Marburg, FRG). Cytotoxicity assay: Target cells (OKT3-hybridoma, P815 and K562 tumor cells) were labeled with ^{51}Cr as desribed previously (Deusch et al. 1986), washed and subsequently used as target cells in a standard 4 hour ^{51}Cr-release assay at a humidified atmosphere (5% CO_2, 37°C). The effector to target ratios (E/T-ratio) are given in the figure. Specific lysis was calculated as described previously (8).

RESULTS

Usage of TCR-τ/δ-V-gene segments in relation to CD8 expression on IEL. To investigate the expression of CD8 molecules on TCR-τ/δ$^+$ IEL in relation to TCR-V-gene segment usage, IEL were stained with anti-CD8 antibodies in combination with antibodies directed to the products of distinct TCR-τ/δ V-gene segments (see methods) and subsequently analyzed by flow-cytometry. Fig 1. shows a representative cell sample depicting the preferential expression of CD8 molecules on V-δ-1$^+$ IEL (71.3% CD8$^+$). In contrast, the fraction of CD8$^+$ cells within the V-δ-2$^+$ and V-τ-9$^+$ IEL subpopulation was substantially lower (45.7% and 55.5%, respectively). Therefore, it appears that CD8 molecule expression on V-δ-1$^+$ IEL, which represent the predominant cell population within TCR-τ/δ$^+$ IEL, is selectively induced.

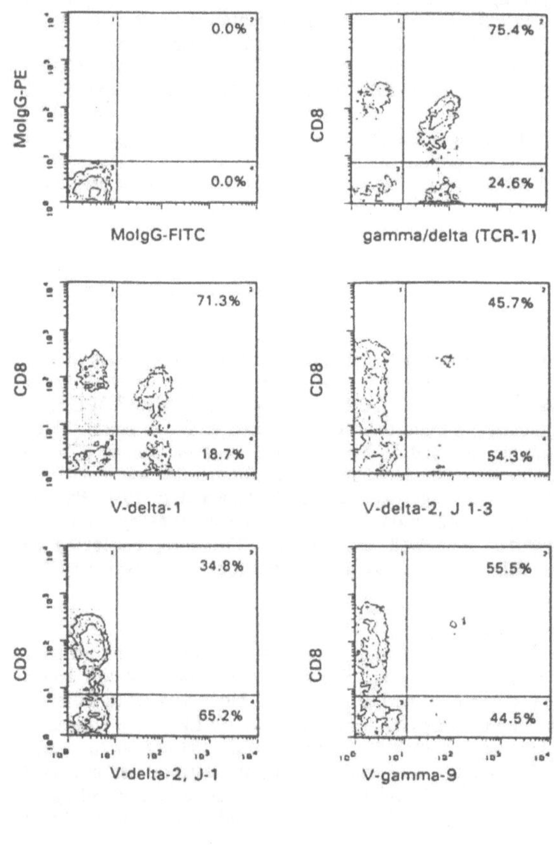

Fig.1 Densitiy gradient purified IEL were stained with anti-CD8-PE in combination with either a pan anti-TCR-τ/δ antibody (anti-TCR1) or V-gene segment specific antibodies as depicted in the figure (see Methods for details) and analyzed for dual color fluorescence by flow cytometry. Percentages were calculated in relation to the total number of cells that stained with the pan anti-TCR-τ/δ antibody or the V-gene segment specific antibodies (anti-V-δ-1, anti-V-δ-2,J1-3, anti-V-δ-2 J1, anti-V-τ-9). Shown is a representative example where the fraction of TCR-τ/δ$^+$ IEL was 74%. The latter figure was taken as 100% for the calculation of TCR-τ/δ$^+$ subpopulations.

The majority of TCR-γ/δ$^+$ IEL does not express the CD5 molecule. Next, we compared the expression of the CD5 molecule TCR-γ/δ+ IEL, LPL and PBL. The presence of this molecule on the surface of peripheral T lymphocytes is believed to indicate their thymic origin. Cell samples were stained with anti-TCR-δ-1 in combination with anti-CD5 antibodies and subsequently analyzed for dual color fluorescence intensity by flow cytometry. The results of this study reveal that the majority (≈59%) of TCR-γ/δ+ IEL do not express the CD5 molecule whereas peripheral blood derived TCR-γ/δ T cells are almost all CD5+. TCR-γ/δ+ LPL appear to represent an intermediate cell population in that the fraction of CD5- cells substantially smaller than that observed for TCR-γ/δ$^+$ IEL (Fig. 2). Hence, these data support the notion that a substantial fraction of TCR-γ/δ$^+$ T cells within the gut epithelium may follow a distinct pathway of differentiation.

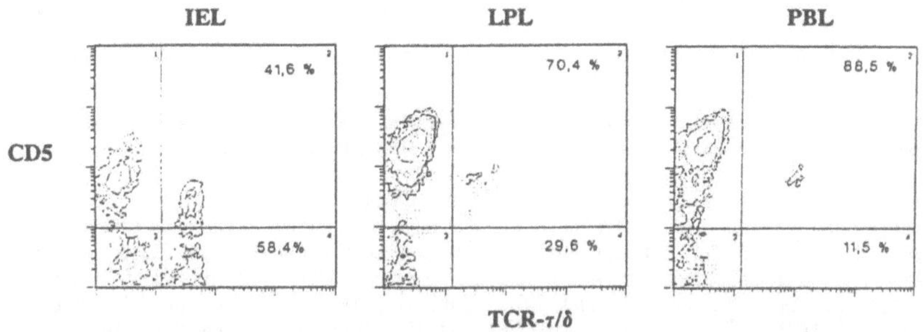

Fig.2 IEL, LPL and PBL were stained with anti-TCR-δ-1-FIT together with anti-CD5-PE and subsequently analyzed for dual color fluorescence. The data shown are representative for four individual experiments. The percentages given were calculated by taking the relative number of TCR-γ/δ cells within IEL, LPL and PBL as 100%.

Nearly all TCR-γ/δ$^+$ IEL are CD7$^+$. Since intestinal mucosal lymphocytes are constantly confronted with all kinds of antigenic materials, we were next interested whether the expression of the CD7 molecule (T-2 blast antigen) was increased on TCR-γ/δ$^+$ IEL compared to TCR-γ/δ LPL and PBL. Fig. 3 gives a representative example of the staining pattern that we observed on TCR-γ/δ IEL, LPL and PBL. It reveals that nearly all TCR-γ/δ$^+$ IEL (≈99.8%) are CD7$^+$ whereas the fraction of these cells within LPL and PBL was substantially lower (81.9% and 71.4%, respectively).

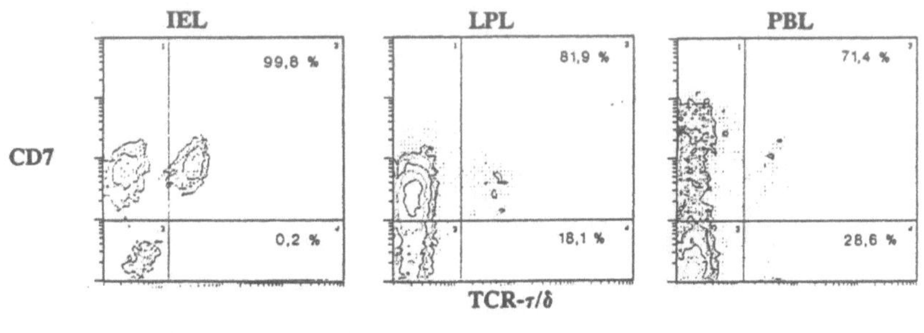

Fig. 3: Freshly isolated IEL, LPL and PBL were stained with anti-TCR-δ-1-FITC and anti-CD7-PE and subsequently analyzed for dual color fluorescence. The percentages given were calculated by taking the total number of TCR-γ/δ cells within each compartment as 100%.

A major fraction of Lamina propria derived CD4+CD8- and CD4-CD8+ T lymphocytes express MHC Class II and CD25 on their surface. To characterize the functional status of IEL, LPL and PBL, T lymphocytes isolated from these compartments were stained with either anti-CD4 or anti-CD8 in combination with antibodies specific for activation markers on human T cells such as MHC ClassII or CD25, respectively. Table I shows that compared to epithelial or peripheral blood derived T cells, a much higher fraction of both CD4+ and CD8+ LPL express MHC class II or CD25 molecules on their surface. However, the differences between IEL, LPL and PBL with regard to CD25 expression were not so striking. However, these data suggest that a substantial fraction of lamina propria derived T lymphocytes reside at this site in a constitutively activated state.

Table I

	% of CD3+ CD3/MHCII	% of CD4+ CD4/MHC II	% of CD8+ CD8/MHC II	% of CD4+ CD4/CD25	% of CD8+ CD8/CD25
IEL	3.9±2.3	4.1±1.6	8.53±0.9	<1.0	<0.2
LPL	36.7±14	54.1±13	58.58±22.0	14.9±2.9	5.4±3.0
PBL	10.6±6.2	9.9±5.6	18.9±8.0	6.1±7.0	3.4±1.7

IEL, LPL and PBL were stained with the combination of antibodies listed in the above table and subsequently analyzed for dual color fluorescence by flow cytometry. The data given are the means ± SD of 10 individual experiments.

A large fraction of Human IEL express both CD16 and CD56.
To investigate the expression of surface molecules that have been shown to delineate natural killer cells, we compared IEL, LPL and PBL for dual color fluorescence employing anti-CD16 and anti-CD56 antibodies. Fig. 4a demonstrates that within IEL the fraction of $CD3^+/CD16^+/CD56^+$ cells is much higher than that observed in LPL and PBL. Similarly, Fig. 4b shows that the frequency of $CD56^+CD16^+$ TCR-τ/δ^+ IEL is higher than LPL and PBL. It can be seen that this difference is due to an increased expression of CD16 (≈10-fold compared to LPL and PBL) on TCR-τ/δ cells rather than CD56, the latter being expressed at equal proportion on TCR-τ/δ^+ IEL, LPL and PBL (Fig. 3b)..

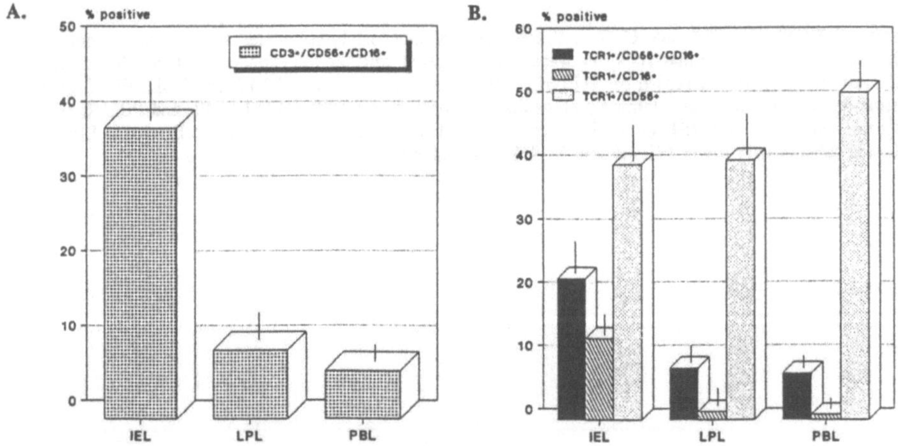

Fig. 4. (A) IEL, LPL and PBL were stained with a combination of anti-CD3-FITC and anti-CD16+CD56-PE antibodies and subsequently analyzed for dual color fluorescence. (B) The same cell samples were stained with a combination of anti-TCR-τ/δ (anti-TCR-δ-1-FITC) and anti-CD16+CD56-PE antibodies.

Human large intestinal IEL do not exhibit natural killer or conventional cytolytic activity. In order to characterize the cytolytic potential of human IEL, freshly isolated IEL were compared with PBL for their ability to lyse 51Crlabeled P815, OKT3 or K562 target cells. As can be seen from Fig. IEL neither exhibit natural killer activity nor anti-CD3 redirected cytotoxicity nor non specific anti-P815 killing. In contrast, PBL showed substantial NK-activity and moderate cytotoxicity against the anti-CD3 hybridoma and P815.

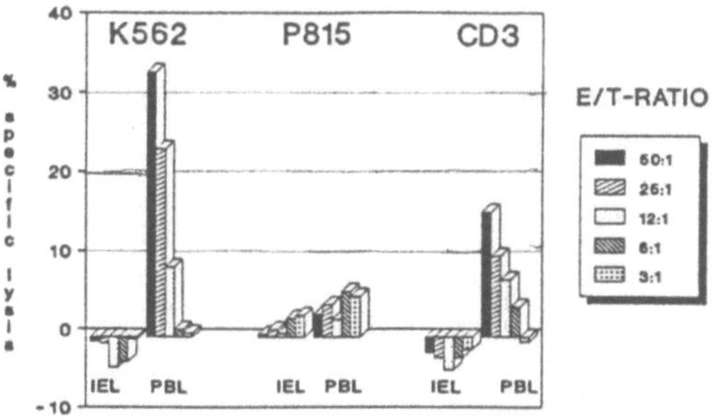

Fig. 5. Freshly isolated IEL and PBL were tested for their cytotoxic activity in a four hour 51Cr-release assay employing OKT3 hybridoma cells (CD3), P815 cells or K562 cells as targets E/T-Ratio denotes effector to target ratio

SUMMARY

Intraepithelial Lymphocytes (IEL) are one of the largest yet least understood populations of lymphocytes exhibiting a number of unique morphological, phenotypical and functional features. In this study, we comparatively analyzed a number of these features on IEL, LPL and PBL. The major observation was the preferential expression of the CD8 molecule on TCR-τ/δ^+ IEL using the V-δ-1 gene segment. Furthermore, the fraction of TCR-τ/δ^+ IEL expressing CD16 alone or in combination with CD56 was much higher than that observed in TCR-τ/δ^+ LPL and PBL. However, the expression of these markers did not correlate with an enhanced cytolytic potential of these cells. A substantial number of TCR-τ/δ^+ IEL do not express the pan T cell marker CD5 suggesting an alternate pathway of differentiation for these cells. Finally, many CD4$^+$ and CD8$^+$ LPL were found to express clas II and CD25 molecules reflecting that they constitutively become activated in the intestinal mucosa. In man, mouse and the chicken TCR-τ/δ T cells appear to localize preferentially in intestinal epithelium and to exhibit a restricted TCR-τ/δ variable gene segment heterogeneity. Moreover, both human and murine IEL express peculiar surface molecules that may serve to direct homing mechanisms and to provide these cells with functional repertoires necessary to protect the organism from invading pathogens. However, to date in man and mouse, apart from information pertaining to their principal lymphokine repertoire and low proliferative capacity towards non specific mitogens and specific antigens (Ebert et al. 1986, own unpublished observations), little is known about the specific function of IEL. Nevertheless, their limited TCR heterogeneity could be explained by specific lymphoepithelial interactions involving hitherto unknown restriction molecules expressed on epithelial cells. One interesting observation is the marked proliferative response to sheep erythrocytes that can be blocked by anti-CD2 antibodies, which. suggests that the surface receptors critical to trigger IEL may be be distinct from those involved in activation processes of peripheral blood or lamina propria derived T lymphocytes (Ebert et al. 1986). At present, we have no explanation for the observed preferential v-gene segment usage and differential expression of CD16 and CD56 on TCR-τ/δ^+ IEL compared to LPL and PBL. However, present and future efforts in our laboratory are aimed to resolve these issues with special emphasis on intestinal lymphoepithelial interactions and their role in the immunological surveillance of the intestinal epithelial surface.

ACKNOWLEDGEMENTS

We thank Mrs A. Wiesel and S. Kromer for expert technical assistance and Prof. J. Siewert for providing the surgical specimens. This work was supported by grant DFG/De-384/

REFERENCES

Asarnow, D.M., Goodman, T., Lefrancois, L., and Allison, J. (1989) Distinct antigen receptor repertoire of two classes of murine epithelium associated T cells. Nature. 341:60

Bucy, P., Chen, C., Cihak, J., Lösch, U., and Cooper, M. (1988) Avian T cells expressing τ/δ receptors localize in the splenic sinusoids and intestinal epithelium. J. Immunol. 141:2200

Deusch, K. Moebius, U., Meyer zum Büschenfelde, K.H., Meuer, S.C. (1986) T lymphocyte control of autoreactivity: analysis with human T cell clones in limting dilution culture. Eur. J. Immunol. 16:1433. SS.

Deusch, K., Lüling, F, Reich, K., Classen, M., Wagner, H. and Pfeffer, K. (1991) A major fraction of human intraepithelial lymphocytes simultaneously expresses the τ/δ-TCR, the CD8 accessory molecule and preferentially uses the V- 231-1 gene segment. Eur. J. Immunol. (in press)

Ebert, E.C., Roberts, A.I., Brolin, R.E., Raska, K. (1986) Examination of the low proliferative capacity ofhuman jejunal intraepithelial lymphocytes. Clin. Exp. Immunol. 65:148

Goodman T, Lefrancois L (1987) EXpression of the τ/δ-T-cell receptor on intestinal CD8+ intraepithelial lymphocytes. Nature 333:855

Itohara, S., Farr, A.G., Lafaille, J.J., Bonneville, M., Tagakaki, Y., Haas, W., and Tonegawa, S. (1990) Homing of the τ/δ thymocyte subset with homogeneous T-cell receptors to mucosal epithelia. Nature 343:754

McI. Mowat, A, (1990) Human Intraepithelial Lymphocytes in: Springer Seminars in Immunopathology 12:165

Tagakaki, Y., DeCloux, A., Bonneville, M., and Tonegawa.S. (1989) Diversity of τ/δ T cell receptors on murine intestinal intraepithelial lymphocytes. Nature. 339:712

A Search for Cells Carrying the γ/δ T Cell Receptor in Mice Infected with *Leishmania major*

M. Lohoff, J. Dingfelder, and M. Röllinghoff

Institute for Clinical Microbiology, University Erlangen/Nürnberg, Wasserturmstraße 3, 8520 Erlangen, FRG

INTRODUCTION

T-cells can be devided into two distinct lineages on the basis of their receptor. Those cells expressing the α/β-heterodimer have been extensively analysed and are known to recognize foreign antigens in the context of MHC-molecules. Several distinct functions of α/β- T-cells including cytolytic and helper functions have been identified. In contrast, little is known about a different T-cell subset expressing the so-called γ/δ-heterodimer. The fact that a high frequency of γ/δ- T-cells is found in epithelial tissues such as the skin or the gut, led to the suggestion that they may represent a kind of primitive immune system for the defense of invading pathogens (Janeway, 1988). Support to this notion came from several publications demonstrating the expansion of γ/δ-T-cells during infections in humans and in mice. Thus, γ/δ-T-cells were found within the peritoneal exsudate cells of mice infected with Listeria monocytogenes (Ohga, et al, 1990) and within the lesion-draining lymph nodes of mice infected with Mycobacterium tuberculosis (Janis, et al, 1989). Along this line, γ/δ-T-cells were identified in human skin chronically infected with Mycobacterium leprae or protozoons of the genus Leishmania (Modlin, et al, 1989).

A subspecies of Leishmania, namely Leishmania major (L.major), is a well examined model system for chronic infections in mice. The different course of the disease in inbred mouse strains has led to the definition of disease-susceptible strains, e.g. BALB/c, which die as a result of the infection, and resistant strains, e.g. C57Bl/6, which overcome the infection and appear to be immune to reinfection (Mitchell, 1982). Given the possible role of γ/δ-T-cells during leishmaniasis in humans, we were interested to test, if γ/δ-T-cells can be found in the lesion-draining lymph node cells (LNC) of mice infected with L.major, and if so, whether or not there is a difference in the expansion of these cells in susceptible and resistant mouse strains. In this report we demonstrate that we were unable to detect γ/δ-T-cells within lymph nodes of L.major-infected mice for up to seven weeks after infection.

MATERIAL AND METHODS

The L.major strain used throughout these studies has been described elsewhere (Solbach, et al, 1986). Female BALB/c and C57Bl/6 mice, 4 -6 weeks of age, were infected with 2 x 10^7 L.major promastigotes in the right footpad. Evaluation of the course of the disease was performed by measuring the increase in footpad thickness, as described (Solbach, et al, 1986).

At different time points after infection, four mice per group and time point were killed, the lesion-draining popliteal lymph nodes were isolated, single cell suspensions were obtained, the cells of the four mice were pooled and tested for the expression of several cell surface markers by immunofluorescence. Antibodies used for cell staining were biotinylated anti-α/β T-cell-receptor (TcR)-antibody (Kubo, et al, 1989), biotinylated anti-CD5 antibody (Ledbetter & Herzenberg, 1989), anti-γ/δ TcR

Current Topics in Microbiology and Immunology, Vol. 173

antibody, kindly provided by Dr L. Lefrancois, Upjohn, Kalamazoo, USA (Goodman & Lefrancois, 1989), FITC-conjugated anti-B220 (Medac, Hamburg, FRG) and FITC-conjugated anti-Thy 1.2 (Becton-Dickinson). Counterstaining of the biotinylated reagents was performed using phycoerythrin-coupled streptavidin (Medac). The anti-γ/δ TcR antibody was counterstained using a phycoerythrin (PE)-coupled goat-anti hamster antiserum which had been absorbed with mouse immunoglobulin before (Medac, Hamburg). In the absence of the anti-γ/δ TcR antibody all cells tested scored uniformly negative with this reagent. As a positive control for the anti-γ/δ TcR antibody we used a T-cell line (T245), kindly provided by Dr A. Elbe, VIRCC, Vienna, Austria, (Stingl, et al, 1987). All samples were analysed using an EPICS-profile 1.

RESULTS AND DISCUSSION

The aim of the present study was to detect and to test for a difference in the systemic expansion of γ/δ-T-cells of resistant and susceptible mouse strains after infection with L.major. Since γ/δ-T-cells have been proposed to represent a first line of defense against invading pathogens (Janeway, 1988) such a difference could have accounted for the different course of the disease in the two mouse strains.

The data given in this report, however, do not support such a notion. We were unable to detect γ/δ-T-cells in the lesion-draining lymph nodes of either of the two mouse strains in appreciable numbers, i.e. above 0.4% which was the level of unspecific staining in the presence of an irrelevant hamster antibody. In addition, we could not find a significant difference in the number of γ/δ-T-cells between the two mouse strains (Tables 1 and 2).

Table 1: Immunofluorescence of LNC from L.major-infected mice

	BALB/c					C57Bl/6				
	days post infection									
	0d	3d	21d	35d	49d	0d	3d	21d	35d	49d
TcR-γ/δ	<0.4	<0.4	<0.4	<0.4	<0.4	<0.4	<0.4	<0.4	<0.4	<0.4
TcR-α/β	83.9	59.3	63.8	60.3	56.0	84.9	58.6	48.3	53.0	32.9
Thy 1.2	87.5	67.0	70.2	61.2	56.0	90.5	72.2	46.4	56.2	38.4
CD5	85.6	63.1	61.7	59.8	59.2	85.8	66.8	50.6	54.7	33.8
B 220	13.5	31.4	33.2	34.6	38.2	13.0	32.1	47.1	46.1	64.5

BALB/c and C57Bl/6 mice were infected with L.major in the right footpad. At the indicated time points, the lesion-draining popliteal LNC were stained for the indicated surface markers. Values give the percentage of positive cells.

Preliminary data obtained using a different staining protocol that included biotinylated protein A and PE-coupled streptavidin (b-pA/s-PE) are compatible with the above results (M. Lohoff, et al, submitted for publication). As in the lesion-draining lymph nodes, we were unable to detect γ/δ-T-cells at greater numbers in the spleens of L.major-infected mice of both strains with the goat-anti hamster reagent, although here the assay system was slightly less sensitive (detection level about 0.9%), due to a higher level of background staining as compared to lymph node cells. Preliminary data of spleen cells stained with lower levels of background staining (<0.3%) using b-pA/s-PE as second step reagents indicate that there may be a detectable number of γ/δ-T-cells (about 0.6%). However, these cells do not change in number during infection.

Table 2: Immunoflourescence of spleen cells from L.major-infected mice

	BALB/c			C57Bl/6		
	days post infection					
	0d	7d	21d	0d	7d	21d
TcR-γ/δ	<0.9	<0.9	<0.9	<0.9	<0.9	<0.9
TcR-α/β	76.4	53.8	56.2	59.0	50.1	49.4
B 220	14.9	38.6	32.0	27.1	41.7	38.9

BALB/c and C57Bl/6 mice were infected with L.major in the right footpad. At the indicated time points, the spleen cells were stained for the indicated surface markers. Values give the percentage of positive cells.

Two types of positive controls were employed. Both staining protocols yielded similarly good staining of T245 cells, a cell-line established from epidermal γ/δ-T-cells. A typical diagram obtained with the PE-coupled anti-hamster antiserum is shown in Fig.1.

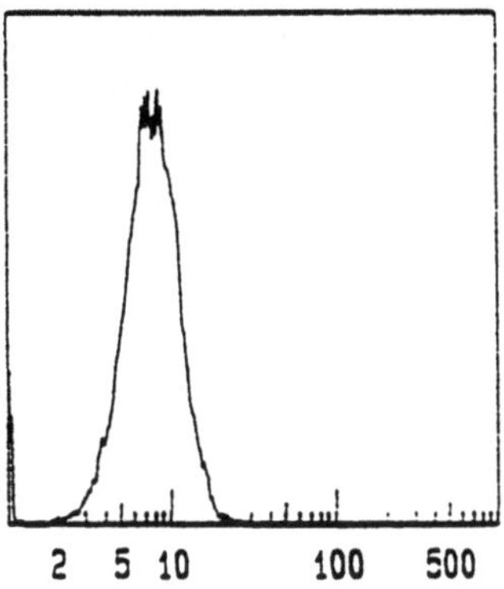

Fig.1: a staining profile of T245-cells stained with anti-γ/δ TcR antibody and phycoerythrin (PE)-coupled goat-anti hamster antiserum

At the same time we also tested for the presence of γ/δ-T-cells within peritoneal exsudate cells of mice infected with Listeria monocytogenes. Here we used b-pA/s-PE as second step reagents and were easily able to detect γ/δ-T-cells (Lohoff, et al, submitted), as was expected from data in the literature (Ohga, et al, 1990). In addition, the same anti-γ/δ TcR antibody supernatant also stained day 12 fetal thymocytes (R. Ceredig, Strasbourg, personal communication). We therefore are confident that the staining protocol used in these studies was sufficient to detect murine γ/δ-T-cells.

We take our data to propose that γ/δ-T-cells are not of decisive importance for the systemic course of leishmaniasis. Of course, we cannot exclude the presence of γ/δ-T-cells at numbers below the detection level of our staining procedure. Hypothetically, such cells could still decide the course of the disease as regulator cells provided that they have a particularly high avidity for their targets. However, we consider such a possibility to be not very likely.

The results shown here have to be discussed in view of two recent reports demonstrating on the m-RNA level the existence of γ/δ-T-cells in mice infected with two other, potentially intracellular pathogens, namely Mycobacterium tuberculosis (Janis, et al, 1989) and Listeria monocytogenes (Ohga, et al, 1990). Both groups tested for γ/δ-T-cells at sites distant from the local invasion of the parasite, i.e. in the lesion-draining lymph nodes, as we did, and in the peritoneum. The fact that γ/δ-T-cells were to be found there, but not in lymph nodes of L.major-infected mice, suggests that a general statement on the expansion of these cells in mice infected with intracellular pathogens at present is premature.

In contrast to the data presented here, γ/δ-T-cells have been detected in the skin of humans chronically infected with Leishmania parasites (Modlin, et al, 1989). Therefore, it appears to be of urgent need to study the local situation in mice infected with L.major. This will help to answer the question whether or not humans and mice differ in their capacity to expand γ/δ-T-cells in response to L.major. If, however, γ/δ-T-cells can be found in the skin of mice infected with L.major, one has to ask, why these cells are unable to home to the lesion-draining lymph nodes in leishmaniasis, whereas the cells can home there in mice infected with Mycobacterium tuberculosis. Studies to test these questions are in progress .

REFERENCES

Goodman, T., Lefrancois, L. (1989) Intraepithelial lymphocytes: Anatomical site, not T cell receptor form, dictates phenotype and function. J. Exp. Med. 170:1569-1581

Janeway, C.A., Jr. (1988) Frontiers of the immune system. Nature 333:804-806

Janis, E.M., Kaufmann, S.H.E., Schwartz, R.H., Pardoll, D.M. (1989) Activation of γ/δ-T-cells in the primary immune response to Mycobacterium tuberculosis. Science 244:713-716

Kubo, R.T., Born, W., Kappler, J.W., Marrack, P., Pigeon, M. (1989) Characterization of a monoclonal antibody which detects all murine α/β- T-cell receptors. J. Immunol. 142:2736-2742

Ledbetter, J.A., Herzenberg, L.A. (1979) Xenogenic monoclonal antibodies to mouse lymphoid differentiation antigens. Immunol Rev. 47:63-75

Mitchell, G.F. Host-protective immunity and its suppression in a parasitic disease: murine cutaneous leishmaniasis. Immunol. Today 5:224-226

Modlin, R.L., Pirmez, C., Hofman, F.M., Torigian, V., Uyemura, K., Rea, T.H., Bloom, B.R., Brenner, M.B. (1989) Lymphocytes bearing antigen-specific γ/δ-T-cell receptors accumulate in human infectious disease lesions. Nature 339:544-548

Ohga, S., Yoshikai, Y., Takeda, Y., Hiromatsu, K., Nomoto, K (1990) Sequential appearance of γ/δ- and α/β- bearing T cells in the peritoneal cavity during an intraperitoneal infection with Listeria monocytogenes. Eur. J. Immunol. 20:533-538

Solbach, W., Forberg, K., Kammerer, E., Bogdan, C., Röllinghoff, M (1986) Suppressive effect of cyclosporin A on the development of Leishmania major-induced lesions in genetically susceptible BALB/c mice. J. Immunol. 137:702-707

Stingl, G., Gunter, K.C., Tschachler, E., Yamada, H., Lechler, R.I., Yokoyama, W.M., Steiner, G., Germain, R.N., Shevach, E.M. (1987) Thy-1$^+$ dendritic epidermal cells belong to the T-cell lineage. Proc. Natl. Acad. Sci. USA 84:2430-2434

Involvement of γδ T Cells in Respiratory Virus Infections

P. C. Doherty[1], W. Allan[1], M. Eichelberger[1], S. Hou[1], K. Bottomly[2], and S. Carding[2]

[1] Department of Immunology, St. Jude Children's Research Hospital, 332 North Lauderdale, Memphis, TN 38101, USA
[2] Section of Immunobiology, Howard Hughes Medical Institute, Yale University School of Medicine and Department of Biology, Yale University, New Haven, CT 06510, USA

INTRODUCTION

Some aspects of the immunobiology of virus-induced respiratory disease are relatively well understood. It is known, for instance, that the adoptive transfer of cloned CD8+ T cells to immunologically naive mice infected intranasally (i.n.) with an influenza A virus greatly enhances virus clearance (Lukacher et al 1984, Taylor & Askonas 1986). The same has been observed for cloned CD4+ T cells, though the kinetics of virus elimination are more protracted (Lukacher et al 1986). In both cases, the effect is mediated by T cells that are conventionally MHC-restricted and are very specific for the infecting pathogen. There is thus no doubt that effector functions mediated by αβ T cell receptor (TCR)-bearing lymphocytes are of central importance in recovery from infection of the respiratory tract with influenza A viruses. Recent studies of the influenza pneumonia model have established that there is substantial involvement of lymphocytes with mRNA for the γδ TCR late in the course of the inflammatory process, after infectious virus has been eliminated (Allan et al 1990, Carding et al 1990 a). The obvious question is whether these γδ T cells contribute to the specific host response, or are simply recruited passively by the αβ TCR+ effectors (Doherty et al 1990).

BASIC EXPERIMENTAL APPROACH

Adult, female C57Bl/6J (B6) mice are infected i.n. with a high dose (100 HAU) of either the mouse-adapted HKx31 (H3N2) human influenza A virus or an influenza B virus (B/HK), or with a low dose (1,000 EID_{50}) of the mouse parainfluenza type 1 Sendai virus. Under these conditions, the three viruses cause severe but non-fatal pneumonitis, characterized by a massive increase in the cellularity of the MLN (>3x) and of the lung lavage population (>7x), with peak counts in the LIE being achieved at about day 7 post infection (Allan et al 1990). In some experiments, recovered mice have been challenged i.n. with 1,000 EID_{50} of the A/PR8/34 (H1N1) influenza A virus, which gives a secondary αβ T cell response in mice primed with the HKx31 virus (Doherty et al 1977), but not in those given B/HK or Sendai virus.

The inflammatory cell populations are then either used for functional studies (Allan et al 1990) or analysed morphologically, often after first adsorbing cells on plastic to remove macrophages. The separated lymphocytes are stained and analysed (Allan et al 1990) by flow microfluorimetry (FMF), or cytospin preparations are made and the slides sent for the determination of TCR distribution in individual cells using <u>in situ</u> hybridization techniques (Carding et al 1989, 1990 a). Unless stated otherwise, the results cited in the text are for primary infection with the HKx31 virus.

Current Topics in Microbiology and Immunology, Vol. 173

KINETICS OF $\alpha\beta$ AND $\gamma\delta$ TCELL INVOLVEMENT

The numbers of cells that can be lavaged from the lungs of mice infected i.n. with the HKx31 influenza virus are increased by day 3 after infection, reach a maximum by day 7, remain high through day 13, and are declining by day 15. The inflammatory process thus continues to be substantial for at least 5 days after infectious virus is cleared from the lung, though there is no evidence of secondary infection with bacteria or mycoplasma (Allan et al 1990, Carding et al 1990 a). Peak counts for plastic-adherent macrophages are found on day 7, while the non-adherent population generally constitutes more than 60% of the total on days 10 to 15. The CD4:CD8 ratio ranges from 1:2 to 1:5 and there are relatively few B lymphocytes present.

The detailed quantitation of TCR distribution in lymphocytes localizing to the lung air spaces of mice with influenza that has been done so far utilizes the in situ hybridization approach (Carding et al 1990 a). The findings are very clear and highly reproducible. The inflammatory process is dominated by $\alpha\beta$ TCR mRNA+ cells until day 7, with $\gamma\delta$ TCR MRNA+ lymphocytes being present at higher frequency on days 10 and 13 (Fig 1). These do not seem to be normal residents that had been shed following virus-induced damage to respiratory epithelium, as few $\gamma\delta$ mRNA+ cells are seen when cryostat sections of lung from uninfected B6 mice maintained in our SPF animal facility were probed.

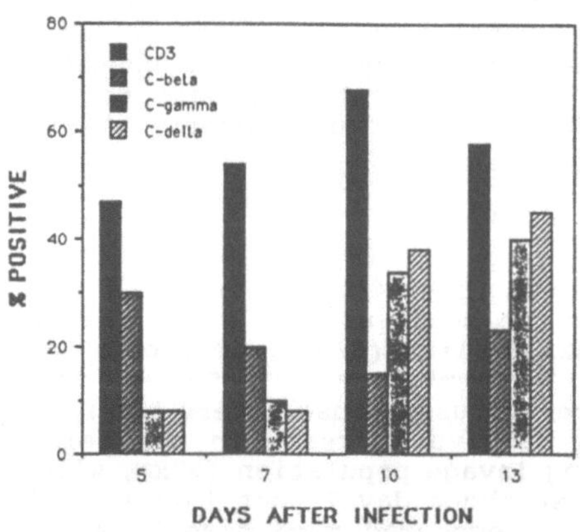

Figure. 1. Distribution of TCR mRNA phenotypes in plastic non-adherent lymphocytes lavaged from the lung of female B6 mice infected i.n. with 100 HAU of the HKx31 influenza A virus. This data is representative of a number of experiments, with there being a good concordance in all cases between the percentages of cells positive with the probes for $C\alpha$ and $C\beta$, $C\gamma$ and $C\delta$ respectively.

Separation by FACS (Carding et al 1990 a) shows that $\gamma\delta$ TCR mRNA was present in few, if any, T cells expressing surface $\alpha\beta$ TCR detected by the H57.597 mAb (Kubo et al 1989). Also, the predominant phenotype of the $\gamma\delta$ TCR mRNA+ set is CD3+4-8- for sorted cell populations (Table 1).

Table 1. Distribution of TCR mRNA in day 10 inflammatory cells separated by FACS

TCR	Sorted population[a]		
mRNA	3+4+8+	3+4-8-	3-4-8-
C-α	67	2	4
V-J-Cγ	7	83	19

[a]The lung lavage cells from mice infected i.n. with the HKx31 virus 10 days previously were adsorbed on plastic to remove macrophages, then stained with the 2C11 hamster anti-CD3± a cocktail of the H129.19 (anti-CD4) and 53.6.72 (anti-CD8) rat mAbs. The second step reagents were TxRd-mouse anti-rat Ig and FITC-goat anti-hamster Ig. The cells were sorted in 2-color mode on a FACStar Plus.

We have also been able to show that a significant proportion of the lung lavage cells are expressing surface $\gamma\delta$ TCR (Goodman and LeFrancois 1989. The initial experiments have utilized the magnetic activated cell sorter (MACS, Biotechnische Gerate, Spezialelektronic, Gladbach, Germany) to deplete Mac-1+ macrophages and CD4+ and CD8+ T cells prior to staining with the GL3 mAb (Goodman and LeFrancois 1989) to the $\gamma\delta$ TCR (Fig 2). This has allowed us to establish the staining profiles and size characteristics of the $\gamma\delta$ T cells, so that we will now be able to repeat the basic kinetic analysis using flow cytometry.

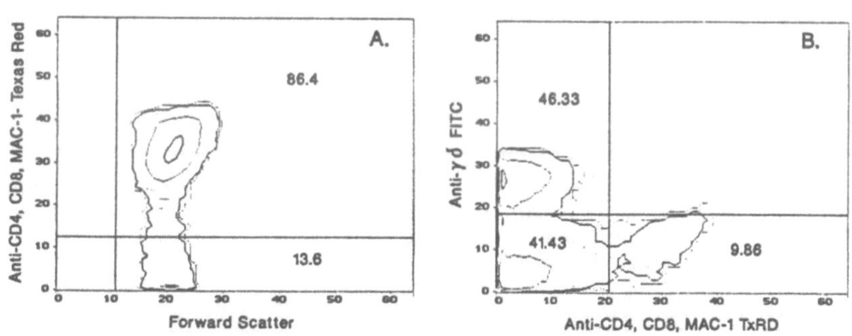

Figure 2. Inflammatory cells lavaged from the HKx31-infected mouse lung on day 10 were stained with a cocktail of mAbs to CD4 (H129.19), CD8 (53.6.72) and Mac-1 (MI/70), followed by biotinylated goat anti-rat Ig, then Strepavidin TxRD, and finally biotinylated MACS microbeads. The staining prior to separation is shown in "A", while the FACS profile for the $\gamma\delta$ TCR+ lymphocytes (GL3 mAb, followed by goat-anti-hamster-FITC) in the flow-through population from the MACS is shown in the top-left quadrant of "B".

Thus, significant numbers of the lymphocytes present in the inflammatory exudate resulting from sublethal influenza infection are $\gamma\delta$ T cells. The $\gamma\delta$ TCR+ subset is at highest prevalence late in the course of the disease process, after infectious virus has been eliminated. We need to know what, if anything, these lymphocytes are doing.

EVIDENCE FOR SPECIFICITY AND MEMORY

Virus challenge studies show quite clearly that the localization of γδ TCR mRNA+ T cells to the lungs of primed mice follows secondary kinetics (Tables 2 and 3). We are currently trying to determine if this reflects true memory for the γδ TCR+ subset, or is simply a consequence of passive recruitment by the primed αβ T cells. The experimental system uses the fact that the HKx31 H3N2 influenza A virus is a recombinant which has the internal proteins of the A/PR8/34 H1N1 virus (Kilbourne 1969). Thus, while there is no cross-neutralization because the surface glycoproteins are different, there is the potential for cross-reactive T cell recognition. Mice are infected with the H3N2 virus or an influenza B virus, left for at least one month and then challenged i.n. with the H1N1 virus.

Table 2. The secondary response at 3 days after i.n. challenge

		Total mRNA+ cells per mouse (x 10^4)			
Priming[a] virus	Challenge inoculum	CD3	Cα	Cγ	Cδ
H3N2	H1N1	5.8	2.2	3.4	3.9
H3N2	AF	0.8	0.7	<.1	0.1
H3N2	Nil	0.4	0.1	0	0
B/HK	H1N1	3.5	4.5	0.3	0.3
B/HK	AF	0.4	0.2	<.1	0.1
B/HK	Nil	0.3	0.4	<.1	0

[a]The mice were primed i.n. with the A/HKx31 (H3N2) or B/HK virus one month prior to i.n. challenge with the A/PR8 (H1N1) virus or allantoic fluid (AF), and the values shown were calculated from the total cell counts per mouse and the % TCR mRNA+ cells.

The pattern for the H1N1 challenge of the B/HK-immune mice on day 3 is characteristic of a primary response, with the cell counts for γδ TCR+ lymphocytes being more than 10-fold higher in the H1N1-H3N2 combination (Cγ and Cδ, Table 2). The gap between the two had dropped to a factor of 2-4x by day 5 (i.n. primed, Cγ, Table 3), with the difference for Vγ2 TCR mRNA+ cells being two-fold or less (Vγ2, Table 3). The priming effect was also more apparent for mice given the initial dose of virus i.n. rather than i.p. (Table 3).

Table 3. Distribution of cells positive for TCR mRNA at 5 days after i.n. challenge of primed mice with an H1N1 influenza A virus

Priming[a]			TCR mRNA (cell count x10^4)			
virus	route	Total[b] cells	CD3	Cγ	Vγ2	Vγ4
H3N2	i.n.	73	13.9	10.2	3.7	5.8
B/HK	i.n.	48	8.1	2.9	1.9	1.4
H3N2	i.p.	41	9.8	4.5	2.9	2.5
B/HK	i.p.	35	5.6	2.5	1.6	2.1

[a]The mice were primed i.n. or i.p. with the HKx31 influenza A virus or the B/HK influenza B virus one month prior to i.n. challenge.

[b]Count per mouse after plastic adherence to remove macrophages: these values were calculated from the % mRNA+ cells.

The "Vγ2" probe used in these experiments (Table 3) does not distinguish between Vγ1 and Vγ2. Expression of Vγ1 paired with Vδ6.3 has been associated with recognition of heat shock protein (HSP) for a panel of T cell hybridomas (Happ et al 1989, Born et al 1990). Primary i.n. infection with both HKx31 and with B/HK (and Sendai virus) causes an increase in the percentage of HSP mRNA+ macrophages, which is again maximal late in the course of the disease (Carding et al 1990 a). At least for the HKx31 model, this is concurrent with high frequency expression of Vγ4 mRNA and precedes the reproducible switch in dominance to lymphocytes with Vγ1/2 mRNA that occurs between day 10 and day 13 after infection. It is thus possible that the greater concordance in the numbers of Vγ1/2 mRNA+ (cf Vγ4 mRNA+) T cells for the HKx31 and B/HK primed mice (Table 3) reflects a secondary response to HSP induced by both viruses. However, by this criterion, it would also seem to be the case that the majority of the γδ T cells recruited to lung in the H1N1→H3N2 challenge are not reactive to endogenous ligands, like HSP (Born et al 1990), that are also induced by immunologically distinct orthomyxoviruses such as B/HK (Tables 2 and 3).

The other indication of viral specificity is that the current PCR analysis (Carding et al 1990 b) is showing a different pattern of γδ TCR V-region usage for Sendai virus and for the HKx31 influenza A virus. The dominant usage in Sendai virus infected mice is Vγ1. All the known Vγ phenotypes are represented in the HKx31 virus-induced inflammatory exudate, with Vγ2 being the most prominent, while Vδ3, 4, 5 and 6 show high levels of expression at different time points.

Evidence of viral specificity, memory and some protective capacity, would make the γδ T cell a worthwhile target for immunopathogenesis and vaccine development studies. Obvious alternatives are that these lymphocytes are simply passively recruited bystanders (Doherty et al 1990),or that they function to promote tissue repair by helping to resolve the inflammatory process. Understanding the latter may be important for dealing with virus-induced immunopathology.

ACKNOWLEDGEMENTS

We thank Anthony McMickle for capable technical assistance. The experiments described here were funded by CA 21765, AI 29579 and GM37759 from the NIH, by the Howard Hughes Medical Institute and by the American Lebanese Syrian Associated Charities (ALSAC).

REFERENCES

Allan W, Tabi Z, Cleary A, Doherty PC (1990) Cellular events in the lymph node and lung of mice with influenza: consequences of depleting CD4+ T cells. J Immunol 144: 3980-3986

Born W, Dallas A, Reardon C, Kubo R, Shinnick T, Brennan P, O'Brien R (1990) Recognition of heat shock proteins and γδ cell function. Immunol Today 11: 40-43

Carding SR, Allan W, Kyes S, Hayday A, Bottomley K, Doherty PC (1990 a) Late dominance of the inflammatory process in murine influenza by γδ Tcells. J Exp Med: 172:1225-1231

Carding SR, Kyes S, Jenkinson EJ, Kingston R, Bottomley K, Owen JT, Hayday AC (1990 b) Developmentally regulated fetal thymic and intrathymic T-cell receptor γδ gene expression. Genes and Development 4: 1304-1315

Carding SR, West J, Woods A, Bottomley K (1989) Differential activation of cytokine genes in normal CD4-bearing T cells is stimulus dependent. Eur J Immunol 19: 231-238

Doherty PC, Effros RB, Bennink JR (1977) Heterogeneity of the cytotoxic T cell response following immunization with influenza viruses. Proc Natl Acad Sci USA 74: 1209-1213

Doherty PC, Allan JE, Lynch F, Ceredig R (1990) Cellular events in a virus-induced inflammatory process: Promotion of delayed type hypersensitivity by CD8+ T lymphocytes. Immunol Today 11: 55-59

Goodman T and LeFrancois L (1989) Intraepithelial lymphocytes: anatomical site, not T cell receptor form dictates phenotype and function. J Exp Med 170: 1569-1581

Happ MP, Kubo RT, Palmer E, Born WK, O'Brien RL (1989) Limited receptor repertoire in a mycobacteria reactive subset of γδ T lymphocytes. Nature 342: 696-698

Kilbourne ED (1969) Future influenza vaccines and the use of genetic recombinants. Bull Wld Hlth Org 41: 643-645

Kubo RT, Born W, Kappler JW, Marrack P, Pigeon M (1989) Characterization of a monoclonal antibody which detects all murine αβ T cell receptors. J Immunol 142: 2736-2742

Lukacher AE, Braciale VL, Braciale TJ (1984) In vivo effector function of influenza virus-specific cytotoxic T lymphocyte clones is highly specific. J Exp Med 169: 814-826

Lukacher AE, Morrison LA, Braciale VL, Braciale TJ (1986) T lymphocyte recovery from experimental viral infection. In: Steinman RM, North RJ (eds) The Rockefeller University Press, new York p 223

Taylor PM, Askonas BA (1986) Influenza nucleoprotein-specific cytotoxic T cell clones are protective in vivo. Immunol 58: 417-420

Current Topics in Microbiology and Immunology

Volumes published since 1986 (and still available)

Vol. 127: **Potter, Michael; Nadeau, Joseph H.; Cancro, Michael P. (Ed.):** The Wild Mouse in Immunology. 1986. 119 figs. XVI, 395 pp. ISBN 3-540-16657-2

Vol. 128: 1986. 12 figs. VII, 122 pp. ISBN 3-540-16621-1

Vol. 129: 1986. 43 figs., VII, 215 pp. ISBN 3-540-16834-6

Vol. 130: **Koprowski, Hilary; Melchers, Fritz (Ed.):** Peptides as Immunogens. 1986. 21 figs. X, 86 pp. ISBN 3-540-16892-3

Vol. 131: **Doerfler, Walter; Böhm, Petra (Ed.):** The Molecular Biology of Baculoviruses. 1986. 44 figs. VIII, 169 pp. ISBN 3-540-17073-1

Vol. 132: **Melchers, Fritz; Potter, Michael (Ed.):** Mechanisms in B-Cell Neoplasia. Workshop at the National Cancer, Institute, National Institutes of Health, Bethesda, MD, USA, March 24–26, 1986. 1986. 156 figs. XII, 374 pp. ISBN 3-540-17048-0

Vol. 133: **Oldstone, Michael B. (Ed.):** Arenaviruses. Genes, Proteins, and Expression. 1987. 39 figs. VII, 116 pp. ISBN 3-540-17246-7

Vol. 134: **Oldstone, Michael B. (Ed.):** Arenaviruses. Biology and Imminotherapy. 1987. 33 figs. VII, 242 pp. ISBN 3-540-14322-6

Vol. 135: **Paige, Christopher J.; Gisler, Roland H. (Ed.):** Differentiation of B Lymphocytes. 1987. 25 figs. IX, 150 pp. ISBN 3-540-17470-2

Vol. 136: **Hobom, Gerd; Rott, Rudolf (Ed.):** The Molecular Biology of Bacterial Virus Systems. 1988. 20 figs. VII, 90 pp. ISBN 3-540-18513-5

Vol. 137: **Mock, Beverly; Potter, Michael (Ed.):** Genetics of Immunological Diseases. 1988. 88 figs. XI, 335 pp. ISBN 3-540-19253-0

Vol. 138: **Goebel, Werner (Ed.):** Intracellular Bacteria. 1988. 18 figs. IX, 179 pp. ISBN 3-540-50001-4

Vol. 139: **Clarke, Adrienne E.; Wilson, Ian A. (Ed.):** Carbohydrate-Protein Interaction. 1988. 35 figs. IX, 152 pp. ISBN 3-540-19378-2

Vol. 140: **Podack, Eckhard R. (Ed.):** Cytotoxic Effector Mechanisms. 1989. 24 figs. VIII, 126 pp. ISBN 3-540-50057-X

Vol. 141: **Potter, Michael; Melchers, Fritz (Ed.):** Mechanisms in B-Cell Neoplasia 1988. Workshop at the National Cancer Institute, National Institutes of Health, Bethesda, MD, USA, March 23–25, 1988. 1988. 122 figs. XIV, 340 pp. ISBN 3-540-50212-2

Vol. 142: **Schüpach, Jörg:** Human Retrovirology. Facts and Concepts. 1989. 24 figs. 115 pp. ISBN 3-540-50455-9

Vol. 143: **Haase, Ashley T.; Oldstone Michael B. A. (Ed.):** In Situ Hybridization 1989. 22 figs. XII, 90 pp. ISBN 3-540-50761-2

Vol. 144: **Knippers, Rolf; Levine, A. J. (Ed.):** Transforming. Proteins of DNA Tumor Viruses. 1989. 85 figs. XIV, 300 pp. ISBN 3-540-50909-7

Vol. 145: **Oldstone, Michael B. A. (Ed.):** Molecular Mimicry. Cross-Reactivity between Microbes and Host Proteins as a Cause of Autoimmunity. 1989. 28 figs. VII, 141 pp. ISBN 3-540-50929-1

Vol. 146: **Mestecky, Jiri; McGhee, Jerry (Ed.):** New Strategies for Oral Immunization. International Symposium at the University of Alabama at Birmingham and Molecular Engineering Associates, Inc. Birmingham, AL, USA, March 21–22, 1988. 1989. 22 figs. IX, 237 pp. ISBN 3-540-50841-4

Vol. 147: **Vogt, Peter K. (Ed.):** Oncogenes. Selected Reviews. 1989. 8 figs. VII, 172 pp. ISBN 3-540-51050-8

Vol. 148: **Vogt, Peter K. (Ed.):** Oncogenes and Retroviruses. Selected Reviews. 1989. XII, 134 pp. ISBN 3-540-51051-6

Vol. 149: **Shen-Ong, Grace L. C.; Potter, Michael; Copeland, Neal G. (Ed.):** Mechanisms in Myeloid Tumorigenesis. Workshop at the National Cancer Institute, National Institutes of Health, Bethesda, MD, USA, March 22, 1988. 1989. 42 figs. X, 172 pp. ISBN 3-540-50968-2

Vol. 150: **Jann, Klaus; Jann, Barbara (Ed.):** Bacterial Capsules. 1989. 33 figs. XII, 176 pp. ISBN 3-540-51049-4

Vol. 151: **Jann, Klaus; Jann, Barbara (Ed.):** Bacterial Adhesins. 1990. 23 figs. XII, 192 pp. ISBN 3-540-51052-4

Vol. 152: **Bosma, Melvin J.; Phillips, Robert A.; Schuler, Walter (Ed.):** The Scid Mouse. Characterization and Potential Uses. EMBO Workshop held at the Basel Institute for Immunology, Basel, Switzerland, February 20–22, 1989. 1989. 72 figs. XII, 263 pp. ISBN 3-540-51512-7

Vol. 153: **Lambris, John D. (Ed.):** The Third Component of Complement. Chemistry and Biology. 1989. 38 figs. X, 251 pp. ISBN 3-540-51513-5

Vol. 154: **McDougall, James K. (Ed.):** Cytomegaloviruses. 1990. 58 figs. IX, 286 pp. ISBN 3-540-51514-3

Vol. 155: **Kaufmann, Stefan H. E. (Ed.):** T-Cell Paradigms in Parasitic and Bacterial Infections. 1990. 24 figs. IX, 162 pp. ISBN 3-540-51515-1

Vol. 156: **Dyrberg, Thomas (Ed.):** The Role of Viruses and the Immune System in Diabetes Mellitus. 1990. 15 figs. XI, 142 pp. ISBN 3-540-51918-1

Vol. 157: **Swanstrom, Ronald; Vogt, Peter K. (Ed.):** Retroviruses. Strategies of Replication. 1990. 40 figs. XII, 260 pp. ISBN 3-540-51895-9

Vol. 158: **Muzyczka, Nicholas (Ed.):** Viral Expression Vectors. 1991. Approx. 20 figs. Approx. XII, 190 pp. ISBN 3-540-52431-2

Vol. 159: **Gray, David; Sprent, Jonathan (Ed.):** Immunological Memory. 1990. 38 figs. XII, 156 pp. ISBN 3-540-51921-1

Vol. 160: **Oldstone, Michael B. A.; Koprowski, Hilary (Ed.):** Retrovirus Infections of the Nervous System. 1990. 16 figs. XII, 176 pp. ISBN 3-540-51939-4

Vol. 161: **Racaniello, Vincent R. (Ed.):** Picornaviruses. 1990. 12 figs. X, 194 pp. ISBN 3-540-52429-0

Vol. 162: **Roy, Polly; Gorman, Barry M. (Ed.):** Bluetongue Viruses. 1990. 37 figs. X, 200 pp. ISBN 3-540-51922-X

Vol. 163: **Turner, Peter C.; Moyer, Richard W. (Ed.):** Poxviruses. 1990. 23 figs. X, 210 pp. ISBN 3-540-52430-4

Vol. 164: **Bækkeskov, Steinnun; Hansen, Bruno (Ed.):** Human Diabetes. 1990. 9 figs. X, 198 pp. ISBN 3-540-52652-8

Vol. 165: **Bothwell, Mark (Ed.):** Neuronal Growth Factors. 1991. 14 figs. IX, 173 pp. ISBN 3-540-52654-4

Vol. 166: **Potter, Michael; Melchers, Fritz (Ed.):** Mechanisms in B-Cell Neoplasia 1990. 143 figs. XIX, 380 pp. ISBN 3-540-52886-5

Vol. 167: **Kaufmann, Stefan H. E. (Ed.):** Heat Shock Proteins and Immune Response. 1991. 18 figs. IX, 214 pp. ISBN 3-540-52857-1

Vol. 168: **Mason, William S.; Seeger, Christoph (Ed.):** Hepadnaviruses. Molecular Biology and Pathogenesis. 1991. 21 figs. IX, 206 pp. ISBN 3-540-53060-6

Vol. 169: **Kolakofsky, Daniel (Ed.):** Bunyaviridae. 1991. 34 figs. IX, 255 pp. ISBN 3-540-53061-4

Vol. 170: **Compans, Richard W. (Ed.):** Protein Traffic in Eukaryotic Cells. Selected Reviews. 1991. 14 figs. Approx. X, 180 pp. ISBN 3-540-53631-0

Vol. 171: **Kung, Hsing-Jien (Ed.):** Retroviral Insertional Mutagenesis and Oncogene. 1991. 18 figs. Approx. X, 180 pp. ISBN 3-540-53857-7

Vol. 172: **Chesebro, Bruce W. (Ed.):** Scrapie and Other Transmissible Spongiform Encephalopathies. 1991. 48 figs. Approx. X, 220 pp. ISBN 3-540-53883-6

MIX
Papier aus verantwortungsvollen Quellen
Paper from responsible sources
FSC® C105338

If you have any concerns about our products,
you can contact us on
ProductSafety@springernature.com

In case Publisher is established outside the EU,
the EU authorized representative is:
Springer Nature Customer Service Center GmbH
Europaplatz 3, 69115 Heidelberg, Germany

Printed by Libri Plureos GmbH
in Hamburg, Germany